ALTERNATIVE
ENERGY SOURCES

ALTERNATIVE ENERGY SOURCES
An International Compendium

Edited by T. Nejat Veziroğlu

Vol. 1	Solar Energy 1
Vol. 2	Solar Energy 2
Vol. 3	Solar Energy 3
Vol. 4	Indirect Solar Energy
Vol. 5	Nuclear Energy
Vol. 6	Geothermal Energy and Hydropower
Vol. 7	Hydrocarbon Conversion Technology
Vol. 8	Hydrogen Energy
Vol. 9	Energy Delivery, Conservation, and Environment
Vol. 10	Energy Economics and Policy
Vol. 11	Index

ALTERNATIVE ENERGY SOURCES
An International Compendium

VOLUME 6
Geothermal Energy and Hydropower

Edited by
T. Nejat Veziroğlu
Clean Energy Research Institute, University of Miami

HEMISPHERE PUBLISHING CORPORATION
Washington   London

DISTRIBUTION OUTSIDE THE UNITED STATES

McGRAW-HILL INTERNATIONAL BOOK COMPANY

New York   St. Louis   San Francisco   Auckland   Bogotá
Düsseldorf   Johannesburg   London   Madrid   Mexico
Montreal   New Delhi   Panama   Paris   São Paulo
Singapore   Sydney   Tokyo   Toronto

ALTERNATIVE ENERGY SOURCES: An International Compendium

Copyright © 1978 by Hemisphere Publishing Corporation. All rights reserved. Printed in the United States of America. No part of this publication may be reproduced, stored in a retrieval system, or transmitted, in any form or by any means, electronic, mechanical, photocopying, recording, or otherwise, without the prior written permission of the publisher.

1 2 3 4 5 6 7 8 9 0   L I L I   7 8 3 2 1 0 9 8

Library of Congress Cataloging in Publication Data

Miami International Conference on Alternative Energy
  Sources, Miami Beach, Fla., 1977.
  Alternative energy sources: an international
compendium.

  "Presented by the Clean Energy Research Institute, University of Miami, Coral Gables, Florida, sponsored by the United States Department of Energy and the School of Continuing Studies, University of Miami."
  Includes index.
  CONTENTS: v. 1-3. Solar energy.—v. 4. Indirect solar energy.—v. 5. Nuclear energy. [etc.]
  1. Renewable energy sources—Congresses.
I. Veziroglu, T. Nejat. II. Miami, University of, Coral Gables, Fla. Clean Energy Research Institute.
III. United States. Dept. of Energy. IV. Miami, University of, Coral Gables, Fla. School of Continuing Studies. V. Title.
TJ163.15.M5   1977   333.7   78-16383
ISBN 0-89116-129-5 (set)
ISBN 0-89116-084-1 (v. 6)

Contents

Preface   vii

Acknowledgments   ix

Conference Committee and Staff   xi

Session Officials   xiii

GEOTHERMAL ENERGY OVERVIEW AND SITING

Harnessing Geothermal Energy in Rotorua, New Zealand
 R. J. Shannon   **2577**

Geothermal Systems in the Hauraki Rift Zone (New Zealand)—An Example for Geothermal Systems over an Inferred Upper Mantle Swell
 M. P. Hochstein   **2599**

The Puchuldiza Geothermal Field
 Pratricio R. Trujillo   **2611**

Geothermal Energy Coupled with Solar Energy Is the Future Energy in Saudi Arabia
 A. A. M. Sayigh   **2623**

Manmade Geothermal Energy
 T. K. Guha, K. E. Davis, R. E. Collins, J. R. Fanchi, and A. C. Meyers, III   **2641**

Hot Dry Rock Energy Project
 R. H. Hendron   **2655**

GEOTHERMAL ENERGY TECHNOLOGY

Pilot Facility for the Experimental Study of Binary-Cycle Conversion Systems—Thermogravimetric Loop
 G. Cefaratti, S. Arosio, G. Sotgia, P. Alia, and G. Morandi   **2683**

The Optimization of Alternative Energy Cycles Using Program GEOTHM
 M. A. Green, P. A. Doyle, H. S. Pines, W. L. Pope, and L. F. Silvester   **2709**

Thermal Fracturing Patterns and Effects of an Imitation Hot Dry Rock by Impinging of Water Jets
    H. Kiyohashi, M. Kyo, and W. Ishihama   **2727**

An Exploration of the Applicability Field of Geothermal Water-fed Heat Pumps
    P. DeMarchi Desenzani and G. Giglioli   **2747**

Geothermal Community Heating Systems
    J. F. Kunze, R. C. Stoker, and L. E. Donovan   **2767**

Conserving Electric Power by Geothermal Refrigeration—Cooling and Freezing
    Edward F. Wehlage   **2785**

Heat Transfer Consideration in Utilizing Solar and Geothermal Energy
    J. W. Michel   **2801**

The Economics of Upgrading Geothermal Steam by Adiabatic Compression
    A. Valfells   **2827**

WATER-BASED ENERGY

Low-head Hydroelectric Power: A Realizable Alternative
    J. I. Mills and G. L. Smith   **2843**

Small Hydropower—Promise and Reality in New York State and the Northeast
    R. S. Brown and A. S. Goodman   **2857**

Energy from Sea and Air by Large-Span Tensioned Foils
    D. Z. Bailey   **2877**

Salinity Power Station at the Swedish West-Coast: Possibilities and Energy Price for a 200 MW Plant
    Allan T. Emren and Sture B. Bergström   **2909**

Heat Extraction from a Salt Gradient Solar Pond
    F. Zangrando and H. C. Bryant   **2935**

Preface

Fossil fuels, particularly oil and gas, which presently provide most of our energy requirements, are rapidly being depleted. Coal has been less actively utilized, since it is less convenient and creates more environmental problems.

Alternative energy sources are available, but are at present either undeveloped technologically or underutilized. Some of these energy sources are renewable, such as solar radiation, hydro, wind, ocean thermal, and salinity gradient energy. Others are depletable, but relatively untapped, such as geothermal heat or synthetic fuels from coal or wastes. Nuclear energy as presently produced depletes uranium more rapidly than energy from breeder reactors. Nuclear fusion is still a hope for the future. Hydrogen complements the nonfossil energy sources and is recyclable. It requires the development of efficient and economical production methods.

The Miami International Conference addressed itself to the present state and the future promises of the main alternative energy sources. It included sessions on solar energy, ocean thermal energy, wind energy, geothermal energy, hydro- and salinity gradient power, nuclear breeders and nuclear fusion, synthetic fuels from coal and wastes, hydrogen production and uses, conservation, environment, economics, and policy matters. There were also two sessions devoted to panel discussions.

The papers recommended by the session chairpersons and co-chairpersons, panel discussions, and the keynote and banquet addresses (which are found in the Introduction), comprise ten volumes of this set. The eleventh volume contains the Index. The reader should be advised that it was difficult to specifically classify some papers when there was overlap in the subject matter. In such cases, we tried to make the best possible choice.

Alternative Energy Sources should serve as a valuable reference collection for future research, development, and planning efforts in this important and growing field.

T. Nejat Veziroğlu

Acknowledgments

The Conference Committee gratefully acknowledges the financial support of the U.S. Department of Energy (Divisions of Solar Energy, Nuclear Energy, and Magnetic Fusion Energy; U.S. Energy Research and Development Administration; and Federal Energy Administration) and the support and services provided by the School of Continuing Studies at the University of Miami.

We also wish to extend sincere appreciation to the keynote speaker, Ishrat H. Usmani, Senior Energy Advisor to the United Nations, and to the banquet speaker, Dennis H. Meadows of Dartmouth College.

Special thanks are due to our authors, lecturers, and panelists, who provided the substance of the conference as published in this eleven-volume compendium.

And last, but not least, our gratitude is extended to the session chairpersons and co-chairpersons for the organization and execution of the technical sessions and for their assistance in the selection of the papers published herein. In acknowledgment, we list these session officials on the following pages.

The Conference Committee

Conference Committee and Staff

CONFERENCE COMMITTEE

Chairperson: T. Nejat Veziroğlu
Clean Energy Research Institute, University of Miami,
Coral Gables, Florida, USA

Co-Chairperson: Harold J. Plass, Jr.
Mechanical Engineering Department, University of Miami,
Coral Gables, Florida, USA

Members: Herbert Hoffman
Oak Ridge National Laboratory, Oak Ridge,
Tennessee, USA

Frank Salzano
Brookhaven National Laboratory, Upton,
New York, USA

Kudret Selcuk
Jet Propulsion Laboratory, Pasadena,
California, USA

CONFERENCE STAFF

Coordinators: Carol Pascalis
Lucille Walter

Special Assistant/
Manuscript Editor: Deirdre Finn

Arrangements: Tony Pajares
James H. Poisant
Lois Rosenzweig

Technical
Assistant: Laxman G. Phadke

Session Officials

SESSION 1A   **SOLAR ENERGY ECONOMICS**

Chairperson: R. Bezdek, U.S. Department of Energy, Washington, D.C., USA
Co-Chairperson: W. Pardo, University of Miami, Coral Gables, Florida, USA

SESSION 1B   **OCEAN THERMAL ENERGY CONVERSION I**

Chairperson: F. E. Naef, Lockheed Ocean Systems, Washington, D.C., USA
Co-Chairperson: W. B. King, University of Miami, Coral Gables, Florida, USA

SESSION 1C   **NUCLEAR FUSION I**

Chairperson: H. S. Cullingford, U.S. Department of Energy, Washington, D.C., USA
Co-Chairperson: J. Davidson, University of Miami, Coral Gables, Florida, USA

SESSION 1D   **COAL CONVERSION TECHNOLOGY I**

Chairperson: P. A. Kittle, Apollo Chemical Corporation, Whippany, New Jersey, USA
Co-Chairperson: J. P. Alexander, University of Miami, Coral Gables, Florida, USA

SESSION 1E   **GEOTHERMAL I**

Chairperson: R. W. Ellis, Florida International University, Miami, Florida, USA
Co-Chairperson: S. Kakac, Middle East Technical University, Ankara, Turkey

SESSION 1F — ECONOMICS AND POLICY I

Chairperson: M. Avriel, Technion—Israel Institute of Technology, Haifa, Israel
Co-Chairperson: T. G. Olsen, University of Miami, Coral Gables, Florida, USA

SESSION 2A — SOLAR COLLECTORS I

Chairperson: A. A. M. Sayigh, University of Riyadh, Riyadh, Saudi Arabia
Co-Chairperson: N. Ozboya, University of Miami, Coral Gables, Florida, USA

SESSION 2B — OCEAN THERMAL ENERGY CONVERSION II

Chairperson: F. E. Naef, Lockheed Ocean Systems, Washington, D.C., USA
Co-Chairperson: H. L. Craig, University of Miami—RSMAS, Miami, Florida, USA

SESSION 2C — NUCLEAR BREEDERS I

Chairperson: J. DeMastry, Florida Power and Light Company, Miami, Florida, USA
Co-Chairperson: L. G. Phadke, University of Miami, Coral Gables, Florida, USA

SESSION 2D — COAL CONVERSION TECHNOLOGY II

Chairperson: D. M. Kennedy, Hydrocarbon Research, Incorporated, McLean, Virginia, USA
Co-Chairperson: K. Akyuzlu, University of Miami, Coral Gables, Florida, USA

SESSION 2E — POWER GENERATION AND TRANSPORTATION

Chairperson: C. S. Chen, U.S. Department of Energy, Washington, D.C., USA
Co-Chairperson: A. Gokhman, University of Miami, Coral Gables, Florida, USA

SESSION OFFICIALS

SESSION 2F	**ECONOMICS AND POLICY II**
Chairperson:	M. Avriel, Technion—Israel Institute of Technology, Haifa, Israel
Co-Chairperson:	R. Zuckerman, University of Miami, Coral Gables, Florida, USA
SESSION 3A	**SOLAR COLLECTORS II**
Chairperson:	H. S. Robertson, University of Miami, Coral Gables, Florida, USA
Co-Chairperson:	R. Zuckerman, University of Miami, Coral Gables, Florida, USA
SESSION 3B	**SOLAR ORGANIZATIONS AND SOME APPLICATIONS**
Chairperson:	C. S. Chen, U.S. Department of Energy, Washington, D.C., USA
Co-Chairperson:	T. Dogan, University of Miami, Coral Gables, Florida, USA
SESSION 3C	**NUCLEAR FUSION II**
Chairperson:	J. DeMastry, Florida Power and Light Company, Miami, Florida, USA
Co-Chairperson:	D. Wells, University of Miami, Coral Gables, Florida, USA
SESSION 3D	**HYDROGEN I**
Chairperson:	F. J. Salzano, Brookhaven National Laboratory, Upton, New York, USA
Co-Chairperson:	R. R. Adt, Jr., University of Miami, Coral Gables, Florida, USA
SESSION 3E	**GEOTHERMAL II**
Chairperson:	S. Kakac, Middle East Technical University, Ankara, Turkey
Co-Chairperson:	T. Khalil, University of Miami, Coral Gables, Florida, USA

SESSION OFFICIALS

SESSION 3F   **ECONOMICS AND POLICY III**

Chairperson: J. O'M. Bockris, Flinders University of South Australia, Bedford Park, Australia
Co-Chairperson: A. Gokhman, University of Miami, Coral Gables, Florida, USA

SESSION 4A   **SOLAR HEATING AND COOLING**

Chairperson: R. D. Scott, U.S. Department of Energy, Washington, D.C., USA
Co-Chairperson: L. E. Poteat, University of Miami, Coral Gables, Florida, USA

SESSION 4B   **SOLAR DESALINATION AND AGRICULTURAL APPLICATIONS**

Chairperson: A. R. Martinez, Frente Nacional Pro Defensa del Petroleo Venezolano, Caracas, Venezuela
Co-Chairperson: H. Hiser, University of Miami, Coral Gables, Florida, USA

SESSION 4C   **NUCLEAR BREEDERS II**

Chairperson: M. C. Cullingford, U.S. Nuclear Regulatory Commission, Washington, D.C., USA
Co-Chairperson: P. Ziajka, University of Miami, Coral Gables, Florida, USA

SESSION 4D   **HYDROGEN II**

Chairperson: F. J. Salzano, Brookhaven National Laboratory, Upton, New York, USA
Co-Chairperson: A. Mitsui, University of Miami—RSMAS, Miami, Florida, USA

SESSION 4E   **ENERGY CONVERSION AND TRANSMISSION**

Chairperson: J. Belding, U.S. Department of Energy, Washington, D.C., USA
Co-Chairperson: J. Davidson, University of Miami, Coral Gables, Florida, USA

SESSION OFFICIALS

SESSION 4F	**ENERGY ALTERNATIVES OVERVIEW**
Chairperson:	R. F. McAlevy, III, Stevens Institute of Technology, Hoboken, New Jersey, USA
Co-Chairperson:	T. G. Olsen, University of Miami, Coral Gables, Florida, USA
SESSION 5A	**SOLAR POWER SYSTEMS**
Chairperson:	D. R. Costello, Solar Energy Research Institute, Golden, Colorado, USA
Co-Chairperson:	J. P. Alexander, University of Miami, Coral Gables, Florida, USA
SESSION 5B	**BIOCONVERSION AND OTHER SOLAR APPLICATIONS**
Chairperson:	H. L. Cr aig, University of Miami—RSMAS, Miami, Florida, USA
Co-Chairperson:	P. Nayak, University of Miami, Coral Gables, Florida, USA
SESSION 5C	**WIND ENERGY I**
Chairperson:	S. Kakac, Middle East Technical University, Ankara, Turkey
Co-Chairperson:	S. Bukkaputnam, University of Miami, Coral Gables, Florida, USA
SESSION 5D	**COAL CONVERSION TECHNOLOGY III**
Chairperson:	R. E. Billings, Billings Energy Corporation, Provo, Utah, USA
Co-Chairperson:	L. G. Phadke, University of Miami, Coral Gables, Florida, USA
SESSION 5E	**ENERGY CONSERVATION**
Chairperson:	F. Singleton, Federal Energy Administration, Atlanta, Georgia, USA
Co-Chairperson:	H. Robertson, University of Miami, Coral Gables, Florida, USA

SESSION 5F — ECONOMICS AND POLICY IV

Chairperson: M. N. Ozisik, North Carolina State University, Raleigh, North Carolina, USA
Co-Chairperson: K. Akyuzlu, University of Miami, Coral Gables, Florida, USA

SESSION 6A — PHOTOVOLTAICS

Chairperson: J. Hirschberg, University of Miami, Coral Gables, Florida, USA
Co-Chairperson: N. Ozboya, University of Miami, Coral Gables, Florida, USA

SESSION 6B — HEAT STORAGE AND TRANSFER

Chairperson: H. Hoffman, Oak Ridge National Laboratory, Oak Ridge, Tennessee, USA
Co-Chairperson: T. Dogan, University of Miami, Coral Gables, Florida, USA

SESSION 6C — WIND ENERGY II

Chairperson: A. A. M. Sayigh, University of Riyadh, Riyadh, Saudi Arabia
Co-Chairperson: H. Hiser, University of Miami, Coral Gables, Florida, USA

SESSION 6D — MISCELLANEOUS SYNTHETIC FUELS

Chairperson: J. Kelley, Jet Propulsion Laboratory, Pasadena, California, USA
Co-Chairperson: T. G. Olsen, University of Miami, Coral Gables, Florida, USA

SESSION 6E — HYDRO- AND SALINITY POWER

Chairperson: L. E. Poteat, University of Miami, Coral Gables, Florida, USA
Co-Chairperson: P. Ziajka, University of Miami, Coral Gables, Florida, USA

SESSION OFFICIALS

SESSION 6F	**PANEL DISCUSSION ON ENERGY POLICIES AND PUBLIC UNDERSTANDING**
Panel Moderator:	J. Shacter, Union Carbide, Oak Ridge, Tennessee, USA
Co-Moderator:	J. Anderson, University of Miami, Coral Gables, Florida, USA
Panel Members:	J. P. Andelin, Committee on Science and Technology, U.S. House of Representatives, Washington, D.C., USA
	J. Ceppos, The Miami Herald, Miami, Florida, USA
	C. Goldstein, Atomic Industrial Forum, Washington, D.C., USA
	A. Mayrhofer, University of South Carolina, Columbia, South Carolina, USA
FINAL PLENARY SESSION:	**PANEL DISCUSSION ON ALTERNATIVE ENERGY SOURCES**
Panel Moderator:	J. E. Funk, University of Kentucky, Lexington, Kentucky, USA
Panel Members:	J. Appleby, Laboratoire d'Electrolyse, Bellevue, France
	L. Bogart, U.S. Department of Energy, Washington, D.C., USA
	J. O'M. Bockris, Flinders University of South Australia, Bedford Park, Australia
	M. Dubey, Lockheed Company, Burbank, California, USA
	D. Gregory, Institute of Gas Technology, Chicago, Illinois, USA
	J. Horak, Oak Ridge National Laboratory, Oak Ridge, Tennessee, USA
	G. H. Lavi, Carnegie-Mellon University, Pittsburgh, Pennsylvania, USA
	E. Wehlage, International Society for Geothermal Engineers, Whittier, California, USA

GEOTHERMAL ENERGY OVERVIEW AND SITING

Harnessing Geothermal Energy in Rotorua, New Zealand

R. J. SHANNON
Ministry of Works and Development
Hamilton, New Zealand

ABSTRACT

The city of Rotorua, is located in a known geothermal resource area, which has been exploited principally to obtain water for space heating.

The replacement of conventional oil and coal fired boilers, with fully automated geothermal systems, has resulted in savings of both manpower and imported fuel oil with less environmental pollution.

Group heating of buildings at the Government Center, from a geothermal source, has set a pattern which could be duplicated for all the buildings in the central city area without undue disruption to existing underground services.

The capital cost of geothermal heating systems compare favourably with those using fossil fuels; but the running costs are much lower.

Other applications of terrestrial heat include clean steam generation for hospital services, air conditioning (heating and cooling) for a large tourist hotel, thermal swimming pools and numerous uses in horticulture, animal husbandry, forestry and industry.

INTRODUCTION

New Zealand like many other nations is facing an energy shortage. The fact that over sixty per cent of its primary energy is imported from overseas is a matter for concern.

Until 1973 there seemed to be no difficulty in meeting the incremental demand for energy from a mixture of coal, oil, natural gas, geothermal and hydro-electricity but the country has been moving into a position of increasing dependence on an imported fuel.

Earlier misgivings over the future of oil were fully substantiated in October 1973 when members of the Organisation of Arab Petroleum Exporting countries announced that they were going to reduce production.

This whole exercise illustrates quite dramatically that oil is no longer a reliable form of energy in terms of overseas supply.

At the same time it has become more expensive and, if the spiralling costs continue, it will adversely affect the country's living standards.

To arrest the trend of reliance on overseas fuel there are positive advantages to be gained in greater development and harnessing of indigenous energy resources of which geothermal energy forms a part.

Although New Zealand has made substantial progress in the last 25 years in the exploration of geothermal energy, in recent years its progress has been hampered by lack of finance, due to greater emphasis being placed on the development of natural gas.

Fortunately government has now seen fit to set aside substantial finance to accelerate the development of geothermal resources, in keeping with its policy that energy must be from a reliable source at prices that the nation can afford.

This country's main thermal region is the Taupo Volcanic Zone which extends in a north-easterly direction from the three major volcanoes (all usually quiet) in the center of the North Island to another on White Island in the Bay of Plenty. The area covered is 240 km long by 45 km wide and within its boundaries hot springs, fumeroles, geysers and abnormal earth temperature gradients occur. Geothermal heat is obtained by tapping the hot water reservoirs which underlie much of this region.

The greatest utilisation occurs in three places - Wairakei, Kawerau and Rotorua - in other places comparatively small scale use is made of geothermal heat.

Estimates of the geothermal potential vary. They range from 1760 MW electrical to more than 2500 MW electrical. This indicates that geothermal resources could provide 25 - 30 per cent of the electricity demand in 20 years if there is the same rate of increase in demand as before 1975. Likewise attitudes to geothermal power vary. There is a body of opinion that favours non-electrical use of geothermal energy. However, since geothermal power stations are ideally suited for base load operation and produce cheaper power than from

any other source,* it seems that the nation could benefit from
a series of geothermal power stations like Wairakei (160 MW)
being built on the Taupo Volcanic Plateau. Already work
has begun on the Broadlands field which is expected to begin
transmitting power from a 150 MW station late in 1982.

Although geothermal development in terms of electric power is
of prime importance, it is nevertheless essentially an inefficient process owing to inescapable thermodynamic
restraints (unless coupled to district heating or industrial
applications).

On the other hand, the direct use of geothermal heat for space
heating and domestic hot water supply, industrial processes,
horticulture or for husbandry, can be highly efficient, since
the losses incurred are not imposed by the laws of thermodynamics, but only by such imperfections as must inevitably
arise from insulation losses, terminal temperature differences
in heat exchangers, etc. These imperfections can be
controlled within economic constraints, they are not
dictated by natural laws.

Furthermore, the sources of high enthalpy natural heat,
suitable for power generation are believed to be less
abundant than those of lower enthalpy fluids, which can be
used for other purposes. In short, geothermal energy is far
too versatile an asset to be used for power generation only.

Out of this comes a pattern of high enthalpy heat for generation of electricity and lower enthalpy heat for non-
electrical use, or possible electrical generation using the
binary system.

Long before the advent of the first European settlers to
New Zealand the Maoris in the Rotorua area used geothermal
heat for warmth, cooking and bathing. Today that tradition
is carried on in the city of Rotorua, for besides being a
tourist and spa center, it is also the district where the
greatest use is made of low enthalpy terrestrial heat for non-
electrical applications. Moreover, with new techniques of
regenerating geothermal effluent by reinjection, it is
likely that this heat can be made a renewable resource for
the centuries to come.

RESERVOIR CRITERIA

Geothermal fluids are found to exist in either the liquid
water or gaseous steam phase in reservoirs. In the Taupo

* excluding hydro

Volcanic Zone hot water is the dominant phase. To utilise geothermal energy as a commercial proposition, a hot water reservoir must generally meet the following standards:-

(a) A sufficiently high temperature for the type of application.

(b) A relatively shallow depth.

(c) Sufficent permability to allow water to flow continuously at a high rate.

(d) Sufficient water to recharge or maintain production over many years.

(3) Relatively non-corrosive and not excessive dissolved solids.

It is fortunate that the central area of the City of Rotorua is built over a geothermal reservoir which meets these standards. Superheated water is found at comparatively shallow depths, and the greater part of the city obtains supplies of low enthalpy heat from wells drilled into the hot water bearing strata.

ROTORUA GEOTHERMAL FIELD

Rotorua Caldera is almost circular in plan, about 16 km in diameter, and has a relief of about 300 m. Lake Rotorua occupies the caldera and the City of Rotorua is located on the southern shore, where there is visible geothermal activity.

The Rotorua Geothermal Field is at present the second most extensively exploited geothermal resource in New Zealand.

Hot water (greater than $100°C$) is obtained in a central zone of the city extending north from Whakarewarewa to Lake Rotorua bordered by relatively narrow zones on east and west where warm water is encountered. See Figure 1.

More than 770 shallow holes within an area of 11 km^2 have been drilled in the field and over 400 of these are known to be in production. Wells generally range from 60 to 120 m in depth, a few exceed 220 m, and the deepest reaches 300 m. Most of the producing wells are drilled to and into rhyolite, and discharge slightly alkaline chloride water. The majority of the wells in which temperatures have been measured pass through a temperature maximum of up to 160°C at about 90-120 m depth, below which temperatures drop within the rhyolite. However, there are exceptions, as one hole near Whakarewarewa has a temperature of 194°C at a depth of 135 m. Well head pressures range from non-artesian to over 780 kilo pascals. Silica and sodium/potassium ratio geothermometry indicate that deep water temperatures of 230-350°C could be expected in source areas beneath the Rotorua field. Deep drilling is required to confirm these expectations, but on present information the Rotorua field might be capable of producing about 100 MW of generated electricity. (Nairn) quotes that in 1965 total flow from all drillholes was estimated at 1.94 ^{3}/sec water discharge and 112 MW heat discharge (relative to 13° C). Since then many more wells have been drilled and heat discharge consequently higher. Some indication of the use of these wells is as follows:

Public buildings	25
Local Bodies	15
Hotels - motels	80
Hospitals	11
Industrial	20
Commercial buildings	30
Domestic and other uses	219

HARNESSING GEOTHERMAL HEAT

The production of geothermal fluids from a reservoir is similar to the production of oil and gas. The practice is to drill through the soft surface layer until firm ground is reached and to line the hole with a casing, the diameter of which will depend on the size of the well required. Some wells have multiple steel casings decreasing in diameter with depth and the bottom of the hole being unlined at production level. Completion of a well requires proper engineering design of casing and cementing materials to prevent subsequent mechanical failures.

Cold water is pumped through hollow drill rods during the drilling operation. This acts as a lubricant and brings up debris formed by the cutting edge, it also acts as a water seal should the drill suddenly meet with a source of high temperature water. When the temperature of the water starts to rise the drill has reached a production zone and is withdrawn.

A small gate valve is then fixed on the head top and a hose is passed through the valve down into the lubricating water and compressed air is blown through the hose. The aerated lubricating water which flows from the blow-off valve is replaced in the tube by hot water forced in by hydrostatic pressure; when the incoming superheated water has risen in the tube to a point where the internal pressure is reduced sufficiently, it flashes into steam and a mixture of steam and water is ejected from the well head. The air hose is then withdrawn and the bore allowed to discharge. Once a bore has been started it must be left operating continuously or it will cool and cease to flow. After the bore has settled a calorimeter test is carried out to determine the heat available.

Heat Exchangers

The hazards associated with uncondensable gases in the geothermal water makes it unsuitable for direct use. It is therefore necessary to extract the heat by means of heat exchangers. These vary in size and complexity, but generally fall into three patterns:

a)   above ground heat exchangers

b)   submersible heat exchangers

c)   down-hole heat exchangers

Above ground heat exchangers are used when the bores are artesian and hot water is discharged with sufficient pressure and temperature to ensure adequate heat transfer. Most of the above ground heat exchangers and sleeve type or shell and tube made from mild steel. With the latter, geothermal water passes through mild steel tubes welded to header plates: and caps can be removed to allow the tubes to be examined and cleared of mineral deposits when required.

Submersible heat exchangers can be inserted in thermal pools or holding tanks where there is flowing hot water. Some are made from mini bore Teflon tubing made up and intertwined in bundles. The overall heat transfer rate of these is high since the tubes have a low fouling factor. As yet this type of heat exchanger is not widely used.

Down-hole heat exchangers are used when the geothermal bore is not artesian and at shallow depth has permability with an ample temperature gradient. Corrosion at the air water interface is minimised by cathodic protection. Where the thermal water is aggressive the outed casing is coated with polythylene. This type of heat exchanger is not widely used in New Zealand.

Control Valves

One of the problems met with when using geothermal water in exchangers is to control the flow so as to prevent boiling or over heating and at the same time ensure that the bore will discharge continuously. This is achieved by a stainless steel control valve located on the discharge side of a heat exchanger. Ball and plunger type control valves are most commonly used and may be operated manually or by electrical or pneumatic controls. Automatic controls are preferable as they provide a continuous balance between supply and demand. A manual bypass bleed valve is often used in conjunction with other valves to ensure that the bore will not stop by maintaining a small continuous flow of thermal water. Sometimes the bypass valve is automatically controlled and wired through a relay to the main control valve so that they operate as master and slave. That is, one valve opens as the other closes and vice versa.

Disposal of Thermal Waters

When geothermal water leaves an exchanger, it is necessary to dispose of the water and uncondensable gases in such a way that they do not cause a nuisance. The discharged water should be passed to a flash tank to allow venting of gases and vapour at high level and the cooled water reinjected into the ground by means of a soak bore. As mentioned by (Burrows), the leaving cooled fluid must be reinjected at a depth sufficiently deep that gravity forces may overcome any resistance to horizontal or vertical disposition. If there is insufficient depth at the point of disposal the hydrostatic head may force the fluid upward to the surface. Furthermore any back pressure is undesirable in the primary heating system.

Not withstanding these comments, Rotorua is a major example for the process of geothermal reinjection with a proven history behind it. This local technique, pioneered in Rotorua, is considered as one of the best methods of disposing of thermal fluid with less damage to the environment.

UTILIZATION

Using the methods described a wide variety of geothermal applications, in diverse industries, is possible. Maximum utilization of geothermal heat is achieved where there is a high temperature source, so that water initially used to produce steam for generation of electricity, may be subsequently used for various other applications. Thus discharged effluent from one industry would be suitable for another down the thermal scale.

Some typical applications are given below:

Water temperature	Use
Above 200°C	Steam for generation of electricity.
180°C	Distillation - paper manufacture.
170°C	Production of heavy water.
160°C	Kiln drying.
150°C	Refrigeration - air cooling.
140°C	Canning industry - drying of crops.
130°C	Sugar refining.
120°C	District heating on large scale.
110°C	Drying cement.
100°C	Cooking.
90°C	Space heating.
60°C	Outdoor agriculture - soil heating.
40°C	Spas - tropical gardens.
30°C	Enclosed agriculture - domestic water heating - fish breeding.

At present in Rotorua most use of geothermal heat is for single purpose application, although these applications vary. Geothermal water is being utilized to generate low pressure clean steam for hospital services and it is providing heating and cooling for a large tourist hotel. Likewise there are many applications of this heat being used for mineral baths, horticulture and soil sterilisation, kiln drying of timber and other uses. But the dominant use of geothermal heat is for the heating of homes and city buildings.

In harnessing geothermal heat for Rotorua the general plan can be divided into five phases:

Phase I	Geothermal heating of all government buildings
Phase II	Generation of clean steam
Phase III	Refrigeration with geothermal heat
Phase IV	Group and district heating
Phase V	Total energy scheme with geothermal heat.

Phase I

Geothermal heating of all government buildings has generally been completed, with the exception of those buildings,

GEOTHERMAL ENERGY AND HYDROPOWER

largely schools, which are outside of the hot water region. The list of buildings at present being heated is given in Table 1. By 1978 the heat load will amount to nearly 24 MW.

Phase II

The generation of clean steam was initiated in 1963, when two old multi-tubular coal fired boilers at Queen Elizabeth Hospital, Rotorua, were replaced with a fully automated geothermal steam generator. This unit uses thermal water drawn from bores in the hospital grounds, at a working pressure of 5.14 bars (60 p.s.i.g.) to produce clean steam at a rate of 454 kg (1000 lb) per hour at a pressure of 2.93 bar (28 p.s.i.g.). Steam from this unit is used for hospital sterilizers, steam ovens, jacketed pans and other cooking equipment in the kitchen, while waste geothermal water from the steam generator is discharged into a large concrete holding vat, where it is cooled before being pumped to thermal baths in the hydrotherapy wing of the hospital, for the treatment of patients suffering from rheumatic diseases. A second geothermal steam generator was installed in 1968. They have given excellent service to date without any major trouble. Other steam generators using geothermal water have been installed in Rotorua and very much larger units are in operation at Kawerau.

Phase III

Refrigeration with geothermal heat was first introduced in 1966 as part of an air conditioning system for a government tourist hotel. This has been described by (Reynolds). Geothermal bore water at a temperature of 143°C passes through heat exchangers to supply secondary hot water at a temperature of 117°C to activate the generator of a 130 ton lithium bromide absorption refrigeration unit. Chilled water from this unit at a temperature of 6.7°C, is piped to the cooling coils for the air conditioning system. Geothermal water from the heat exchangers is passed to the hotel swimming pool before being discharged to waste.

Phase IV

There are no large district heating schemes in Rotorua. However, seven Government buildings in the city center are linked to a central geothermal heating plant, so forming a group heating scheme on a modest scale.

When the Government Departmental Building was erected in 1962 it was decided to use geothermal heat and to oversize

the heat exchangers so that subsequent buildings could use
this heat source. Today the total heat load of the district
heating scheme is 1011 kW, incorporating the following
buildings:

Building	Floor Area	Heat Load		
Departmental building	4000 m^2	439.61 kW	(1.50 M Btu/hr)	
Maori Land Court	1700 "	140.69 "	(0.48 M Btu/hr)	
Main Court House	1500 "	123.09 "	(0.42 M Btu/hr)	
Probation Office	316 "	23.44 "	(0.08 M Btu/hr)	
Labour Department	624 "	70.33 "	(0.24 M Btu/hr)	
Government Life Insurance	1450 "	190.49 "	(0.65 M Btu/hr)	
Tourist Department	260 "	23.44 "	(0.08 M Btu/hr)	
	9850 m^2	1011.07 kW	3.45 M Btu/hr	

The primary heating system is duplicated. There are two
heat exchangers, each served from a separate producing bore
and discharging effluent into separate soak bores. Thus
they can be run individually or in tandem to suit the demand.

A standby bore is available, also an oil fired boiler can
boost the system in the event of exceptional peak loads.
This boiler is rarely used. A general arrangement of the
system is given in Fig. 2.

Thermal water is taken to the heat exchangers in 100 mm
diameter supply pipes, over a distance of 80 m. These pipes
are encased in rigid high temperature polyurethane foam,
having an outer casing of P.V.C. The pipes are laid in
concrete ducts.

The heat exchangers are large shell and tube type, made from
mild steel, having two passes. Controls comprise a temperature controller with immersion detector, which modulates a
motorised MH stainless steel valve to maintain a flow
temperature of 82°C to the secondary system. The thermal
water enters each heat exchanger at 104°C and discharges at
76°C through 125 mm insulated pipe into an open vented flash
tank and allowed to drain into a soak bore.

From the heat exchangers in the central plant room a L.T.H.W.
heating system pipework is site reticulated to seven buildings
in the Government Center. The Secondary water serving these
buildings is carried in 100 mm flow and return pipes laid in
concrete ducts. These pipes also have rigid polyurethane

foam insulation with outer P.V.C. casing. The design temperatures are 82°C flow and 71°C return at 1.7°C outdoor temperature. Pumps in the central plant room circulate the water to the headers in the various buildings where there are individual heating controls.

Since the heating system is running continuously during the winter months the buildings retain considerable heat, consequently the system can tolerate substantial lowering of the secondary flow and return temperatures. Domestic hot water services for the main Department building is supplied through H.W. calorifiers heated with geothermal water.

Technical Data

Total heat load	1 MW	3.45 m Btu/hr
Producing Bores	2/100 mm/64 m	depth/118°C
Pressure	200 kPa	14 psig
Output	10 litres/sec	130 g.p.h.
Soak Bores	2/100 mm/85 m	depth
Heat Exchangers	2/Shell and tube/two pass	
Diameter	1.5 m	5 ft
Length	2.75 m	9 ft
No. of tubes	110/50 mm	
Primary T.D.	27.8°C	50°F
Secondary T.D.	11.1°C	20°F
Primary Control	s/s Satchwell MH valves - auto	
Secondary Control	Satchwell controllers 3-way valves	
All site		
reticulation	100 mm Insapipe	

Using the principle of group heating, as at the Government Center, it would be possible to heat the whole of the central city area. There are just over 100 separate blocks of buildings within the hot water region - excluding those used for government buildings. Allowing half a mega watt heating load for each city block, the heating requirements amount to 50 MW.

At present it is estimated that only one-third of this is being utilized, including district heating of residences in Fenton Park.

Outside of the hot water region lies some light industry with an anticipated load of 20 MW. To service these areas it would be a practical proposition to provide district heating by reticulating thermal water over varying distances up to 1.25 km. The transmission of geothermal water over this distance has already been carried out in heating the Rotorua Boys' High School, as mentioned by (Kerr, Furness et al).

Summarising the present loads:

Government buildings	24 MW
Other city buildings	16 MW
Other applications	8 MW
	48 MW

Future heating loads:

Group heating of city buildings	24 MW
District heating of industry	20 MW
	44 MW

Whilst the heating of buildings is important, one must not neglect the other applications to which thermal heat is put. (Linton) states that it is in the field of organised and professional horticulture that the greatest use is made of geothermal heat - and the range is very considerable, from orchids to mushrooms. Indeed, one of the most enterprising users of geothermal heat was an individual who grew some 200 pineapples each year in his own backyard - while another grew bananas.

Phase V

At present no total energy system exists in Rotorua, since a preliminary requirement is water temperature in excess of 200°C to produce steam for electrical generating units.

This is most likely to take place at the Forest Research Institute, which is located at Whakarewarewa, where the highest water temperatures are recorded. The Institute already uses substantial quantities of geothermal heat for heating, hot water, kiln drying of tree cones and generation of steam for laboratories. Deep well drilling is to commence, and if this proves fruitful, it is envisaged that the following multi-purpose use of geothermal heat will be made at the establishment, which will then be self-supporting on geothermal energy.

Production bore 200°C

Generation of electricity (1 MW)

Clean steam generation (150 kW)*

Kiln drying*

Absorption Refrigeration for chilled water circuits and air-conditioning (250 ton)

Space heating of Institute* and hot water supply (5 MW)

Group heating of residences and primary school (200 kW)*

Heating of glasshouses

Heating of fish ponds

Re-injection 30°C

Items marked with an asterisk are already in existence and the last two items are due to be commenced in 1978.

FACTORS INFLUENCING GEOTHERMAL SYSTEMS

The two main factors influencing geothermal systems are mineral deposits in pipes and corrosion.

It has been observed that one bore may require frequent cleaning during production, while another bore within 30 m may have run for 15 to 20 years with no signs of diminishing production. The usual mineral deposits are calcite and silica. Calcite tends to deposit inside the bore casing or near the well head and this can gradually restrict bore discharge. Separators at the well head are of some assistance. Silica on the other hand, tends to remain in the water until the water is discharged at atmospheric pressure. It can form at leaking glands or joints.

Hydrogen sulphide, carbon dioxide, ammonia, chlorides, steam wetness and the presence of air in pipes are the most significant factors influencing corrosion. However, no trouble has been experienced with corrosion within pipe

lines and vessels while the system is under pressure.
External corrosion, mainly due to hydrogen sulphide, is a
problem, as high concentrations of the gas can damage
electrical equipment and certain metals. Generally non-
ferrous metals are to be avoided. Fortunately mild steel
and close grained cast iron have sufficient corrosion res-
istence to give economic life, as do austenitic stainless
steels.

ECONOMIC ASPECTS

The main cost with most geothermal heating systems is the
capital investment in drilling of bore holes, pipelines for
collecting and transporting the fluids, heat exchangers and
controls, distribution piping and other associated heating
equipment. Labour and maintenance costs as a rule are
rather insignificant compared to the capital cost. Provided
the geothermal heat source is near to the utilization point
the cost of geothermal heating schemes compares favourably
with other systems using oil and coal. This is indicated
by the cost taken out for the Government District Heating
Scheme with relation to the heat raising plant.

Item	Geothermal	Coal fired plant	Oil fired plant
Capital cost/kW	$17-95	$16-9	$13-7
Annual running cost	$2625	$4415	$9411
Annual running cost /kW	0.216 cents	0.635 cents	1.527 cents

(Shannon) states that had the geothermal plant been relocated
nearer the heat source, the capital cost would have been
comparable to that for an oil fired system.

As it is, the break even point for geothermal/oil is 1.35
years and 3.35 years for coal. The total cost of the
geothermal equipment ($21,000) is amortized in 3.6 years
in comparison with oil. This shows the importance of
converting existing oil fired systems to geothermal.

At the Forest Research Institute it is estimated that over
$100,000 per annum is saved by using geothermal energy for
heating, as against using imported fuel oil. At the Rotorua
Public Hospital the saving is about $250,000 per annum and

will rise to nearly $750,000 per annum when additional geothermal systems are introduced next year. These figures vary with the load factors. If coal is used as an alternative fuel the savings would be about half of the costs given - but still a very substantial amount.

With reference to geothermal energy for residences, it is no longer a viable proposition to sink production and soak bores for single homes, as the cost would be in the vicinity of $4,000 to $6,000. Hence it is the practice to link a number of houses together from a common heat source and charge an annual rental, which works out to be competitive with traditional systems.

Sometimes it is possible to reduce the capital investment by compromising on lower efficiency of energy utilization, and this is quite often fully justified.

The most effective measure for obtaining minimum energy costs is to increase the annual load factor for the production plant of the geothermal installation as at the Government Center in Rotorua, where more buildings are being brought into the group heating scheme.

At present the Rotorua City Council make no charge for geothermal energy. However, they propose to introduce a fee for annual inspection and licence, which for most establishments, will not exceed $10 per annum. Likewise, government legislation is to amend the charges laid down in the Geothermal Energy Act. This Act relates to all energy derived from the earth's natural heat-excluding water at temperatures up to $70^{\circ}C$. It is known that there will be no charge for geothermal heat up to 1 tera joule per annum. Rates beyond this amount have yet to be determined, but indications are that it will not be excessive, so as to encourage the use of geothermal energy.

CONCLUSION

In a time of escalating energy costs our survival depends on the realistic utilization of our natural assets. Geothermal heat is an indigenous wealth of considerable potential, and when harnessed has been shown to be an economic proposition resulting in substantial savings of imported fuel oil.

The favourable costs in using terrestial heat can be even improved by controlling waste heat and introducing new techniques to reduce maintenance expenses.

Moreover, the use of geothermal heat does not contaminate our environment to the same degree as fossil fuels.

As indicated, further development of this resource is possible with multi purpose applications and with group and district heating in Rotorua and other suitable areas.

If we are to be a source of influence in the world we need to see to it, as far as in us lies, that we are not found wanting in meeting the challenge of our day and generation.

ACKNOWLEDGEMENTS

The writer acknowledges the assistance given by colleagues and staff of the Ministry of Works and Development and wishes to thank the New Zealand Geological Survey staff for their assistance and the Commissioner of Works for permission to present this paper.

REFERENCES

1) Nairn I.A., 1974 Geothermal Resources Report of N.Z.G.S. (Part D)

2) Burrows W., 1974 Utilization with reinjection. Trans. of International Society for Geothermal Engineering.

3) Reynolds G., 1970 Cooling with Geothermal Heat. U.N. Symposium on the Development and Utilization of Geothermal Resources, Pisa.

4) Kerr, Furness et al., 1961 Recent developments in New Zealand in the utilization of geothermal energy for heating purposes. U.N. Conference on New Sources of Energy, Rome.

5) Linton A.M., 1974 Innovative geothermal uses in agriculture. International Conference on Multi-purpose use of geothermal energy. Oregan Institute of Technology, Klamath Falls, Oregon.

6) Shannon R.J., 1975   Geothermal heating of government
                          buildings in Rotorua, New
                          Zealand. U.N. Symposium on
                          Geothermal Energy, San Francisco.

TABLE 1

GOVERNMENT BUILDINGS USING GEOTHERMAL HEAT

Building	Load in millions of Btu/hr
Government Center Buildings	3.45
Post Office	1.20
Telephone Exchange	0.54
M.W.D. Workshop with extensions	2.05
Forest Research Institute (stages 1, 2 & 3)	13.70
Forest Research Institute (stage 4)	3.00
Forest Training Center Accommodation	2.30
Polynesian Pools (former Ward Baths)	2.00
Blue Baths	3.00
Rotorua Public Hospital	12.20 *
Queen Elizabeth Hospital	4.34
Nurses Home	1.00
N.Z.R. Building	1.00
Broadcast Building	0.55
Waipa Hostel	1.20
Rotorua Community College	2.00
Rotorua Boys' High School	2.00
Primary Schools (Rotorua and Glenholme)	0.80
Rotorua International Hotel	2.00
Police Station	0.30
	58.63 M. Btu/hr

* In 1978 a further 22.4 M. Btu.hr load will be harnessed for Rotorua Public Hospital 22.40

81.03 M. Btu/hr

Total load by 1978   23.75 MW

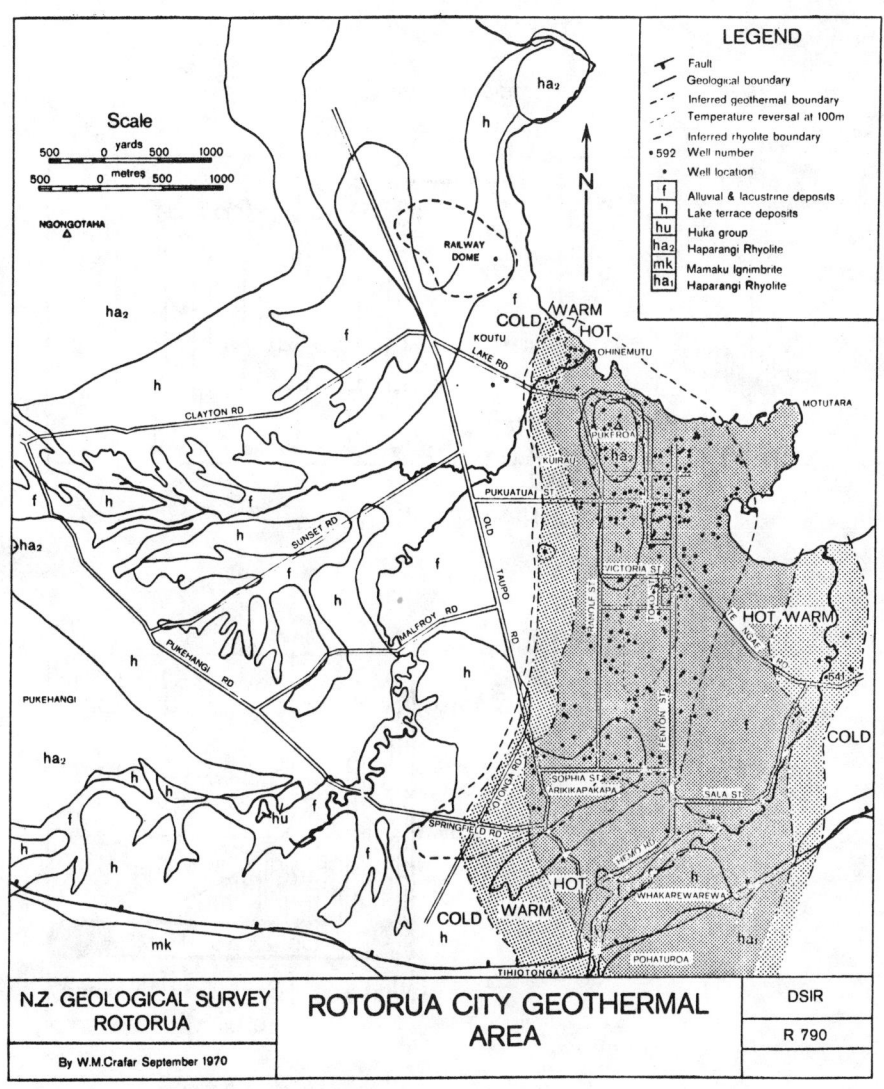

Fig. 1 Rotorua City Geothermal Area

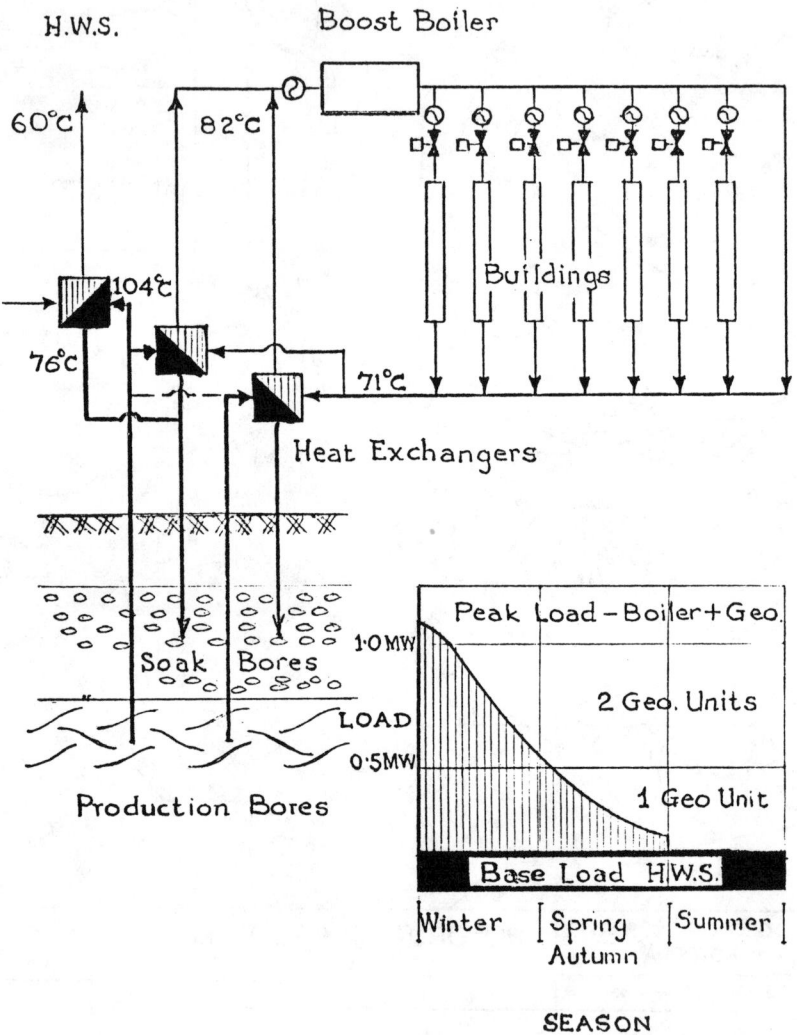

Fig. 2 Geothermal Heating Government Center

Rotorua International Hotel, Rotorua, New Zealand
(World's First Geothermally Air Conditioned Hotel)

Geothermal Systems in the Hauraki Rift Zone (New Zealand)— An Example for Geothermal Systems Over an Inferred Upper Mantle Swell

M. P. HOCHSTEIN
Geology Department, University of Auckland
New Zealand

ABSTRACT

Numerous hot springs occur over the central part and the outer flanks of the Hauraki Rift. Geophysical studies have shown that the rift valley is about 25 km wide and at least 220 km long. The energy discharged by individual thermal areas is rather small and of the order of 10×10^6 J/s. The areal extent of the near-surface reservoirs is quite large beneath the median springs and exceeds 6 km^2 in at least one case. The mineralization of the springs is caused in part by leaching of hot basement rocks; Na-K-Ca equilibrium temperatures of greater than 200°C are common. It is likely that an anomalous heat flux, about twice the normal flux, is the source of all the heat discharged over the whole rift zone. This flux is probably caused by a cushion of anomalously hot upper mantle rocks beneath the crust.

INTRODUCTION

Numerous hot and warm springs can be found in the northern part of the North Island of New Zealand. The springs occur over the flanks and the central active strip of the Hauraki Rift which has been detected only recently [1]. The geological structure of the rift is similar to that of other well-studied continental rifts, like the Rhine Graben [2] and the Lake Baikal depression [3]. The global tectonic setting of the Hauraki Rift, however, differs from that of continental rifts. The southern part of the rift overlies an inferred active subduction zone which is part of the Indian-Pacific plate boundary; the northwest dipping subducted Pacific plate lies about 200 km beneath the southern end of the Hauraki Rift. It is possible that the Rift belongs to the group of 'back arc rifts' or back arc basins as defined by Karig [4].

Continental rifts and back arc rifts are the result of extensional forces within the crust which also cause widespread shallow earthquakes (focal depths < 20 km). Rifting is accompanied by some arching of the crust which in turn is the result of secular upwelling of hot deeper Upper Mantle material [5]. These rocks form a large cushion which also extends beneath the flanks of the rift. Basalts extruded over the flanks of the rifts indicate that sporadic melting occurs within these anomalously hot rocks. Andesitic volcanics together with basalts can often be found over back arc rifts; these rocks occur also to either side of the Hauraki Rift (see Fig.1), whereas basalts are dominant over continental rifts [2].

Because of the hot Upper Mantle rocks (T about $1000^{\circ}C$) beneath the crust, all active continental rifts and back arc rifts are regions with high heat flux. The flux is usually not much greater than twice the normal flux [6, 7]. This anomalous flux heats up deeply infiltrating meteoric waters; some of the heat is transported to the surface by convective cells. Discharge of thermal waters in the form of numerous thermal springs can therefore be found not only over the central part of these rifts but also over the flanks. The heat stored in the rocks beneath these rifts constitutes a large energy reservoir. The potential of these resources has not been fully recognized. It is the purpose of this paper to draw attention to these resources using the Hauraki Rift as an example.

DESCRIPTION OF THE HAURAKI RIFT ZONE

The Hauraki Rift is at least 220 km long; the southern extension of the rift is concealed by a sheet of Pleistocene ignimbrites, the northern extension of the rift is still unknown. The width of the central rift valley is about 25 to 30 km; the rift might be even wider if the Coromandel Depression, which lies to the east (see Fig.1), is also part of the rift zone. The structure of the Hauraki Rift is known from geophysical studies along numerous east-west trending profiles, some of which are shown in Fig.1. Only the southern part of the rift valley is above sea-level, the central and the northern part are concealed by the ocean.

The structure of the southern part has been described by Hochstein and Nixon [1]. A section of the rift based on the interpretation of gravity data along the southernmost profile D is shown in Fig. 2. It can be seen that the Hauraki Rift is made up from west to east by a fault-angle depression, a median horst, and a graben. Quaternary and Tertiary sediments which fill the two depressions reach a maximum thickness of about 3 kilometres. Horst and graben are bounded by steeply dipping normal faults. Almost half of all thermal springs shown in Fig. 1 occur in the southern part of the rift which is also the most active region as indicated, for example, by median active faults within the rift and by a bunching of earthquake foci with focal depths of less than 12 km. The thermal springs in the rift valley lie either over the boundary faults or the median basement horst.

The structure of the central part of the Hauraki Rift is rather similar to that of the southern part. Measurements across the Hauraki Gulf indicate that the median horst widens and that the depth of the graben decreases to the north of profile B B' (Fig. 1); the graben can hardly be recognized north of profile A. The fault-angle depression is broader; the thickness of the sediments within the depression also deceases to the north of profile B. The northern extension of the depression has not been traced yet. The seismicity over the central part of the rift is significantly smaller than that in the south although a few earthquake swarms have been observed [8] which seem to be connected with some tectonic activity beneath the Coromandel depression. Similar swarms also occur in the southern part of the rift; swarm activity has been reported for some continental rifts. Minor thermal activity has been observed over the flanks of the rift.

The structure of the northern part of the rift, i.e. that part lying to the north of profile A in Fig.1, is hardly known. It is possible that the uplifted basement along the northern extension of the rift axis is the result of crustal arching although no significant rifting has taken place. Some tectonic activity is indicated by crustal earthquakes; a few earthquakes have a magnitude similar to that of the largest quakes (magnitude 5) in the sourthern part of the rift. The most impressive thermal activity of the whole rift occurs around Ngawha (locality 1 in Fig. 1) where temperatures of $236°C$ have been observed at the bottom of a 600 m deep well [9].

THERMAL ACTIVITY OVER THE HAURAKI RIFT

All thermal springs occur at low altitudes, most springs in the rift valley lie near sealevel. The visible discharge is rather small and heat discharge rates rarely exceed $2 \times 10^6 J/s$. The total discharge rate of some thermal areas in the rift valley might be greater since subsurface seepage has been observed.

Resistivity measurements over the Kerepehi Springs (Feature 9 in Fig.1) have shown that thermal fluids occur over an area of at least 5 km^2 (Fig.3). Since the greywacke basement occurs at a depth of only 200 m, the subsurface extent of the hot mineralized water at Kerepehi cannot be explained in terms of discharge of hot water along a single fissure. It is likely that the hot fluids ascend within the basement in the form of a large column of hot water; i.e. the Kerepehi Springs, despite their low discharge rates, are the surface manifestations of a hot water system.

Similar hot water systems occur in the southern rift valley at Lower Waitoa and at Manawaru where warm water is discharged over an area of greater than 2 km^2. It has been estimated that the total heat discharged by any of these thermal systems is between $\overline{10}$ and 20×10^6 J/s. The discharge rate at Ngawha is greater, namely about 30×10^6 J/s [9]. These natural losses are still about one order of magnitude lower than those of the active thermal systems in the Taupo Volcanic Zone of New Zealand where discharge rates between 100 and 600×10^6 J/s have been measured. Most of the other thermal springs appear to be isolated springs, except for some wider spread discharges at Lake Waikare and at Great Barrier Island. The heat discharged by each of these thermal areas is less than 10×10^6 J/s.

The thermal activity and the surface manifestations of thermal discharge features over the greater Hauraki Rift Zone are therefore not impressive. This also applies for thermal discharges over other continental rifts, i.e. Rhine Graben and Lake Baikal depression, using the description given in Waring [10] for comparison. If one remembers that in each case the heat is supplied by hot Upper Mantle rocks at the bottom of the crust, the moderate thermal activity over these rifts can be understood. A more intense discharge of heat occurs in parts of the East African rifts where Upper Mantle melts have apparently intruded the crust beneath the rift valley. Such intrusions, however, are absent beneath the southern part of the Hauraki Rift as indicated by the rather undisturbed magnetic anomalies over the rift valley [1].

CHEMISTRY OF THE SPRINGS

Information about the chemistry of some of the springs shown in Fig.1 can be found in the report by Petty [11]. In this paper only the chemistry of the springs from the southern part of the rift is discussed in some detail since here a complete set of data of springs over each flank and the central rift valley is available. The chemistry of these springs is summarized in Table 1; all analyses listed are based on recent studies by the Geology Department, University of Auckland.

The mineralization of all thermal waters discharged over the rift is rather low (total solids usually less than 1 g/kg). Only the springs at Ngawha, Great Barrier Island, Hot Water Beach and at Te Aroha discharge waters with higher mineralization (total solids greater than 3 g/kg). Three of these springs occur at the foot hills of rather thick volcanics and it is likely that the higher mineralization is caused in part by leaching of the deeper section of these rocks. Most of the other springs occur over higher standing basement rocks (greywacke) which are either covered by a thin layer of recent sediments or outcrop nearby; springs 4, 5 and 11 are an exception since thicker Tertiary sediments can be found at these localities.

The springs discharge neutral sodium bicarbonate waters. All constituents in the springs are enriched in comparison with those found in non-thermal groundwater of the region. The isotopic

composition of the thermal springs indicates that all thermal
waters are heated-up groundwater. Table 1 shows that the highest underground temperatures occur beneath the median springs in
the rift valley. High temperatures at depth are indicated by the
high SiO_2 concentrations and the high Na-K-Ca equilibrium temperatures [12]. All thermal waters are diluted by heated-up groundwater. Using realistic values for the $\delta^{18}O$ composition of the
basement rocks, it was found that the ratio between diluting waters and undiluted fluids might be as large as 10 : 1 in the case
of the median springs (P. Blattner, pers.comm.). The ratio is
probably greater for the springs on the flanks. The Na-K-Ca equilibrium temperature, however, is not much effected by such dilution and it can be inferred that minimum temperatures between
220 to 250°C occur at not too great depths beneath the rift valley; minimum temperatures between 150 and 215°C occur elsewhere
beneath the other thermal springs (see Fig.1).

HYDROLOGICAL SETTING OF THE THERMAL SYSTEMS BENEATH THE RIFT
VALLEY AND TEMPERATURE STRUCTURE OF THE RIFT

An estimate of the approximate depth at which these high temperatures occur can be obtained for the rift valley springs in the
southern part of the rift. The rift valley constitutes a large
hydrological basin and significant horizontal groundwater movement occurs within the Quaternary sediments of this basin which
dilute any thermal fluids which rise to a level of about - 300 m
beneath the fault angle depression and the graben (Fig. 2).
Thermal fluids which ascend within the median greywacke ridge are
less affected; a similar protection against dilution is given by
a high standing greywacke ridge in the Lake Waikare-Ohinewai Area
(feature 10 in Fig. 1). It has been inferred by Hochstein and
Nixon [1] that the thermal fluids discharged over the median
ridge in the southern part of the rift valley come from a depth
of at least 3 km but no more than 5 to 6 km. If most of the leaching of the basement minerals takes place at these depths, a
temperature gradient of 70 ± 30°C/km can be inferred for the upper crust from the Na-K-Ca equilibrium temperatures.

This is only a rough estimate but it implies that all the thermal
activity over the greater Hauraki rift zone can be explained by
an anomalous heat flux of the order of $14 \times 10^{-2} W/m^2$, which is
about twice that of the normal flux. The inferred temperature
gradient and heat flux are compatible with those observed over
other continental rifts [6, 7]. Both parameters might vary significantly between adjacent parts of the rift.

THE HAURAKI RIFT AS AN ENERGY RESERVOIR

It can be assumed that temperatures of about 250°C at depths of about 3 to 5 km occur not only beneath the rift valley but also beneath the flanks of the Hauraki Rift. It can also be assumed that the temperatures are not significantly disturbed by convective heat transfer, although numerous convective cells occur at shallower depths. Using an average thermal capacity of 2×10^6 J/m^3°C it can be inferred that the energy stored in 1 km^3 of rock at depths between 3 to 5 km is about 5×10^{17} J. The energy stored beneath the southern part of the rift valley is therefore of the order of 25×10^{20} J. Whether part of this energy can be extracted will depend on the development of suitable dry rock fracturing and heat extraction technology. Since it can be inferred that the energy stored beneath other continental rifts and back arc rifts is similar to that stored in the Hauraki Rift, these rifts constitute large energy reservoirs and are therefore targets for 'dry-rocks heat extraction' studies.

On a smaller scale, extraction of heat from the shallower systems within the rift valley appears to be feasible. It can be predicted that temperatures greater than 200°C prevail within the upper 1-2 km in the hotwater systems of the Hauraki rift valley. Such temperatures have already been observed at depths of 0.5 km at Ngawha, a hot water system lying over the northern extension of the Hauraki Rift and which should be included in the rift valley systems. Although these systems appear to have reservoirs which are similar in volume to that of the more active hotwater systems in the Taupo Volcanic zone, the small natural heat output of these systems and their small buffer capacity, given by the low porosity (.03) of the basement rocks, will restrict the energy which can be extracted from these systems.

REFERENCES

1. HOCHSTEIN, M.P.; NIXON, I.M. (in press) : Geophysical study of the Hauraki Depression North Island, New Zealand. New Zealand Journal of Geology and Geophysics.

2. ILLIES, J.H. 1969 : An intercontinental belt of the world rift system. Tectonophysics 8 : 5-29

3. FLORENSOV, S.R. 1969 : Rifts of the Baikal Mountain Region Tectonophyscis 8 :443-56

4. KARIG, D.E. 1974 : Evolution of arc systems in the western Pacific. Ann.Rev.Earth Planet, 2 :51-75

5. OSMASTON, M.F. 1969 : Genesis of ocean ridge median valleys and continental rift valleys. Tectonophysics 11 : 387-405

6. LUBIMOVA, E.A. 1969 : Heat flow patterns in Baikal and other rift zones. Tectonophysics 8 : 457-67

7. WERNER, D; DOEBL, F. 1974 : Eine Geothermische Karte des Rheingraben-Untergrundes. In : Approaches to Taphrogenesis. Inter-Union Commission on Geodynamics Scientific Report No. 8, 460 pp. Schweizerbarth, Stuttgart.

8. EIBY, G.A. 1966 : Earthquake swarms and volcanism in New Zealand. Bulletin Volcanologique 29 : 61-74

9. THOMPSON, B.N. ; KERMODE, L.O. 1965 : New Zealand Volcanology, Northland - Coromandel - Auckland. New Zealand Department of Scientific and Industrial Research Information Series Nr. 49, 103 pp.

10. WARING, G.A. 1965 : Thermal springs of the United States and other countries of the world - a summary. U.S.Geological Survey Professional

Paper 492, 383 pp.

11. PETTY, D.R. 1972 : Springs of the Auckland Region. New Zealand Geological Survey Report NZ 65 57, 56 pp.

12. FOURNIER, R.D.; TRUESDELL, A.H. 1973 : An empirical Na-K-geothermometer for natural water. Geochemica et Cosmochimica Acta 37 : 1255-75

Table:1. Chemistry of Hot Springs across the Hauraki Rift and Eastern and Western Ranges
(all constituents in mg/kg)

LOCALITY	Tmax °C	pH cold	Li	Na	K	Ca	Mg	F	B	Cl	SO₄	CO₂ (free)	SiO₂	Na-K-Ca equ T°C	distance to centre depᵒⁿ (km)
Western Foothills of Western Ranges															
Ohinewai	27	n.d.	n.d.	190	20	26	0.1	1.8	16	285	n.d.	nil	107	183	36
Lake Waikare(-3m)	≥71	8.3	n.d.	260	21	6	0.1	1.8	17	370	16	nil	110	186	30
Hauraki Rift, Western Boundary Springs															
Miranda(-61m)	56	8.4	n.d.	140	2.2	5	2	1.2	0.4	160	n.d.	nil	63	n.a.	10
Walton	31	8.1	nil	30	8	14	5	0.4	3	21	<1	120	75	n.a.	8
Hauraki Rift, Median Springs															(undiluted)	
Kerepehi	55	7.1	0.8	220	30	8	9	0.4	3.5	78	<1	280	205	212 (215-220)	2
Lower Waitoa	54	8.3	0.6	220	43	37	15	0.3	4	57	<1	2500	175	220 (225-230)	1
Manawaru	40 / 46*	8.2	0.4	135 / 163*	31* / 37*	10* / 10*	6	0.3	3	28* / 30*	<1	420	141 / 157*	237 (245-255)	1
Hauraki Rift, Other Discharges															
Ngatea (-350m)	30	8	n.d.	131	10	8	3	0.9	n.d.	200	<2	80	151	n.a.	3
Okauia(-10m)	39	7.7	0.3	100	11	24	9	0.4	3	18	<1	690	70	171	3
Hauraki Rift, Eastern Boundary Springs															
Te Aroha (-70m)	85	8.3	5	3160	40	8	4	n.d.	132	580	390	abun-dant	115	(126?)	9
Sheehan	23	7.6	nil	17	5	24	6	0.1	2	14	<1	410	68	n.a.	7
Waiteariki	34	6.7	nil	64	17	69	58	0.2	2	21	<1	3450	120	204	7
Eastern Foothills of Eastern Ranges															
Hahei	28	6	n.d.	36	9	n.d.	n.d.	n.d.	n.d.	51	10	45	56	n.a.	40
Hotwater Beach(-18m)	55	7.5	8.7	1250	70	136	4	1.8	n.d.	1985	4	40	76	168	40

* corrected for dilution

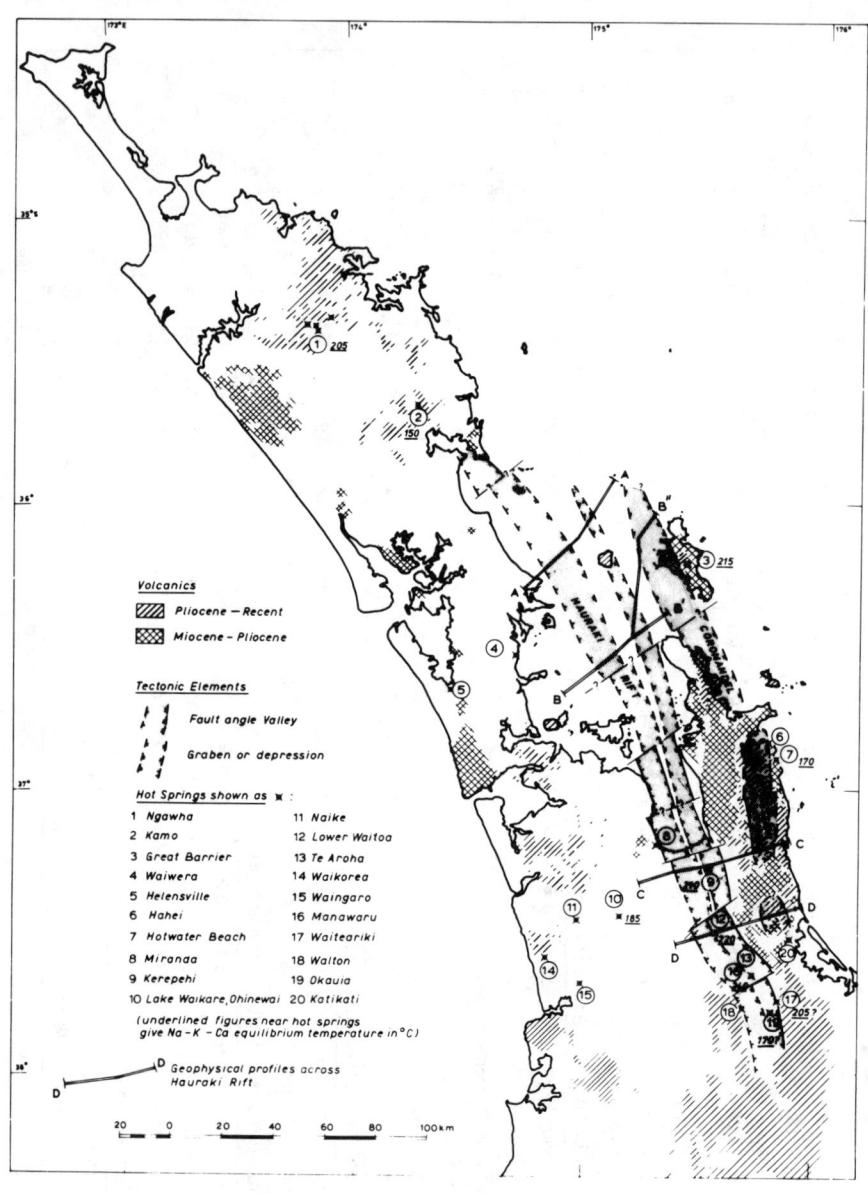

Fig. 1 Map showing the locality of hot springs over the Hauraki Rift, New Zealand.

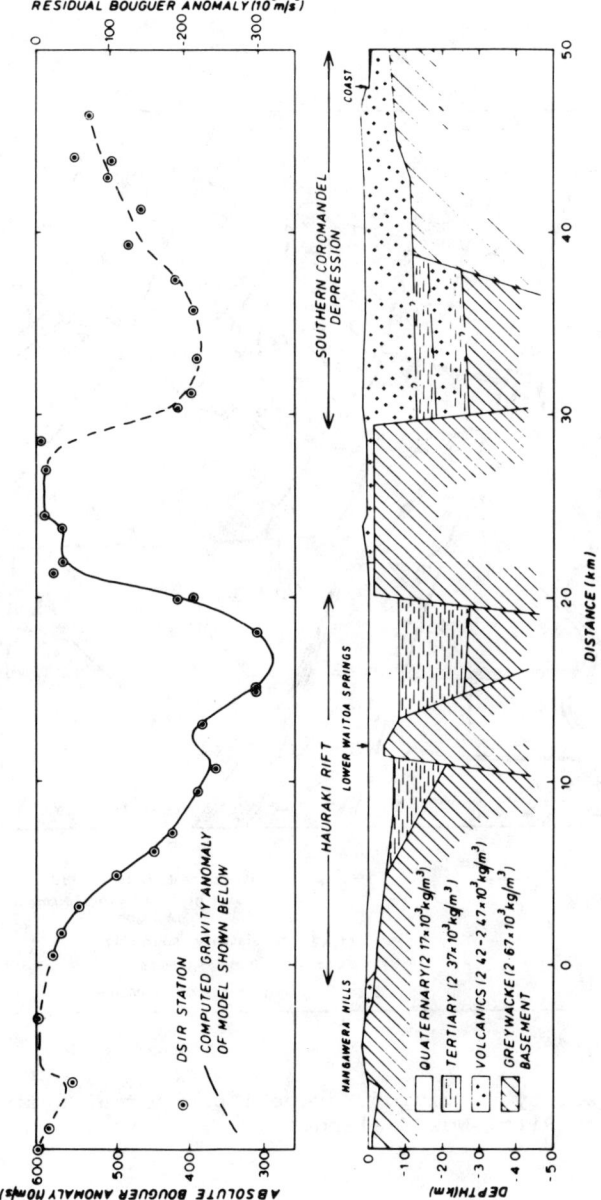

Fig. 2 Section across the southern part of the Hauraki Rift along profile D, for locality of profile, see Fig. 1.

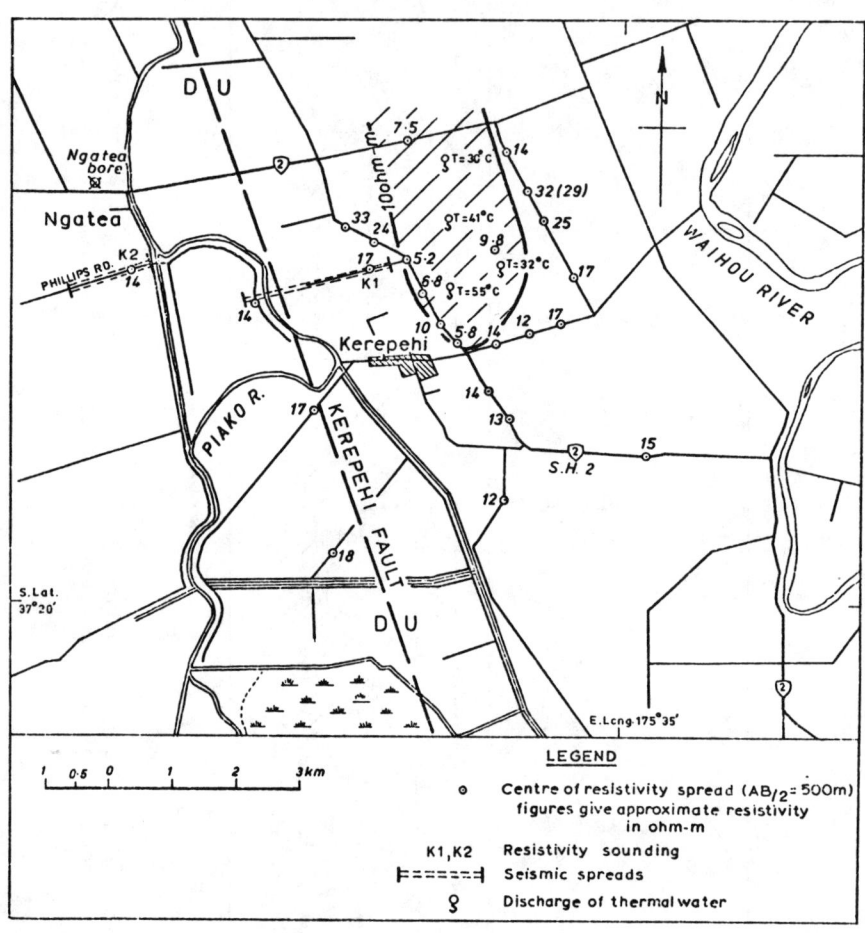

Fig. 3 Resistivity survey of Kerepehi geothermal system, Hauraki Rift, New Zealand.

The Puchuldiza Geothermal Field

PATRICIO R. TRUJILLO
Comité Geotérmico-CORFO
Santiago, Chile

ABSTRACT

The Puchuldiza Geothermal Field is located in the high Andes range at about 150 kms towards de NE of Iquique port. It his in a volcano-tectonic depression surrounded by Plio-Pleistocene andesitic volcanoes. As a basament it has a continental sequence argillaceous and tuffaceous sandstones with conglomerate intercalations, have tentatively been asigned to lower to Middle Cretaceous that may compel the acuifer at depths. Geoelectrict studies have found a resistivity anomaly with values under 10 Ohm. m. in a surface of about 30 km2, running along the main structures.

Geochemical studies of the hot springs, allow us to stablish a mininum temperature between 200-205°C for the Geothermal fluid at depth. The principal producing aquifer actually tapped has a temperature of 174°C measured with an amerada and it is located between 640 and 800 m the geochemical analysis of the well fluids indicate a temperature of 230°C situation that is suggesting that we are tapping an aquifer that has a miture with colder fluids.

Geothermal exploration in Puchuldiza began at the middle in 1969 and was interrupted at the end of 1972 Surveys included geological, geophysical and geochemical studies as well as an account of hot springs and thermal survey by means af 30 m. gradient holes. The studies were reinitiated in October 1974, the surface geology, geophysics and geochemestry were completed by the middle of 1975. Work that allow us to select five drilling sites, actually finished.

INTRODUCTION

The Puchuldiza Geothermal fields located 100 kms East of the Pisagua port in the high Andes Ranges near latitude 19° 22', longitude 68° 58' at on elevation of 4300 above sea level.

Fig 1 Methodical exploration began at El Tatio early in 1969 as a result of on agreement between the United Nations Development Program and the Chilean Goverment. Geological, geophysical and geochemical studies were carried out through this agreement until 1972, when suspended. The geothermal investigations were reinitiated in 1975 by CORFO whose operations are financed by the I Region Goverment and included five exploratory wells with depths between 428 and 1013 m from 1976 to present.

At present in November 1977 well N° 5 was completed to 1013, and late this month well N° 4 will be opened; well production studies will be completed by June 1978, All geothermal information available at that date will be evaluated by a foreign consulting to decide on investments and development programs in a very near future.

STRATIGRAPHY AND STRUCTURE

Geology

The oldest rocks that crop out in the geothermal area of Pu - chuldiza correspond to the Churicollo Formation, a continental sequence of argillacous and tuffaceous green greyish sandstone with conglomerate intercalations. These deposits, of fluvial and lacustrine character, have tentatively assigned to the lower to Middle Cretaceous. The generally argillaceous matrix of the sediments and the high degree of compaction make this formation practically impermeable. This formation was affec = ted by compressive movements referred to the Paleocine Laramide orogenesis. Periods of extenssion followed the Laramide phase and also the late Eocene-early Oligocene phase (1). This extenssion produced north - south normal faults which defined tectonic depressions. These tectonic landscape were filled by thick volcanic deposits and fluvial lacustrine sediments.

The oldest known ash flows of Puchuldiza area the Ulayane Ig - nimbrite, which unconformably overlies the Mesozoic units. This formation consists of 160 m. of densely welded and partialty vitreous ryolitic tuffs. This formation appears to be strongly fractured and leached and might constitute a good aquifer at depth. It is overlayed by the Chojña-Chaya Formation, 60-80 m. of well stratified green sandstones with a few thin intercalations of fine conglomerates. (Fig. 2).

The Chojña-Chaya formation is overlain by the Condoriri Formation, which consists of dacitic ignimbrite and pumice tuffs with some augite-oliviene basaltic andesites on top. This formation seems to interfinger towards the west with the Puchuldiza formation, which in its lower part consists of augite bearing andesites and in its upper part of andesisic ignimbrites with fiamme. An erosional disconformity separates the Puchuldiza formation fron the overlying Lupe Formation(2).

The Lupe is formed by sandstones and conglomerates of fluvial characteristics in its lower part. The upper part consists of pumice tuff breccias and pumice lapilly piroclastic sediments. (table 1)

The late Cenozoic formation consists manly of andesite, dacite and ryodacitic lava flows that builded the actual Strata volcanoes.

Structures

Cenozoic formations are gently folded, structures are oriented N 30 - 50 W. The folding phase has been asigned to the middle to late Miocene compressive phase. After this compression more tectonic extension resulted in a nearly north-south system associated with north-est-southeast and almost east-west fault system. Major vertical displacements of blocks were produced along the north-south fault system. The Puchuldiza geothermal area is located in a block which has sunk 500 m. with respect to the horst of the Cerro Condoriri. These north-south faults have been out by the wrench faults of the northwest-southeast and east west system to which the Quaternary volcanic centers of the area are related.

A Pleistocene-Holocene fracture system characterized by northeast-southwest faults has affected the Plio-Pleistocene volcanoes and produced in part the surficial hydrothermal activity in Puchuldiza.

Geophysics

Geophysical exploration mainly resistivity, and resistivity soundings were undertaken in order to evaluate the characteristics of the reservoir in the subsurface. The low resistivity values of the rocks with hot salty water have been used to determine the limits of the field and to obtain the pattern of the movements of the geothermal fluids in Puchuldiza

(Fig.3)
With geoelectrical prospections using the Schlumberger array with AB/2= 250,500 and 1.000 was used with DC commuted every 10 seconds, unable us to determine a resistivity anomaly of about 30 km2, with 10 Ohm. m. has been found in the Puchuldiza graben, and within it. The east and west limits of the anomaly abovementioned concour totally with the N-S normal faults of Puchuldiza and Tuja respectively.

The isoresistivity contourlines towards North are stretching out by the side of Natividad mountain. These is due to the topografhic efects and high resistivity values that the Piocene Pleistocene lava flows have. The southern part of the anomaly defined by the 10 Ohm.m. contour line remains open; fact that determines, that the Puchuldiza fault does not establish an effective barrier for the geothermal fluids.

A dipole survey carried out by Risk et al in 1970 found anomaly with values under 7 Ohm.m located in Tahipicollo hill with aproximately N-S direction that has the shape of 1 km width trench, this low resistivity body is inside the 10 Ohm. m contour line determined with the AB/2= 1.000 Schlumberger array.

The geological setting of the zone, the temperature gradients measured in this area, and the geochemical data available allow us define, that this low resistivity anomaly detected in the Puchuldiza field corresponds to high temperature saline water reservoirs.

Geochemestry

Geothermal geochemical analyses were undertaken by (4). The hot water discharged by the thermal springs corresponds to an aproximately neutral solution (mean PH 7.3) with the follo - wing main components: Na Cl, K Cl, Ca Cl2, B and SiO2. Its also presents some concentrations of Li, Rb and Cs.

The SiO2 content and Na/K ratio gives us a minimun undergrond temperature between 200 and 205°C, in the Tatio field using these same indicators we determined minimun temperature of 190°C (5). A significant temperature (260°C) was later tapped by drilling in Tatio, this situation is considered as indicative of the waters, that may be also ocurring in Pu - chuldiza.

Recent geochemical analysis of the wells 1,2 and 3 waters are indicating temperatures of 200°C considering the SiO_2 content and 230°C using Na-Ca-K as geothermometers. Temperatures measured in the wells with an amerada thermometer are lower 178°C the highest in well 2. This is indicating that the fluid tapped with the exploratory wells corresponds to a diluted geothermal fluid.

These temperature determinations are in accordance also with the contents of Ca, Mg and other elements whose solubilities and concentrations in the fluids, are related with the temperature of these.

HIDROTHERMAL ACTIVITY

The Puchuldiza thermal springs emerge in two main areas an eastern area inclunding Puchuldiza (s.s) and Tuja, located 3 Km. west of Puchuldiza. Both areas are covered with white sinter and evaporites precipitated from the hot spring. Most of the thermal manifestations are flowing springs and clear hot water pools, also some superheated fumaroles and a few geysers are present. The temperature of the hot spring commonly exeed the 85, 5°C, the boiling point for 4.300 m.a.s.l.

Most of the springs emerge at low levels along the Puchuldiza river and its small tributaries. The hot springs reach the surface through fissures mainly north east-southwest orientation and some north-south. Yellowish white hidrothermal altered rocks extend along the traces of these faults.

REFERENCES

1. Charrier Vicente 1970 Liminary and geosinclinal Andes major orogenic phases and synchronical evolutions of the Central Andes: Buenos Aires Solid Earth Problems Conf. p. 451-470.
2. Lahsen A. 1975 Evaluacion del Sistema Geotérmico de Puchuldiza. Santiago. Chile unpub. rept. Com. Geot. CORFO.
3. Marinovic et al 1975. Estudio de Resistividad de Puchuldiza, Santiago Chile unpub, rept. Com. Geot. CORFO.
4. Cusicanqui H. 1975 Estudio Geoquímico del Campo Geotérmico de Puchuldiza, Santiago Chile, unpub, rept, Com.Geot. CORFO.

5. Ellis A. 1968 Geochemestry of The Tatio Geothermal Field, unpub. rept. Com. Geot. CORFO
6. Trujillo P. 1970 Manifestaciones Termales de Puchuldiza, Santiago Chile, unpub. rept. Com. Geot. CORFO.

CAPTION TO FIGURES

Figure   1. Location Map
             2. Geological Map
             3. Geophysical Map

TABLE 1

PUCHULDIZA STRATIGRAPHY

AGE		FORMATION	LITHOLOGY
QUATERNARY	PLIOCENE	Alluvial Deposits / Alluvial Fans	Unconsolidated clastic materials
		Morraine Deposits	Sands gravels and heterogeneous chaotic blocks.
		Pleistocene Volcanoes	Hornblende and biotite Andesites
		——————— U N C O N F O R M I T Y ———————	
		Mauque Series	Sandstones, breccias, tuffs and piroclastics.
		Lupe Formation Upper member	Tuffaceous breccias and piroclastics sediments
		Lower member	Sandstones conglomeradic breccias with pumice levels
		——————— U N C O N F O R M I T Y ———————	
TERTIARY	MIOCENE	Puchuldiza Formation Upper member	Mauve andesite Ignimbrites with "fiamme"
		Lower member	Dark augite Andesites
		Andesitic Intrusives	Radial an anular andesitic dikes
		Cenderiri Formation Upper member	Dark olivene and augite Andesites
		Lower member	Dacitic Ignimbrites, with abundant pumice in its base.
		Chojña Chaya formation	Green sandstones with conglomeradic intercalations
		——————— U N C O N F O R M I T Y ———————	
CRETACEOUS		Churicolle Formation	Sandstons and coarse conglomerates

TABLE 2

CHEMICAL COMPOSITION OF WELLS 1,2,3 OF PUCHULDIZA IN MG/L

Sample	pH 18°C	Na	K	Ca	Mg	Li	Cs	Rb	Cl	SO_4	B	HCO_3	CO_3	CO_2	SiO_2
Well 1 (550m)	6,24	1525	70	41	2,5	6,4	0,38	0,38	2174	179	70	449	0	340	201
Well 1 Surf.	6,32	1560	70	41	2,7	6,5	0,40	0,42	2198	205	71	453	0	330	203
Well 2 310 m	6,23	1425	183	47	0,7	10,2	1.75	1.50	2253	113	74	350	0	270	261
Well 2 515 m	6,25	1412	192	43	0,9	10,0	1.75	1.50	2320	108	76	311	0	210	267
Well 2 Surf.	8,33	1660	231	18	1,0	11,9	1.96	1.70	2674	120	88	236	4,0	1,2	340
Well 3 220 m.	6,22	1180	162	47	1,9	8,8	1,40	1.21	1861	129	66	256	0	200	188
Well 3 421 m	6,21	1165	162	43	1,8	8,6	1,40	1.21	1826	129	65	295	0	230	180
Well 3 Surf.	8,46	1325	183	26	2,2	9,4	1.56	1.30	2070	144	73	251	5,6	1,1	237

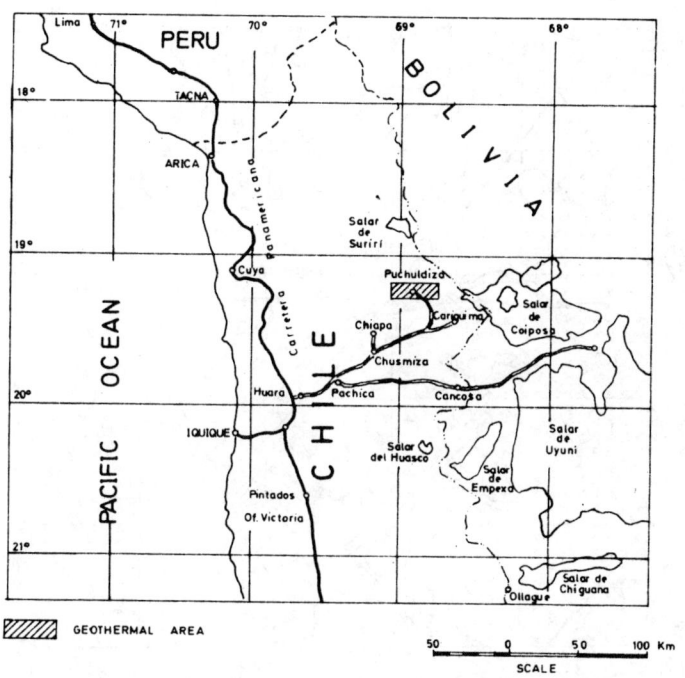

Figure 1. Location Map

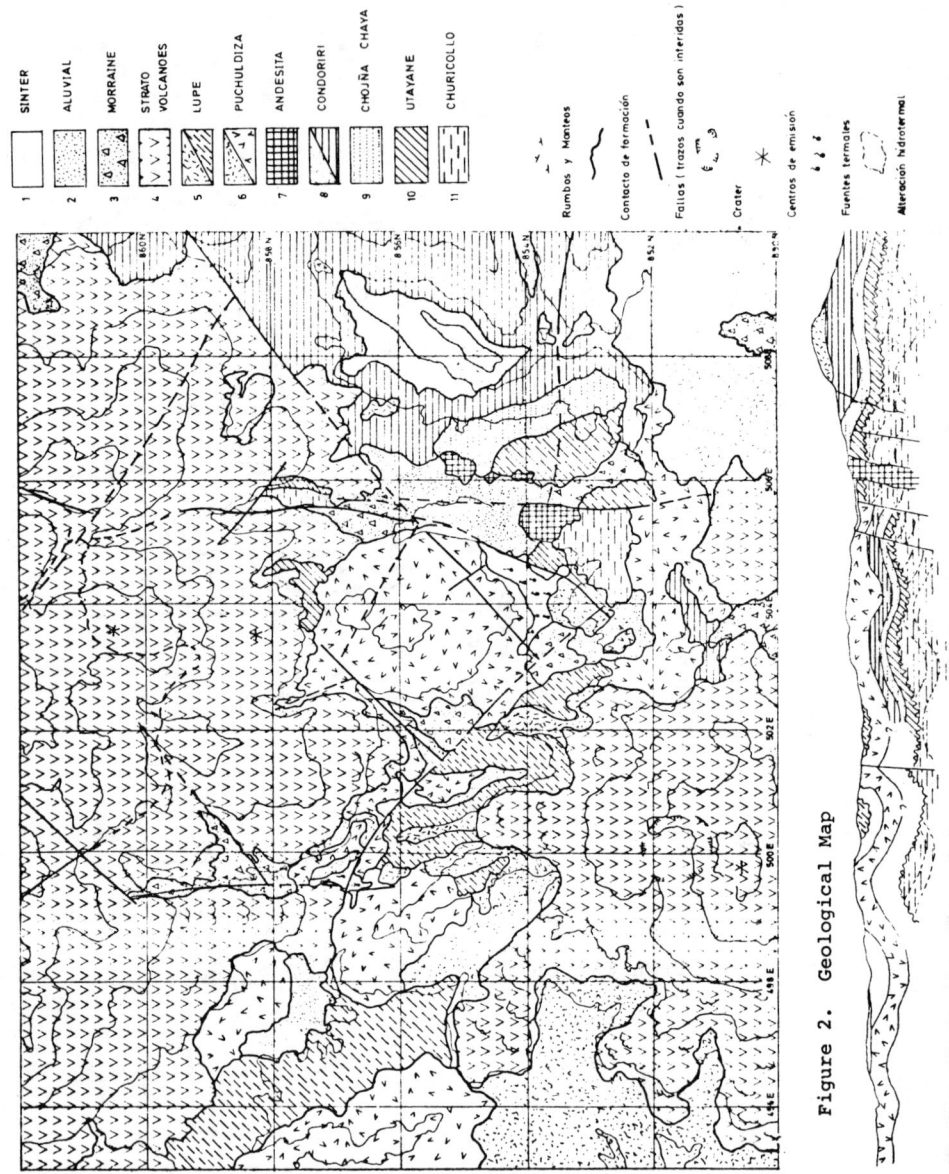

Figure 2. Geological Map

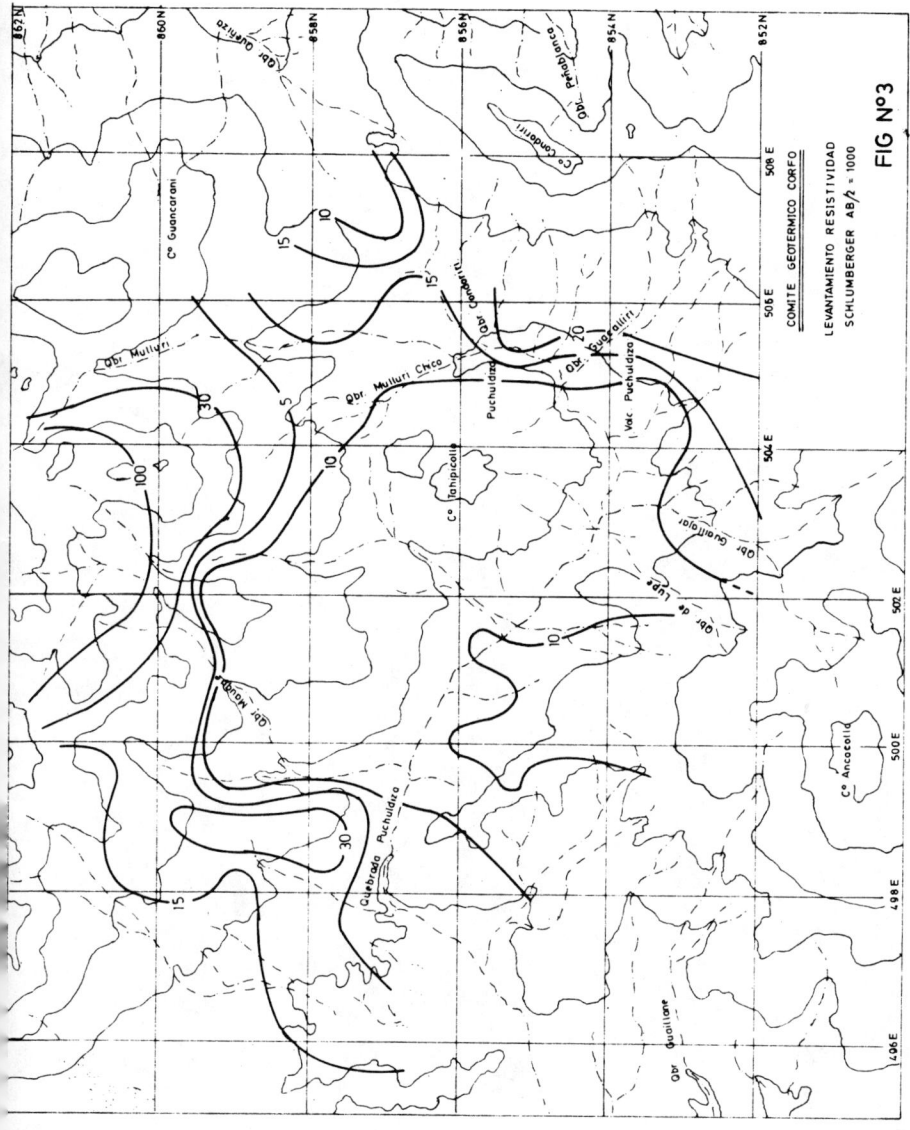

Figure 3. Geophysical Map

Geothermal Energy Coupled with Solar Energy Is the Future Energy in Saudi Arabia

A. A. M. SAYIGH
College of Engineering
University of Riyadh
Riyadh, Saudi Arabia

ABSTRACT
The paper discusses the potential of geothermal energy and solar energy in Saudi Arabia. The various utilizations of such sources in producing electricity, fresh water for drinking and agricultural use. The geothermal areas were located in the Kingdom and various contour maps are presented. Solar energy intensities and some possibilities of utilizing such abundant energy by using flat plate collectors or concentrators were also discussed. A discussion about the coupling of geothermal energy and solar energy to produce more power in most parts of Saudi Arabia and some broad lines about the economics of such systems are also mentioned.

INTRODUCTION
The history of geothermal energy and its conversion to electricity dates back to 1904, at Larderello, in Italy, where an experimental plant of 250-kW was installed. The work gradually commercialized to produce 365.5 MW. Then there was a gap with no geothermal energy activities until 1960 when small plants were established in New Zealand and the United States. By the year 1973, each of the following countries was producing electricity from geothermal energy: Italy - 390.5 MW, New Zealand - 170 MW, United States - 396 MW, Japan - 33 MW, USSR - 29 MW, Iceland - 3 MW and Mexico - 75 MW. The total is 1059.5 MW which is supposed to reach 1876.5 MW by the end of this year with the addition of countries like El Salvador, Guadeloupe, Philippines, Taiwan and Turkey. It was predicted in United States that in 1985, geothermal energy could produce 132000 MW of electricity which is equivalent to 20% of 1985 US electrical power needs. In Saudi Arabia geothermal energy was first realized during the late thirties as result of ARAMCO oil explorations and until this date has not been utilized, [1].

The country of Saudi Arabia consists of mostly desert except for few eastern and western strips. The country has no rivers and relies totally on fossil waters or ground water. Climatically the country can be divided into three regions,

a- Extremely arid region which is represented by the northwest coastal area along the Red Sea and Rub'al Khali

(Empty Quarters) where annual mean rainfall is below 50 min.

b- Semi-arid region which is represented by the Assir Region further south-west which is affected by monsoon rain where annual mean rainfall is about 300 mm.

c- Arid zone, represented by the whole country with the exception of the two other regions, [2].

Temperatures can reach $50^{o}c$ in some parts of the Kingdom and it can drop as low as $-10^{o}c$. Apart from the coastal areas, the relative humidity never exceeds 40% and in most areas it is about 25%. As for the sunshine hours the daily average is about 10 hours while the average yearly intensity of solar radiation on a horizontal surface is 550 gm. cal/cm^2/day,[3,4,5]. Therefore the lack of adequate water for agricultural use, and in some areas for human use, is still a real problem. Digging wells in some regions at a cost of one million Saudi Riyal per well ($ 1 = 3.5 SR) has produced brackish water, and has added to the cost of the well the cost of refining such water and makes it almost beyond a viable proposition. This cost can be reduced by using geothermal energy and solar energy to distill the water and then transport it if necessary to remote areas by the use of solar pumps.

This paper deals with the nature of ground water available in Saudi Arabia, the amount of solar energy and their applications in generating electricity, producing portable water whether for drinking or whether for agriculture.

FOSSIL AND GROUND WATER
Saudi Arabia is rich in fossil water which is meteoric origin infiltrated underground and moved downward several thousand years ago. This water undergoes a very slow recharge, much lower than the rate of discharge, [2]. The country consists of 28 sedimentary formations, 21 only produce water. The well water from such formations varies in depth some at 1250 ft while others at 3500 ft.

The formations which represent prolific aquifers in Saudi Arabia are: Saq. Tabuk, and Wajid in the paleozoic; Minjur, Biyadh, and Wasia in the mesozoic; and Umm Er Radhuma, Dammam, and Neogene in the cenozoic. There are also a few other aquifers, such as Jilh and Dhruma, which are not considered prolific due to their limited geographic extent and/or variation of hydrogeologic and chemical characteristics. These waters have originated from precipitation, they are the result of infiltration to underground of rain water some 10,000 to 40,000 years ago. Their temperature was almost equal to the ground temperature at that time, and their salt content was very small say between 10 to 30 ppm t.d.s. If the temperature of these waters is very high now, it is because they have moved downward following the dip of the formation in which they are accumulated. Hence, the main factor affecting the increase of temperature is the depth in which they are found. Therefore, the reason for the temperature of hot waters is the geothermal gradient. The average value of this gradient is $33°c/km$. How, ever, there are anomalies in Saudi Arabia which bring this value to about $150°c/km$ [6]. Hot waters, of interest, from an energy viewpoint are those which are accumulated in great depth and under high pressure. If they have a static water level below ground surface, they will not be of interest, since they will require energy for exploitation. This is why these two parameters, high temperature and high pressure, are the prerequisite for their usage. Figure 1 shows the area of these wells and their pressure contour. As a result, some examples of wells producing hot water under high pressure are described below.

Qiba, Tannumah and Al-Kharj have wells with reasonably high pressures and temperatures, [7,8].

One of the most promising acquifer which can supply the whole of Saudi Arabia with water and energy is Wasi acquifer and the HAWP well is an example from such acquifer. It is located about 20 km South-East Hofuf. It was drilled recently for observation purposes to determine the hydraulic relationship between Wasia and other formations. Table I shows the main characteristics of this well. There are many other wells drilled by ARAMCO in the same aquifer, especially in the Uthmaniyah area, producing five times more water than the HAWP. Figures 2,3 and 4 show some characteristics of Wasia aquifer.

SOLAR ENERGY
Saudi Arabia is one of the few countries in the world which enjoys prolong daily sunshine, about 10 hours per day. If Riyadh is taken as the average town in Saudi Arabia, then figure 5 shows the various metereological data for it, [9]. Using flat plate collectors with honeycomb structure, a temperature of $150°c$ can be obtained in most part of the year if the collector was fixed at an angle of $30°$ with the horizontal and facing south. The collector receives 25000 KJ/m^2 per day. If a choice of two tilts was used such that the collector is placed horizontally for the months April, May, June, July and August and placed at $30°$ for the rest of the months, then the amount of daily solar radiation received by such collector will be 27000 KJ/m^2, [10,11]. Using a parabolic concentrator a temperature of $300°c$ can easily be obtained. Using a heliostat mirrors and a central receiver, with the previous amount of solar radiation, one can easily deduce that Saudi Arabia

is one of the best country for utilizing solar energy.

GEOTHERMAL UTILIZATION

Using figures 1 and 2, the Lith area and Wasia aquifer with pressure contour 73.5 psi and above and with temperature $80°c$ and above, geothermal energy utilization in Saudi Arabia is a reality and an economical process. The conversion system which has already been proven since 1960 in New Zealand and is in operation now in Japan, Mexico and New Zealand, is the Flash Steam System, [12]. This is a simple system as shown in figure 6. The wellhead product is fed into a flash separator where the vapor fraction is increased by an isenthalpic expansion to a low pressure. The steam is then used to drive a standard axil-flow multistage turbine to generate electricity. The overall thermal efficiency is about 10%. This is very practical when the salinity of the water is low $\leqslant$ 3% dissolved solids and Wasia water is well within this limit, 3515 ppm.

A more durable system is the binary cycle system which is basically a Rankine cycle with an organic working fluid. Figure 7 shows this system. A heat exchanger is used to transfer a fraction of the brine enthalpy to vaporize the secondary working fluid. The expansion through the turbine to a lower pressure which is fixed by the temperature of the rejected heat supplies the power generation. At the moment there is one plant in the world which uses the binary system. This plant was built in 1967 in the USSR [13], and uses Freon-12. The gross power generation is 680 kWe and utilizes wells with $80°c$, while the net power output is 440 kWe. Another way of utilizing geothermal energy and the binary system is in refrigeration as shown in figure 8, [14]. The production of water at $4°c$ is certainly possible.

If we take the calculations of A.D.K. Kaird, [15] which uses one million pounds per hour, with the lowest reservoir temperature about $150°c$, wellhead temperature is $93°c$, pressure is 11.5 psia, the lowest temperature is $71°c$, and if the ratio of fresh water to well fluid is 0.26 with 90% plant availability per million lb/hr of fluid supplied, then the water production is 0.62 million gallons per day (m.g.d) and the power production rate will be 1.2 MW. This can easily be met by the geothermal resources of Saudi Arabia. Another example is sighted by A.L. Austin et al, [13] as shown in figure 9 which is for binary system with flashing flow and sink temperature of $49°c$. From these two examples, it is essential to raise the reservoir temperature to a level of $200°c$ or above. Using flat plate collectors with two tilts in order to get the maximum solar intensity and with honeycomb structures, it is possible to reach $200°c$, if the inlet temperature is $90°c$. However for more temperature a system of parabolic collectors will give $300°c$ or above and hence power generation of 17 MW per one million pounds per hour of groundwater can be acbieved. Figures 10 and 11 shows a combination of solar and geothermal energy to produce power and fresh water.

CONCLUSION
In studying the wells which has been excavated for water in Saudi Arabia, one relizes that in the eastern province and the south west area, water is plentiful with high temperature reaching 93°c and high pressure reaching to 250 psia. Coupling this potential with solar energy, the country's power requirement, which was estimated to be 1000 MW per day, can be met easily. Either using binary system, then Rankine engine can be used with appropriate choice of fluid, [16] or using flash system, then a multistage turbine can be used as shown in figures 11 and 12. As for distilling geothermal brive, it is technologically similar to distilling seawater with the advantage of no feed preheating. As for pretreatment it is also less costly due to the absence of calcium and magnesium scale-forming agents and other organic and nitrogenous pollutants,[17],[18].It is at least 30% cheaper to produce fresh water from geothermal brive than using seawater and in using a system similar to figure 11, 28 MW can be generated per day and 1.7 mgd can be produced from a well producing 1.5×10^6 lbs/hr and the cost of electricity will be 2.8 mills per kWh. Therefore having one hundred wells of this kind, with the proper combination of solar and geothermal energy, the country's energy requirements will be met twofold.

REFERENCES

1. Otkun, K. and Sayigh, A.A.M. "General Aspects of Geothermal Energy in Saudi Arabia, International Solar Energy Conference, Comples, University of Petroleum and Minerals, 2-5 Nov., 1975.

2. Otkun, G., Butain, N.A., and Hussein, A.H. "Recognition and Dating of Fossil Waters in Saudi Arabia" 25th International Geological Congress, Sydney, Australia, August, 1976.

3. Sayigh, A.A.M. "Solar Energy Availability Prediction from Climatological Data," Solar Energy Conversion, 2nd Course, International College on Applied Physics, Catania - Italy, August 1975.

4. Sayigh, A.A.M. "The Energy Prospects in the Arab World," International Conference in Mechanical Engineering with Emphasis on Energy, March 1976, University of Engineering and Technology, Lahore, Pakistan.

5. Sayigh, A.A.M. "The World Energy Situation and the Islamic countries" Islamic Solidarity Conference in Science and Technology, Riyadh University, Riyadh, Saudi Arabia, March 1975.

6. Otkun, G. "More About Paleocene Karst Aquifer in Saudi Arabia," IAH Karst Hydrogeology Congress, Huntsville, Alabama, September, 1975.

7. Sayigh, A.A.M. "The Role of Solar Energy in Producing Fresh Water for Agricultural Use in Saudi Arabia," First Agricultural Conference of Muslim Scientists, College of Agriculture, University of Riyadh, Saudi Arabia, April, 19-22, 1977.

8. BRGM "The Study of Wasia Aquifer - Al - Hassa Area" Ministry of Agriculture and Water Resources, Saudi Arabia, 1976.

9. Sayigh, A.A.M. "Summer Night Cooling in Saudi Arabia" Solar Cooling and Heating A National Forum, Miami Beach, Florida, Dec., 13-15, 1976.

10. Sayigh, A.A.M. "Solar Flat Plate Collector - Facts and Expectations," UNIDO, Expert Group Meeting in Evaluating Existing Solar Technology, 14-18 Feb., 1977, Vienna, Austria.

11. Sayigh, A.A.M. "The Technology of Solar Heat Flat Plate Collectors" Solar Energy Conversion - 4th Course, Trieste-Italy, 6-24 Sept., 1977.

12. Fucho, R.L. and Westphal, W.H. "Energy Shortage Stimulates Geothermal Exploration" World Oil, Dec., 1973, p.37-41.

13. Austin, A.L. and Lundberg, A.L. "Electric Power Generation from Geothermal Hot Water Deposits," Mechanical Engineering, Dec., 1975, p.18-25.

14. Wehlage, E.F. "Geothermal Energy Needed Effective Heat Transfer Equipment" Mechanical Engineering, August, 1976, p.27-33.

15. Laird, A.D.K. "Water From Geothermal Resources" Geothermal Energy, Resources, Production, Stimulation, Ed. Paul Kruger and Carel Otte. Stanford University Press, 1973, p.177-196.

16. Ishikawajima - Harima Heavy Industries Co., Ltd., "Solar Rankine Engine" Comples, Dhahran, Saudi Arabia, Nov. 2-6, 1975.

17. Laird, A.D.K. "Rankine Research Problems in Geothermal Development" R & D Progress Report No.711, August 1971.

18. Green, M.A. and Laird, A.D.K. "Comparison of Elementary Geothermal - Brive Power - Production Processes" Prepared for the U.S. Atomic Energy Commission under contract W-7405-ENG-48., August 2, 1973.

TABLE I   CHARACTERISTICS OF THE HAWP WELL

Characteristics	Value
Coordinates	25°13'E - 49°35'E
Total depth	1560 m
Water producing formation	Wasia (Cenomanian)
Discharge (U-855)	About 55 l/sec
Static water level	About 100 m above ground
Temperature at well head	93°c
Total dissolved solids (U-855)	3515 ppm

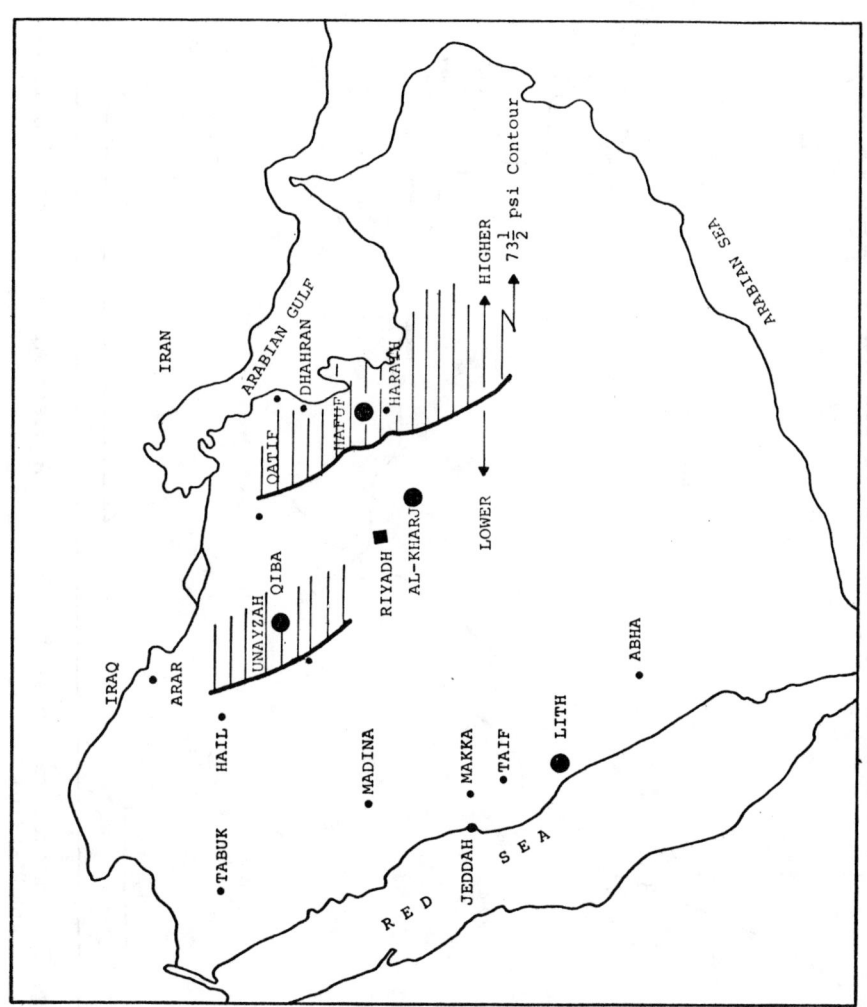

Fig. 1 The Contours of High Pressure Ground Water

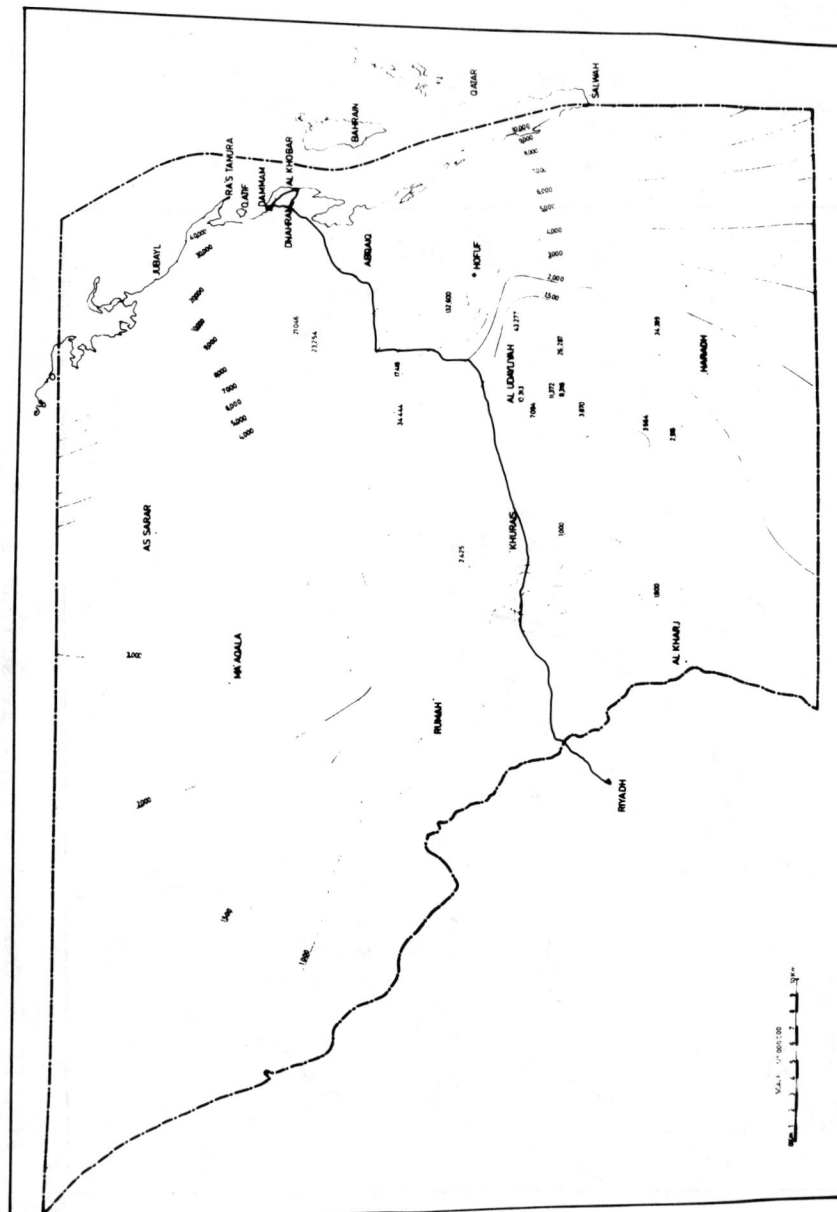

Fig. 2 Temperature Map in °C of Wasia Aquifer Al Hassa Area (Modified After BRGM-1976)

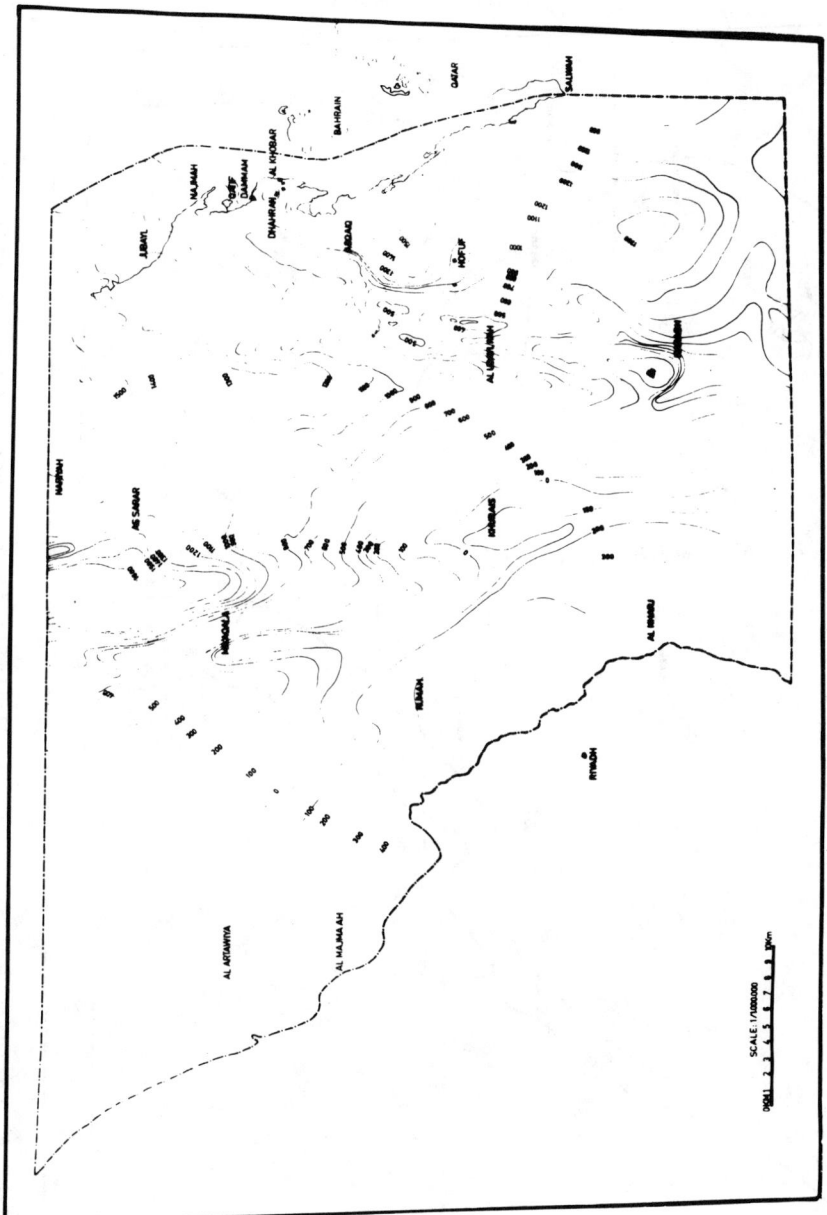

Fig. 3 Structure Contour Map Top of Wasia Formation After BRGM-1976

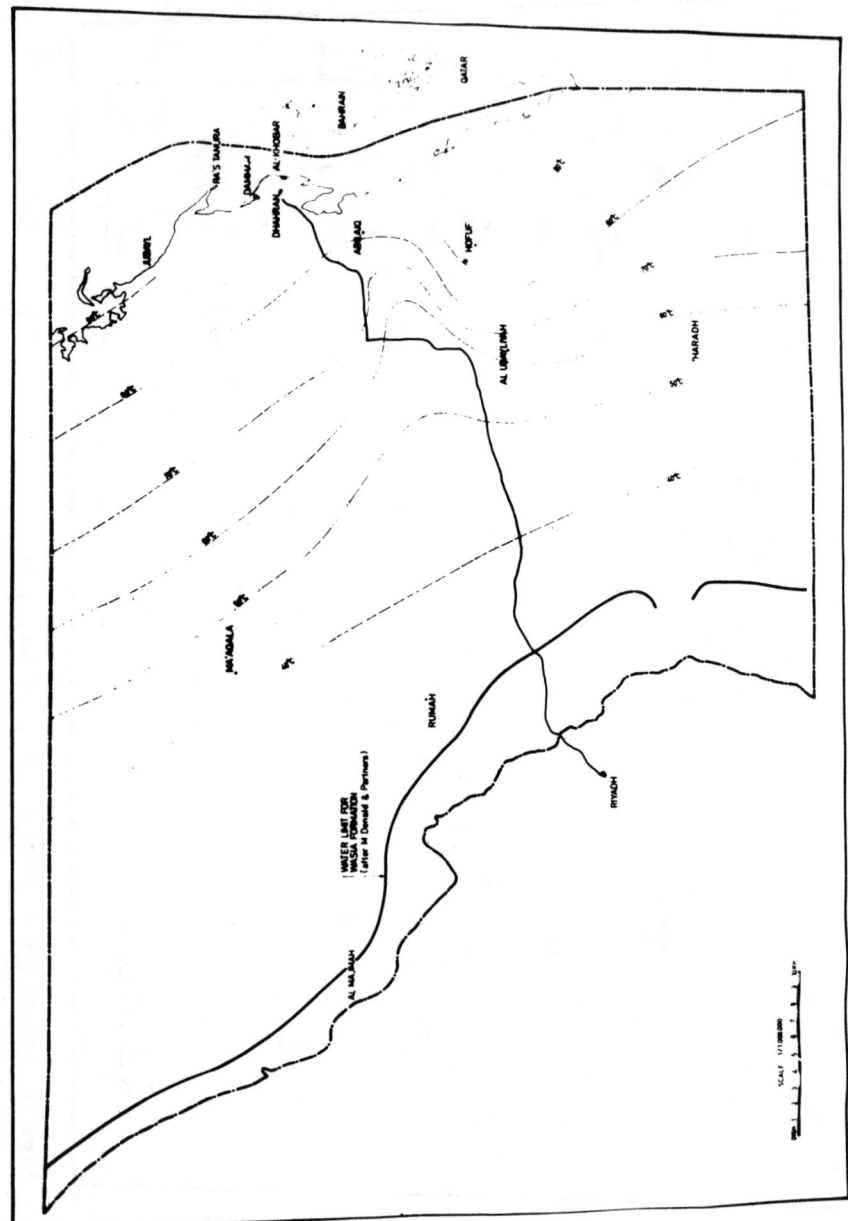

Fig. 4 Salinity Map in p.p.m. of Wasia Aquifer Al Hassa Area (Modified After BRGM-1976)

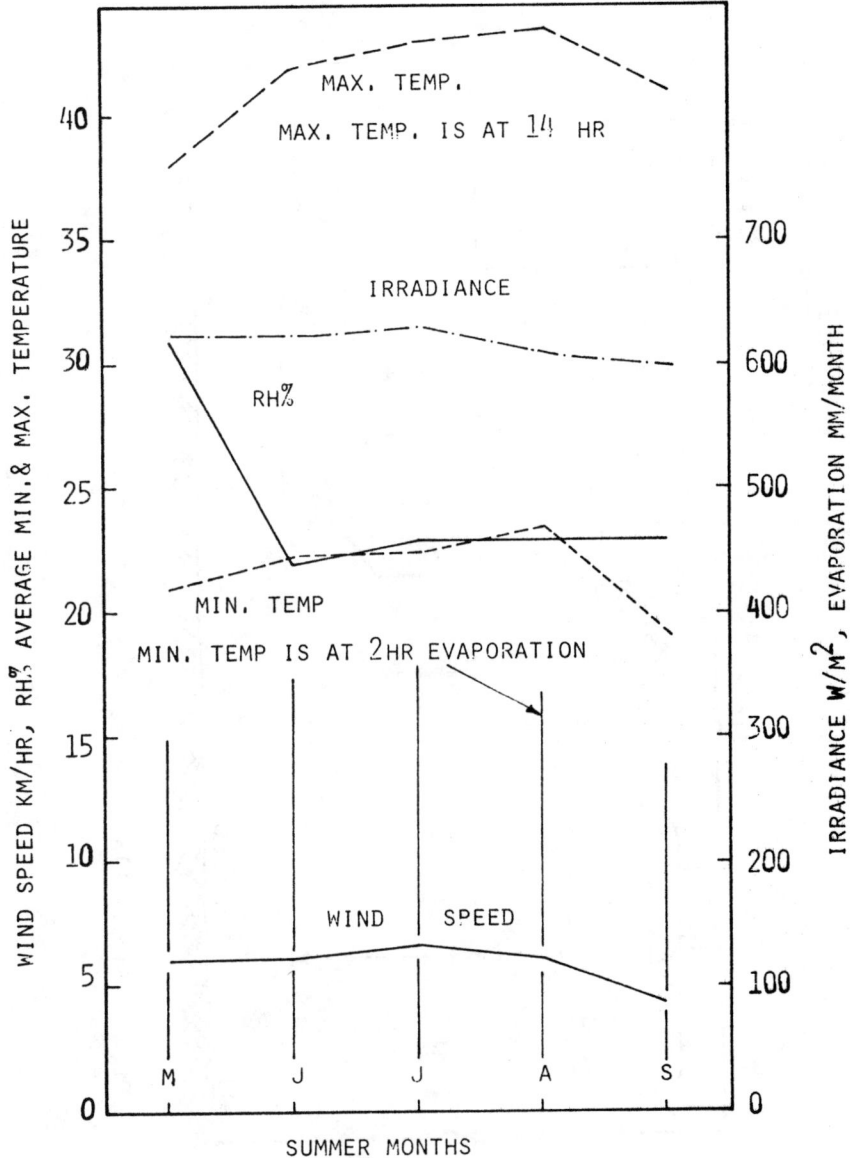

Fig. 5 Some Meteorological Data in Riyadh

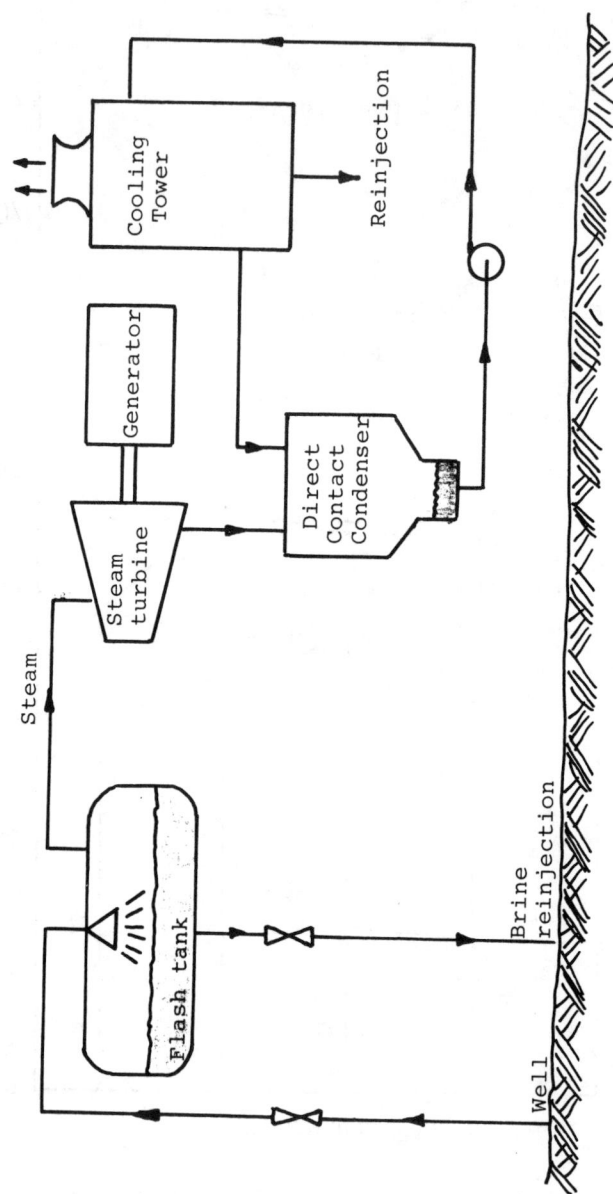

Fig. 6 Simple Flashed-Steam Cycle

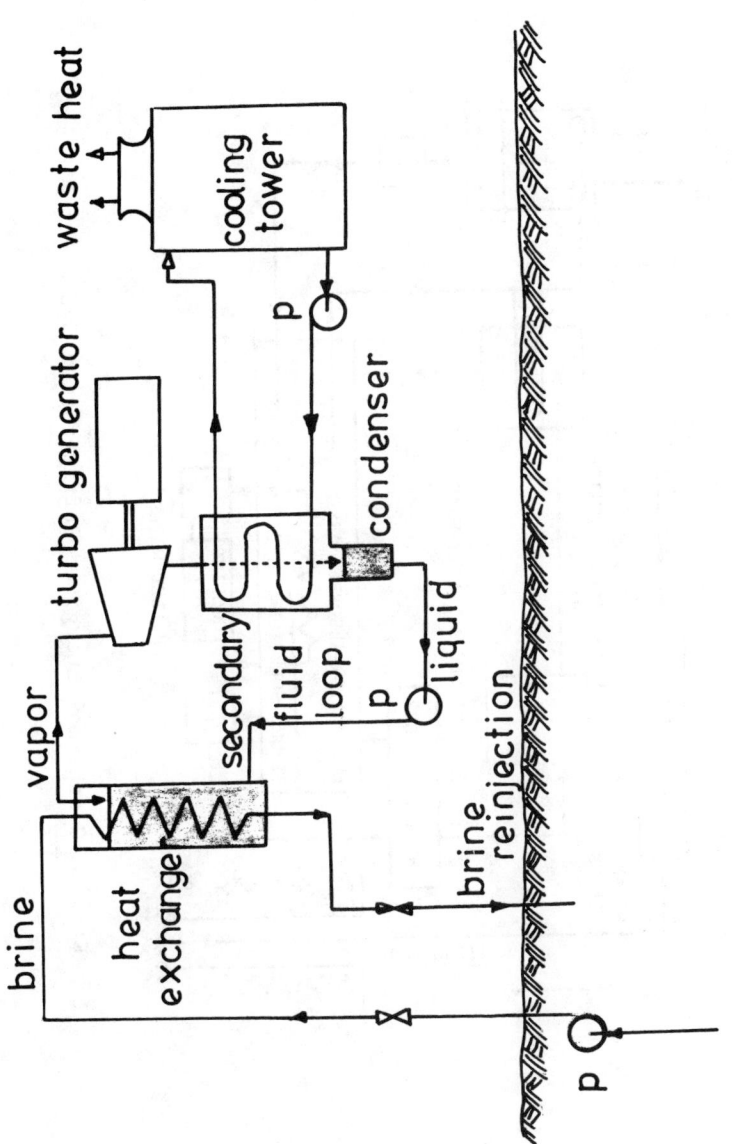

Fig. 7   Isobutane Binary Cycle

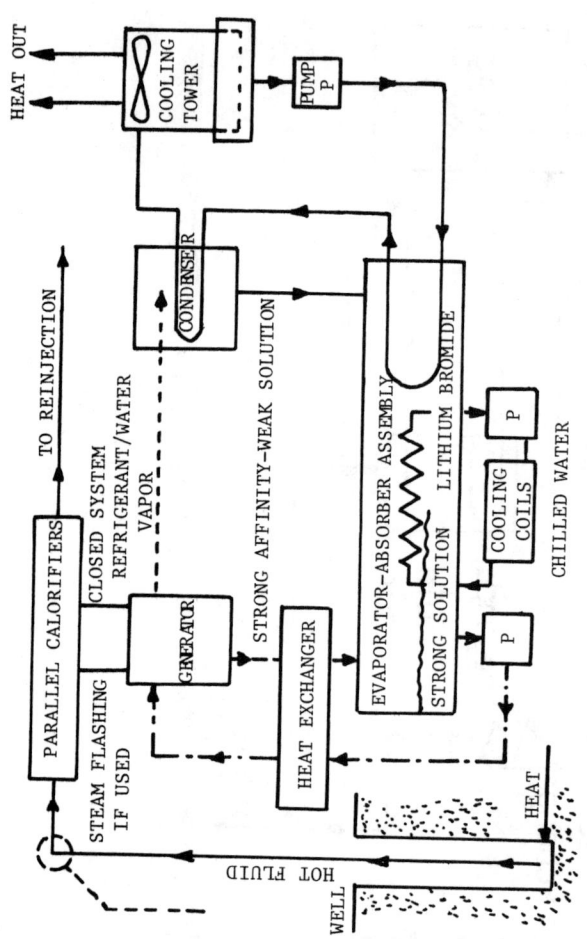

Fig. 8 Geothermal Energized Refrigeration Cycle For 4°C

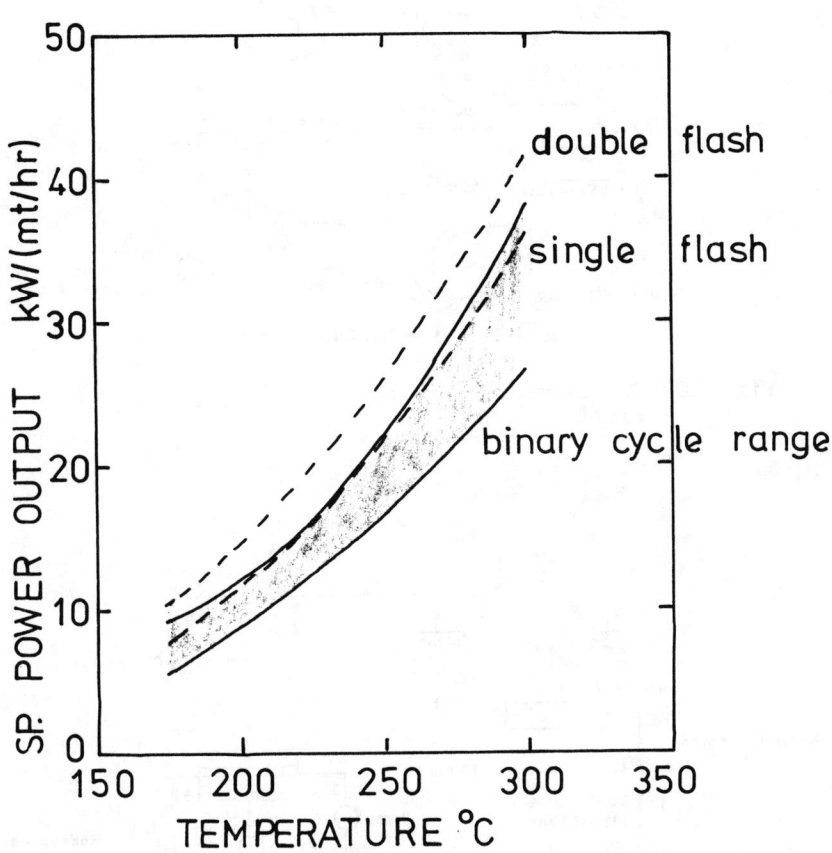

Fig. 9 Relative Performance Characteristics of Geothermal Energy Conversion Concepts

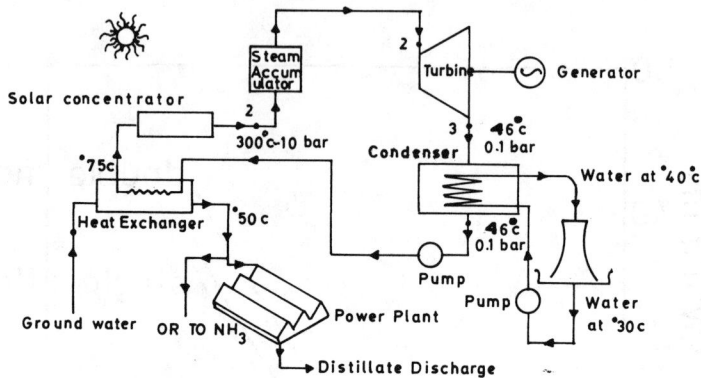

Fig. 10 Geothermal Supplemented Solar System Power Plant

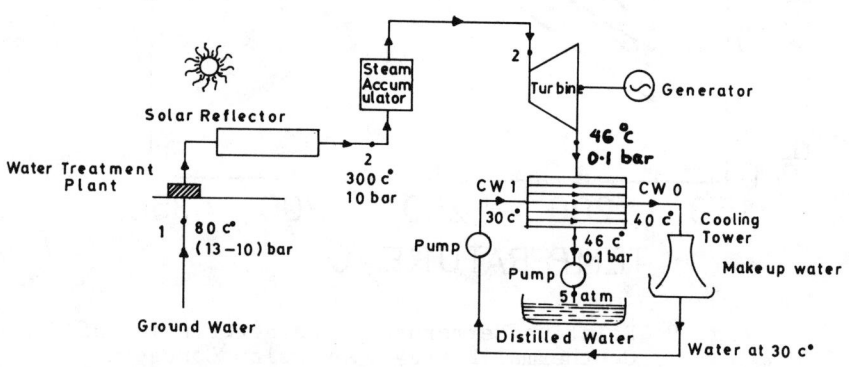

Fig. 11 Flow Diagram of Solar-Assisted Geothermal Engine

Manmade Geothermal Energy*

T. K. GUHA and K. E. DAVIS
Subsurface, Incorporated
Bellaire, Texas 77401, USA

R. E. COLLINS, J. R. FANCHI, and A. C. MEYERS, III
University of Houston
Houston, Texas 77004, USA

ABSTRACT

The ever increasing demand for energy, coupled with the steady decline in fossil fuel reserves, has led the modern scientist on a quest for alternate energy sources. More efficient use of our energy resources is a concurrent objective in this quest.

One of the major problems associated with the efficient use of energy is the storage from thermal energy systems including solar and waste heat, i.e., engine exhaust, furnace exhaust heat, heat discharge from various chemical reactors, heat losses to the atmosphere, etc. A potential solution to this problem is the injection of hot fluids, through wells, into deep permeable aquifers or impermeable caverns.

At present a feasiblity study is being made by the University of Houston and Subsurface, Inc. Funding is from ERDA through the Energy Foundation of Texas. The study encompasses: an investigation of the efficiency of underground heat storage with reference to thermal losses and pumping requirements utilizing computer simulation; a literature survey of potential operational problems; and a geological survey to determine applicable areas within the United States.

A primary application envisaged for geothermal storage wells is for large scale solar collector-concentrator systems to create load following for base line electric power systems. High temperature, high pressure storage wells, 650° and 2700 psi, can be utilized for storage for steam-electric systems. To date the study has shown that thermal losses may be less than 1%. Eighty percent (80%) of the continental United States is amenable to this method of storage.

---

*This paper was prepared with the support of U.S. Energy Research and Development Administration Contract #EG-77-C-04-3974. However, any opinions, findings, conclusions, or recommendations expressed herein are those of the authors and do not necessarily reflect the views of ERDA.

INTRODUCTION

The rapid increase in demand for energy, as well as dwindling reserves of fossil fuels, has led the modern scientist on a quest for alternate energy sources. The objectives in this quest are to substitute new energy sources for existing ones and achieve more efficient use of present energy sources.

One of the major problems associated with the efficient utilization of energy sources is the storage of heat, in particular storage of waste heat of various industries and electric power plants, and storage of thermal energy from large solar collectors. There have been a number of proposals for underground storage of thermal energy using waste heat from industries and electric power plants [1. Meyer & Todd, 1973; 2. Meyer & Hausz, 1975; 3. Meyer, 1976; 4. Dooley et al., 1977] as well as, for the underground storage of solar energy [5. Collins, 1974; 6. Davison et al., 1975; 7. Collins & Davis, 1976; 8. Tsang et al., 1977].

In this paper we describe preliminary results of a feasibility study of deep underground storage of thermal energy. This study is being carried out by the University of Houston and Subsurface, Inc. on a contract from ERDA through the Energy Foundation of Texas.

THERMAL ENERGY SOURCES

It has been estimated that the total energy consumption in the United States is approximately 73.1×10^{15} BTU/yr. [9. U. S. Department of Interior News Release, 1975] of which approximately 26% is used in commercial power plants, 39% in industries and the remaining 35% for transportation and domestic commercial uses.

Electric power plants, fueled either by fossil fuels or nuclear power, are a major source of waste heat, as they run at an efficiency of only 31%. To generate 1 KWH of electricity approximately 10,000 BTU's of energy are required, at a 69% waste heat loss to the environment. In the United States there are 63 nuclear power plants in operation and another 148 plants are either under construction or permitted for construction, to produce a total of 208 Megawatts of electricity in the very near future [10. Parks & Thomas, 1976]. We can safely say that double this amount of energy will be lost to the environment.

Most of the energy in a petroleum refinery is derived from the combustion of fossil fuels. These combustion operations are 72.8% efficient with 27.2% of the total energy consumed lost to the environment [11. Prengle, 1974]. Chemical manufacturing plants are growing in the United States and the product demand growth rate to 1980 will be 6.4%. A significant amount of heat loss to the environment from such plants has been described in a recent survey.

Primary metal production from ore or other natural raw materials and metal-alloy production consumes a major quantity of energy. For these industries an 11% energy loss rate to the environment was suggested by Prengle, 1974. In Figure 1 we have summarized the geographical locations of these various sources of waste heat within the continental United States.

Another potential source of energy in the United States is solar energy. Figure 2 shows a U.S. Weather Service map of the total insolation i.e. direct beam plus sky brightness of the United States in Langlys (1 gm cal/cm^2-day). From Figure 2 we can see that approximately 80% of the United States (southern and western states) is favorable for collecting solar energy on a daily basis. The rest of the country is suitable for solar collection on a seasonal basis, during the summer and part of spring and fall.

In summer, because of the earth-sun relationship, the number of hours of daylight are greater and the solar angle with respect to the normal to the earth's surface is smaller at noon. During winter, in the midwest and west, the air is cooler, drier, and clearer, thus a greater percentage of the available sunlight reaches the surface. The net result is that a solar tower central receiver collector system located in the northern midwest can collect approximately 80% of the energy it would collect if the same system were located in the southwestern United States. If the tower system were optimumly sized for the location and latitude, this percentage could be increased slightly.

STORAGE SYSTEMS

In previous papers [12. Collins, 1974; 13. Collins and Davis, 1976] it has been pointed out that deep underground storage of thermal energy, using a working fluid at high temperature and pressure, can be achieved with relatively small conduction losses for cyclic injection and withdrawals if the system is sufficiently large. In particular,

$$(1) \qquad \frac{\text{Heat lost per cycle}}{\text{Total Heat Stored}} \sim \frac{3\kappa\tau}{R^2}$$

where κ is thermal diffusivity of the rock, τ is the period of injection-withdrawal and R is the equivalent radius of the cyclicly heated region. Thus only <u>large</u> storage systems should be considered

Furthermore, if the working fluid is to remain in the liquid phase, a high pressure is required in the storage region. This in turn dictates a deep storage zone if only the earth overburden pressure is to confine the fluid. The pressure in the storage region must be less than the fracture pressure. A rough rule of thumb, incorporating an appropriate safety factor, assigns 0.5 psi per foot of depth as the fracture pressure.

Storage systems meeting the constraints dictated by these considerations can be realized by using water as the working fluid and injecting through a well into a porous aquifer. Also, oil may be used as a working fluid and injecting through a well into a cavern, as say a solution cavern in a salt dome.

Currently we are designing computer simulators to study details of thermal losses and pumping requirements for deep aquifer storage (superheated water, 650°F, 2700 psi) and deep cavern storage. These studies will also address other aspects of such geothermal storage systems, in particular the solution and transport of minerals, in the case of aquifer storage, and thermo-mechanical stresses on earth and well components. These studies will be reported at a later date. Here we address the general question of geological feasiblity for four recognized types of underground thermal storage:

(a) deep aquifer storage
(b) shallow aquifer storage
(c) solution cavity storage
(d) excavation cavity storage

Deep Aquifer Storage

This system requires a permeable aquifer (saline) of appropriate porosity, permeability, thickness and depth. In order to accomodate high fluid injection rates a permeability on the order of 500 md, and a thickness on the order of 100 feet or more is desirable. Also, in order to sustain a high pressure, 2700 psi, a depth of at least 5500 feet is required. Figure 3 shows differentiated basement rock depth of the United States. The areas where the thickness of the sedimentary column is over 5500 feet are potential regions for this type of storage system.

Shallow Aquifer System

The shallow aquifer system is a term applied here for convenience where the depth ranges from 3000 feet to 5500 feet. This system is similar to the deep aquifer system except that the temperature and pressure of the injected fluid will be comparatively less, depending upon the depth. This thermal storage system is suggested where no other alternative is available. Figure 3 shows these suitable regions in the continental United States.

Solution Cavity System

Artificially created solution cavities in salt formations are made using water to dissolve the salt. A significant characteristic of most rock salt is its very low porosity and permeability, or the apparent lack of it. Salt is semi-plastic and tends to close small fractures and openings made in it. It is this property which makes massive salt formations such ideal storage sites. Today, underground storage facilities in salt store LPG, crude oil, and petroleum products in a safe and most economical manner. Caverns in salt can also be used for thermal storage using some fluid other than water as the working fluid. In the United States there are a large number of salt formations of considerable thickness. Figure 4 shows the distribution of salt formations and locations of sedimentary basins.

Excavation Cavity System

Recent advances in rock mechanics and drilling and mining techniques now permit the use of deeply buried rock cavities for storage of water or steam. Economics, environmental effects and safety are favorable to this thermal storage system with power output being extremely satisfactory [14. Dooley et al., 1977]. Rock types suitable for constructing underground excavations are crystalline and massive rocks. The most favorable rock types are massive granite-like rocks, massive intrusive rocks, and many coarsely crystalline metamorphic basement types (e.g. gneisses). These rock types have the requisite compressive strength of approximately 20,000 psi. Other rock types favorable for this system are massive siliceous limestone, massive well-cemented sandstones, massive dolomites, marbles and high grade metamorphics such as quartzite, massive schist and slate. Basalt and other extrusive volcanic rocks are less favored for this storage system. Figure 5 shows the distribution of crystalline formations suitable for excavation cavity storage systems in the continental United States.

GEOLOGICAL FEASIBILITY OF THERMAL STORAGE

Based upon the physical requirements of these four types of underground thermal storage systems we have attempted to identify those areas of the continental United States in which each type of storage system might be feasible. In making these preliminary assessments a reconnaissance survey was done from published data. We made an overview differentiating the continental crust on the basis of basement depth and thickness of the sedimentary column (Figure 3). This step was followed by identification of different geological formations on the basis of type of storage system. (Figures 4 and 5) and finally a composite map could be prepared (Figure 6) to show regions of the continental United States suitable for each of the four types of underground storage systems. In drawing this composite map regions affected by tectonics, seismicity and volcanic activity have been excluded.

On the basis of these preliminary studies it is evident that underground thermal storage of energy is possible in most parts of the continental United States. The next phase of this geological study will include detailed hydrological and rock property evaluation of areas selected as potential construction sites for storage systems.

CONCLUSION

On the basis of our present study we conclude the following:

Waste heat, as well as solar energy, must be stored to meet the increasing demand for energy and for a better environment. Several major problems associated with storing heat can be solved through the use of underground thermal storage systems.

Four different storage systems (deep aquifer, shallow aquifer, solution cavity and excavation cavity) are favorable for storing heat. The geology is such that one or more of these systems can be feasible in every area of the continental United States where significant sources of waste heat exist or where large scale solar concentrator-collector systems are practical. Approximately 80% of the United States is geologically feasible for underground storage of high temperature and high pressure fluid, with the exception of the West Coast and areas of mountain intrusions.

Studies now underway using computer simulation will eventually provide detailed design and operational criteria for aquifer and cavern storage systems. These will provide guidelines for in-depth geological evaluation of particular sites for such storage systems.

REFERENCES

1. Meyer, C. F. and Todd, D. K.; "Conserving Energy with Heat Storage Wells", Environmental Science & Technology, Vol 7, No. 6, (1973).

2. Meyer, C. F. and Hausz, W., "A New Concept in Electric Generation and Energy Storage", Presented at Frontiers of Power Technology Converence, Oklahoma State University (Oct. 1975).

3. Meyer, C. F.; "Status Report on Heat Storage Wells", Water Resources Bull., Vol. 12, No. 2 (1976).

4. Dooley, J. L., Frost, G. P., Gore, L.A., Hammond, R. P., Rawson, D. L. and Ridgway, S. L.; " A Feasibility Study of Underground Energy Storage Using High-pressure, High-Temperature Water", ERDA Contract No. 5(04-3)-1243, RDA-TR-7100-001, National Technical Information Service, U.S. Department of Commerce, 5285 Port Royal Road, Springfield, VA. 22162 (Jan. 1977).

5. Collins, R. E., "Geothermal Storage of Solar Energy", in "Governors Energy Advisory Council Report, Project NT-3," A. E. Dukler, Executive Director. The University of Houston; Houston, Texas (Nov. 1974).

6. Davison, R. R., Harris, W. B. and Martin, J. H., "Storing Sunlight Underground", Chemtech (Dec. 1975).

7. Collins, R. E. and Davis, K. E., "Geothermal Storage of Solar Energy for Electric Power Generation," Proc. Int. Conf. on Solar Heating and Cooling, Nejat Veziroglu, Ed. University of Miami, Dec. 1976, (in Press).

8. Tsang, C. F., Goranson, C. B., Lippman, M. J. and Witherspoon, P. A., "Modeling Underground Storage in Aquifers of Hot Water from Solar Power Systems", presented at International Solar Energy Society (American Section) "Solar World" meeting Orlando, Florida (June 1977).

9. Parks, J. W. and Thomas, D. C., "Plans for operating enrichment plants and the effects on uranium supply, "Presented at Uranium Industry Seminar, U. S. ERDA, Grand Junction (Oct. 1976).

10. U. S. Department of Interior, Energy Perspectives (Feb. 1975).

11. Prengle, W. H. (Jr.), Crump, J. R., Fang, C. S., Grupa, M., Henley, D. and Wooley, T., "Potential for Energy Conservation in Industrial Operations in Texas" in "Governors Energy Advisory Council Report, Project S/D-10," The University of Houston, Houston, Texas (Nov. 1974).

12. Op. Cit. Collins, R. E. (1974).
13. Op. Cit. Collins, R. E. and Davis, K. E. (1976).
14. Op. Cit. Dooley et al (1977).

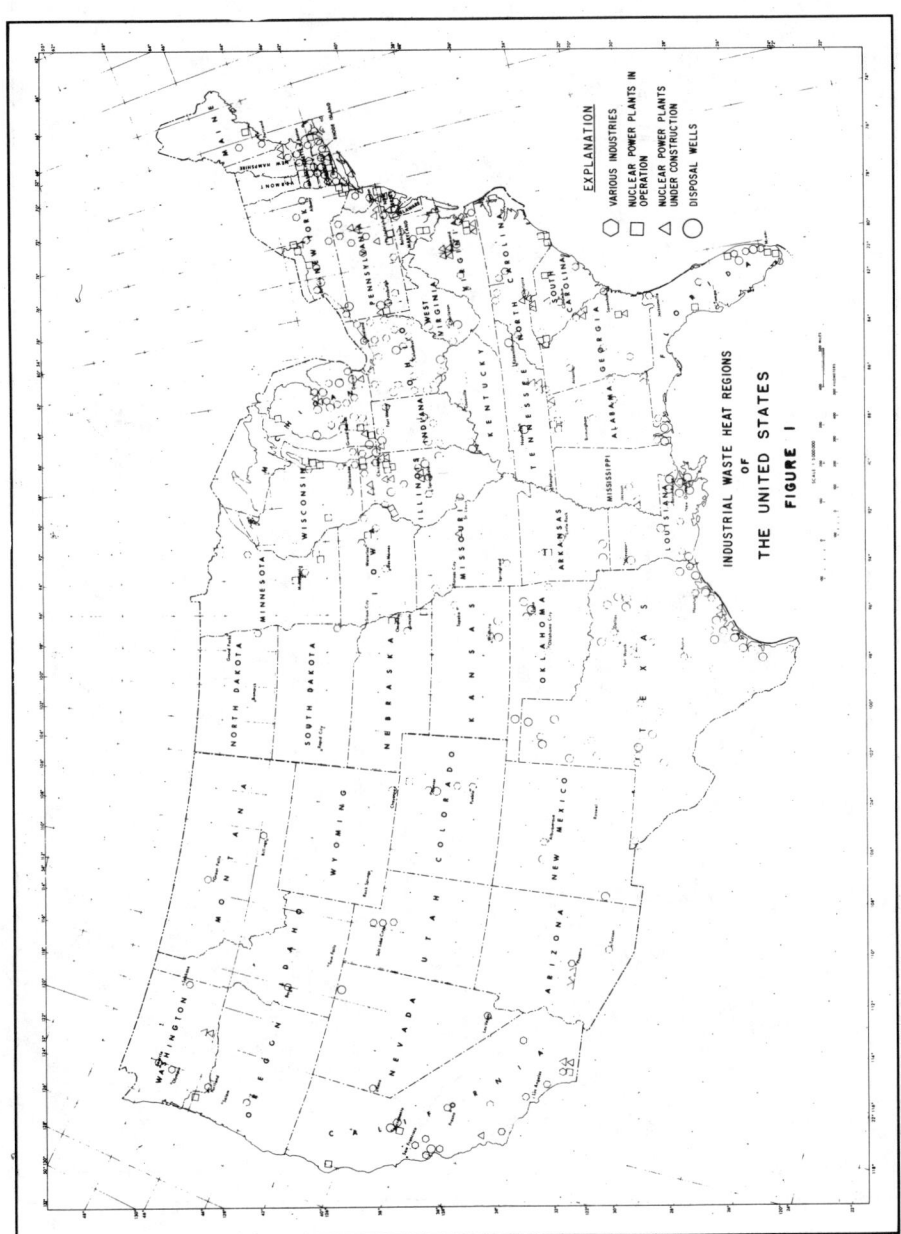

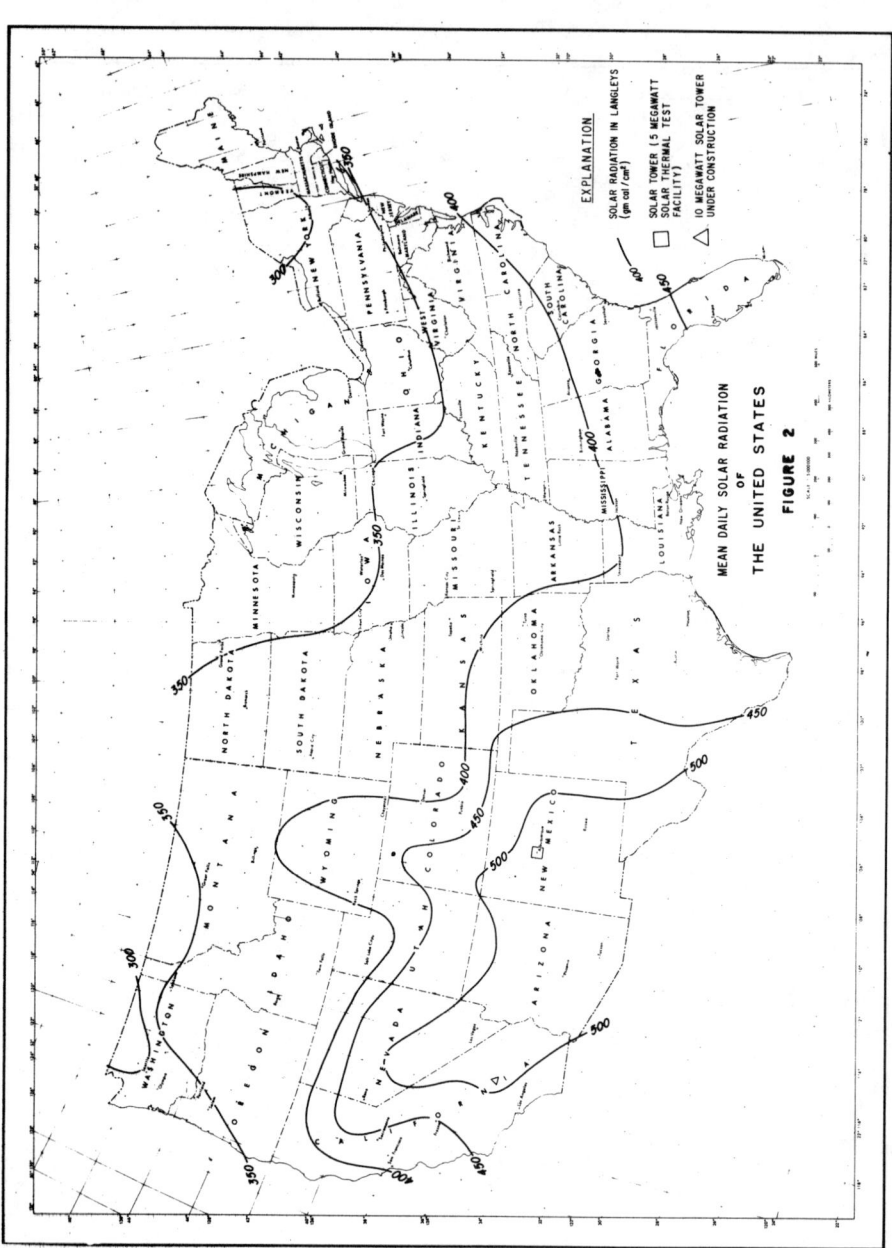

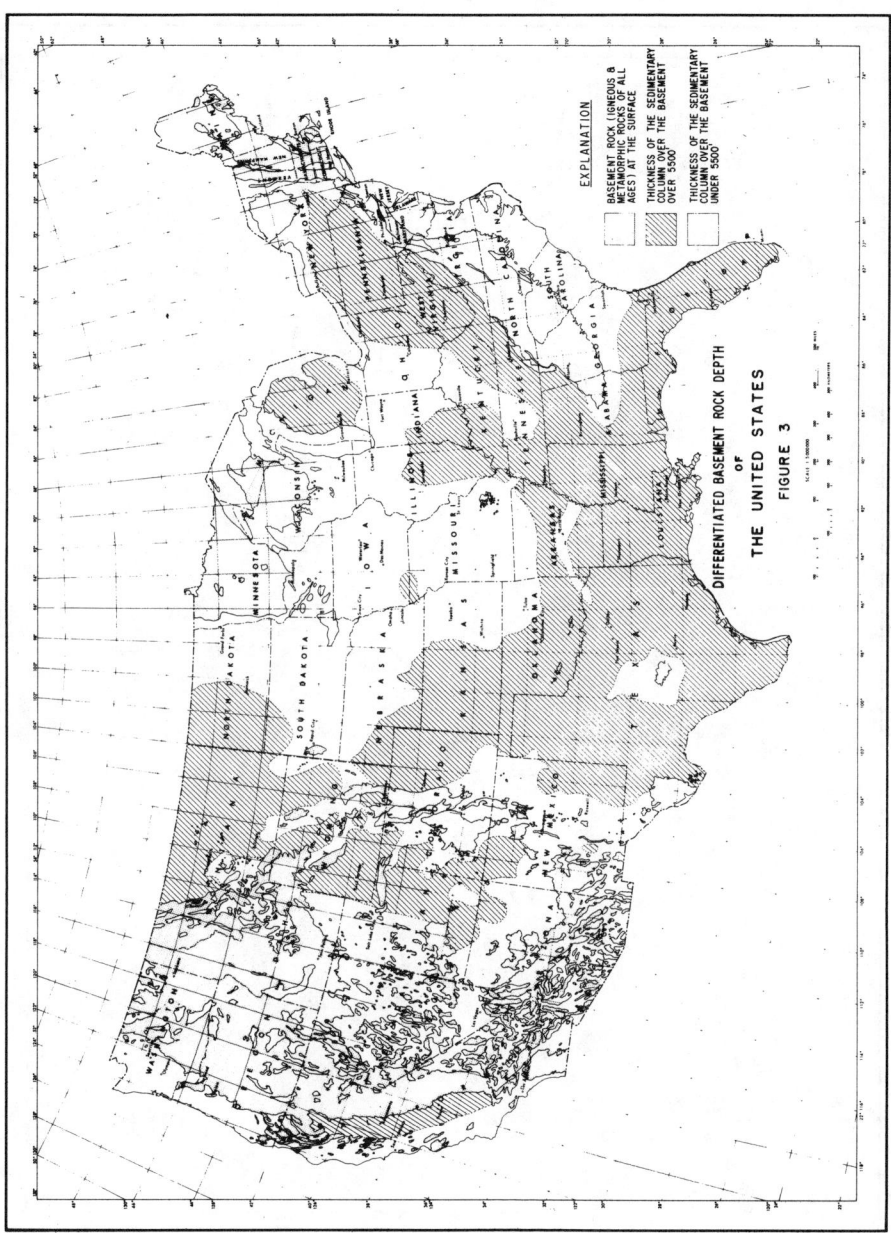

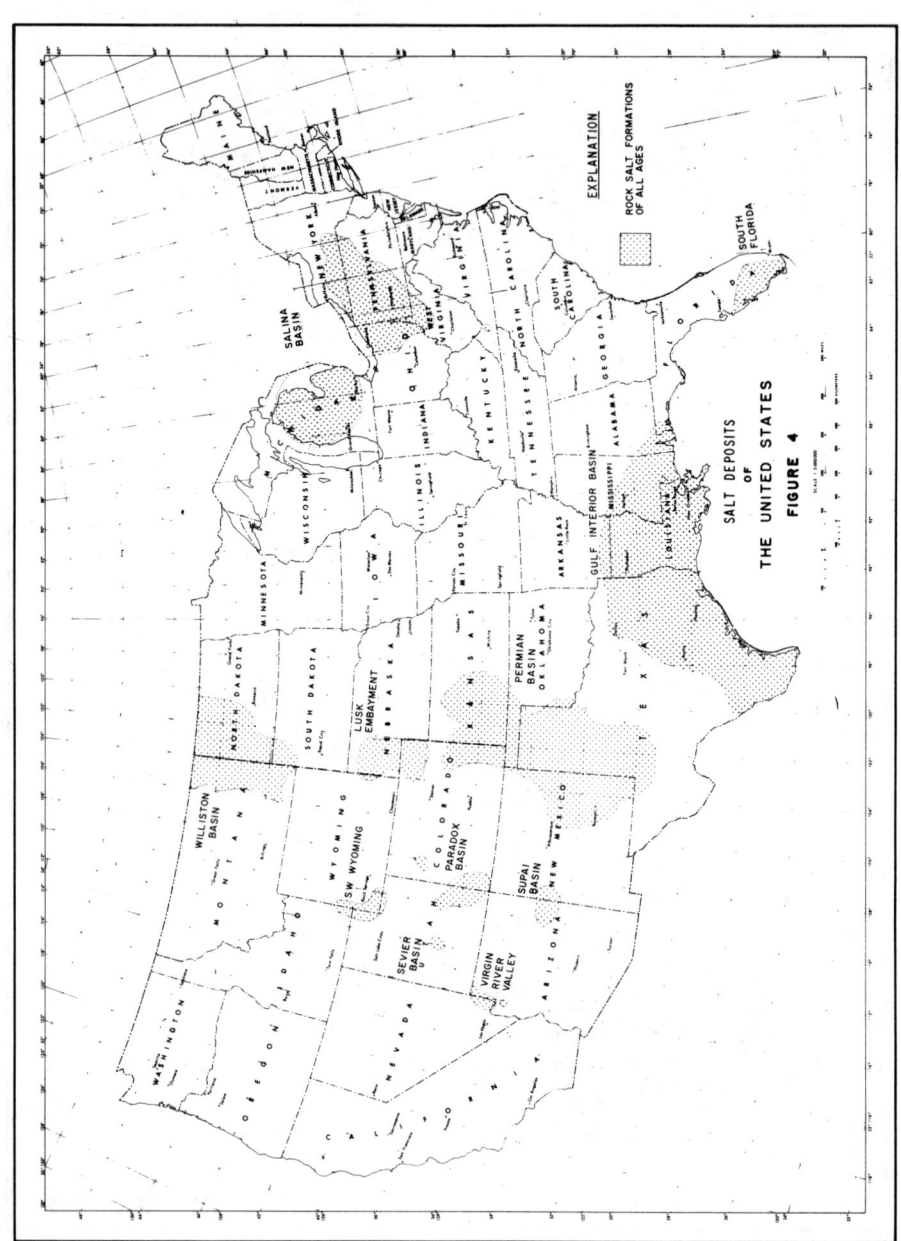

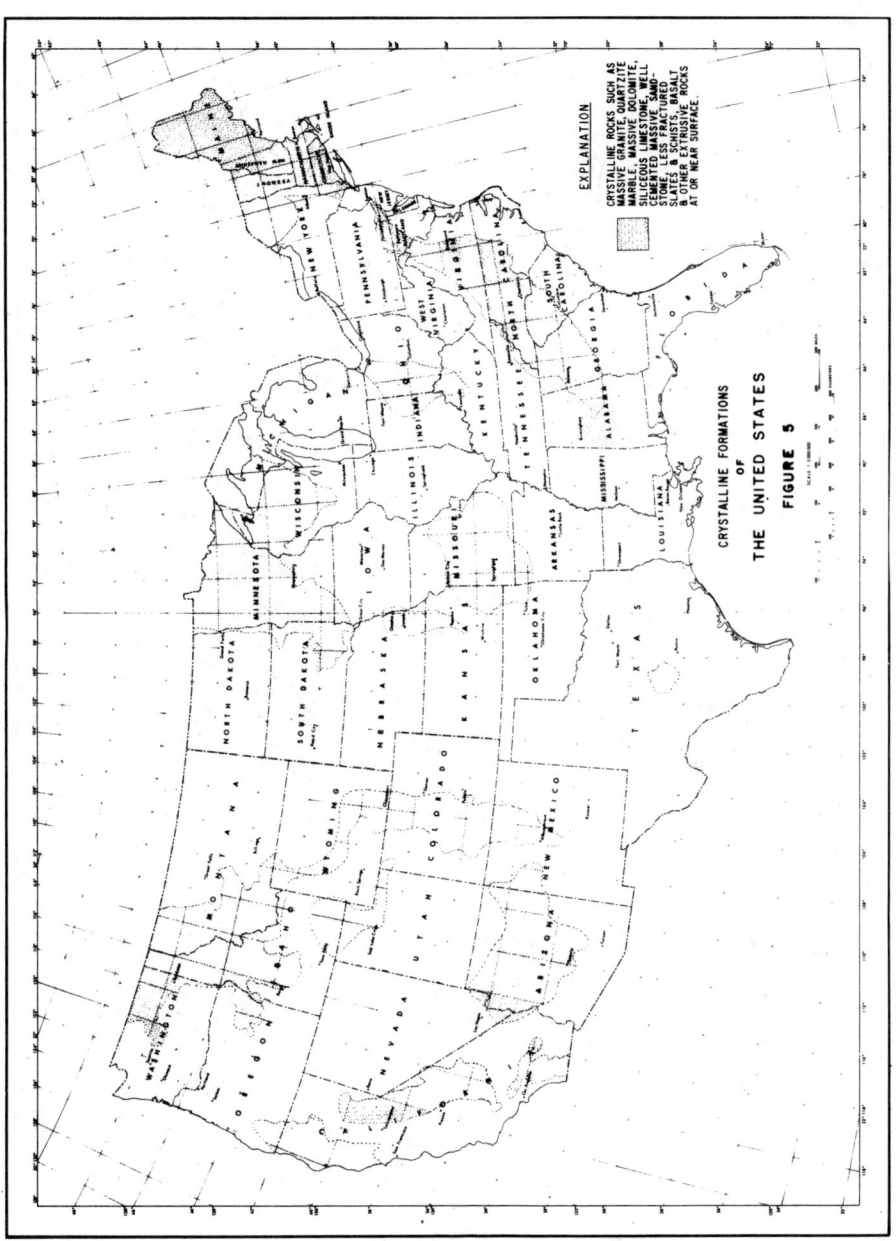

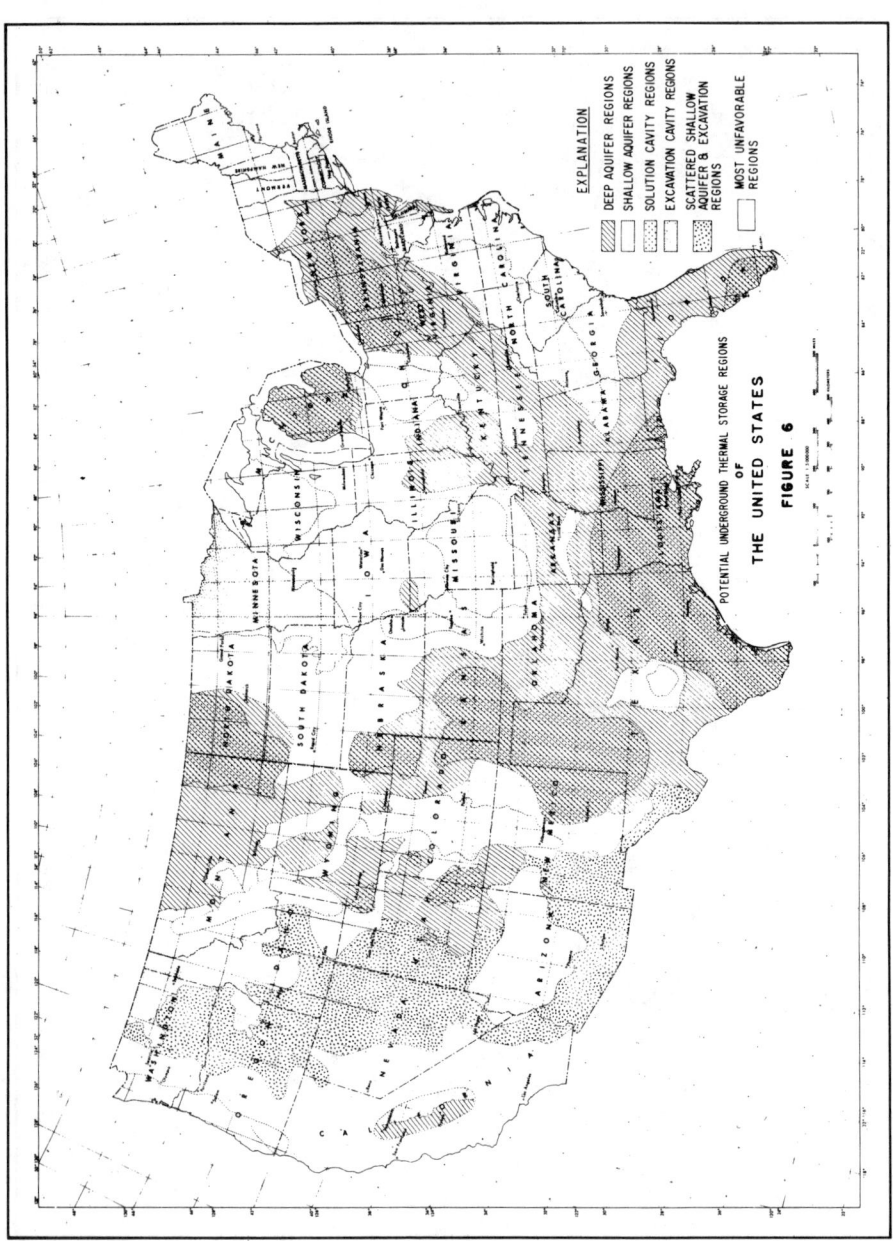

POTENTIAL UNDERGROUND THERMAL STORAGE REGIONS OF THE UNITED STATES

FIGURE 6

Hot Dry Rock Energy Project*

R. H. HENDRON
University of California
Los Alamos Scientific Laboratory
Los Alamos, New Mexico, USA

ABSTRACT

A proof-of-concept experimental project by the Los Alamos Scientific Laboratory endeavors to establish the feasibility of exploitation of the thermal energy contained in the earth crust where such energy and a transporting fluid have not been juxtaposed in nature. A region of high heat flow and apparently unfaulted basement rock formation was selected. Two boreholes, drilled to a total depth of about 3 km (10,000 ft) and penetrating about 2.5 km (7500 ft) into the Precambrian formation, to a rock temperature of 200°C, have been connected at depth by a hydraulically fractured zone to form the heat extraction surface. Energy was extracted at a rate of 3.2 MW(t) with water temperature of 132°C during a 96-h preliminary circulating test run performed late in September 1977. This paper traces the progress of the project, summarizes procedures and salient events, and references detailed reports and specialized topics.

INTRODUCTION

Geothermal energy has, of course, been utilized in a number of locations and for different uses depending on the intensity and quantity of the source. In Iceland and New Zealand large quantities of hot water are profitably used for heating and for power generation. Lardarello, Italy, the Geysers, California, and areas of New Zealand are examples of a high intensity energy source (dry steam) being utilized for power generation.

An innovation of the geothermal scheme was patented by three staff members at the Los Alamos Scientific Laboratory (LASL) [1,2]. The idea was to drill two holes into the hot earth crust and make a flow passage with large surface area between the two dry wellbores. Injecting water into one hole and extracting the same water, now heated, from the second wellbore provides hot pressurized water for power generation or thermal processes. Three attractive features of this scheme are: with deep drilling the locations of energy sources are probably more widespread and numerous [3,4] than known or expected for hydrothermal sources which depend on the combination of underground water and high-temperature regions; extracting or transporting pressurized water (rather than steam) permits more energy to be extracted in a comparable hole diameter; environmental impact of operations is

---
*Prepared for DOE under Contract No. W-7405-ENG-36.

expected to be minimal. There are, of course, many details and perplexing problems, technical and economic, to be solved to attain useful application of this concept. The program objective at the LASL is to develop methods and technology toward that application. This paper is a brief summary of the sequence of operations, the technical events encountered, current status, and ideas on applications and resources.

PROOF OF CONCEPT OPERATIONS

The LASL is located on the eastern flank of a dormant volcano in north central New Mexico. The volcano erupted some million years ago and subsided into its own empty magma chamber forming what is now the Valles Caldera. As a result of this activity there is still a large amount of heat remaining in the formations underlying the general area and within a few kilometers of the surface. The proximity of the Valles Caldera and LASL was obviously fortuitous and preliminary activities were begun toward a development program.

Preliminary Operations

Available geologic studies suggested that a large unfaulted mass of basement granite close to the surface lay to the west and southwest outside the caldera rim [5,6]. Permission was obtained from the Santa Fe National Forest to conduct heat flow investigations on Forest lands. These investigations were begun in 1971 with a series of holes to depths of about 30 m (100 ft) in which temperature gradient measurements were made. Results were encouraging when considered with the available geologic, geophysical and hydrologic information. A smaller array of deeper holes, to about 200 m (600 ft) produced values of about 5 heat flow units (HFU).*

The next steps were to address some of the specific problems of basement rock characteristics and the realities of drilling and fracturing in the hopefully unfaulted granitic formation. Thus a Granite Test-1 (GT-1) exploratory hole was drilled in 1972, west-southwest of the caldera. The hole was completed at 785 m (2576 ft) after drilling about 150 m (500 ft) into the basement rock. The permeability was low and at this relatively shallow depth the temperature of 100°C was also encouraging. This experience indicated that hard rock drilling problems were not great and moderate pumping pressures had achieved an hydraulic fracture of the granite.

Drilling of GT-2

With such encouraging preliminary information and specific funding from the AEC, the next and more extensive phase of operations was undertaken. A site for a deeper granite test was selected (Fig. 1) in a more accessible and serviceable location about 1.5 km (1 mile) south of the GT-1 site and 1.5 km west of the outermost ring fault of the Valles Caldera (about 13 km

---

*HFU = $\mu cal/cm^2$-sec. The worldwide average heat flow is 1.5 HFU.

from the center of the caldera). The 2650-m (8700-ft) elevation and 36° latitude of the Fenton Hill site presents interesting and challenging winter operating conditions.

Drilling of GT-2 began in February 1974. The target was 1360 m (4500 ft) and 200°C -- based on the preliminary data. However, as drilling progressed the early heat flow data was shown to be misleading. After confirming the 5 HFU measurements in the volcanic and sedimentary formations, a lower value of 3.7 HFU was determined for the Precambrian basement rock. Casing was set into granite (encountered at 733 m [2404 ft]) to 773 m (2535 ft) after considerable wildcat drilling difficulties, such as swelling clays and faulted limestone formations. Drilling continued to 2042 m (6700 ft) using 24.4-cm (9-5/8-in.) diam, full-face cone rock bits with tungsten carbide buttons. Hydrologic and fracturing experiments and core samples were encouraging and it was determined to proceed with experiments at the site. Drilling resumed and GT-2 was deepened to 2932 m (9619 ft) where an equilibrium rock temperature of 197°C was recorded (Fig. 2). The hole had drifted west-northwest and from 1 to 4° from vertical without directional correction. A 185-m (608-ft) liner was set 11.6 m (38 ft) off bottom to provide a good seat for packers required for fracturing experiments. Hydraulic fracturing experiments were then conducted in the open hole section at bottom through tubing landed in the liner. The rig was released in December [7,8]. Later, with a work-over rig more fracturing was done through perforations in the liner [9]. A near vertical, 122-m (400-ft) radius fracture was thought to have been created near the bottom of the hole.

Drilling of EE-1

In May 1975, concurrently with the workover rig and experiment/developmental program in GT-2, another drilling rig began drilling EE-1 (Energy Extraction-1) about 77 m (252 ft) north-northeast of GT-2 [10]. Drilling proceeded to 1957 m (6420 ft) where 27.3-cm (10-3/4-in.) casing was set and cemented to facilitate additional fracturing experiments. The drilling had drifted much as GT-2 had. Directional drilling techniques were then used to direct the hole to intercept the fracture from the GT-2 borehole. The directed path ultimately formed a 205° spiral, counter-clockwise from an initial northwest heading to a northeast heading (Fig. 3), and 965 m (3176 ft) to total depth of 3062 m (10,047 ft). Flow between the two boreholes was established on October 14, 1975. A final 19.4-cm (7-5/8-in.) diam casing was cemented into EE-1 to a depth of 2926 m (9600 ft).

A series of experiments ensued to gain information of the downhole configuration, including flow characteristics, fracture location, dimensions and behavior, and in situ rock properties. Flow experiments quickly revealed that the connection was not adequate for energy extraction. The impedance (pressure drop per unit volume of flow) was much too high. Seismic ranging tests with a source in one hole and detector in the other indicated a closest distance between holes of about 7 to 10 m (22 to 31 ft). An attempt to improve the flow was made by leaching silica from the rock with injected sodium carbonate solution; no improvement was detected. Final confirmation that the EE-1 drilling had not intercepted a fracture from GT-2 came when two independent surveys with newly-developed, temperature-hardened gyroscopic instruments were obtained. The spiraling EE-1 had turned short of the GT-2 borehole and fracture. Fracturing efforts in this borehole seemed to produce fractures but no significant improvement of flow path was achieved. Figure 4 shows actual drill paths as established by information described under Instrumentation and Logging.

Redrilling of GT-2

The accumulated evidence of ranging experiments, repeated borehole surveys, attempts at fracture identification or mapping, and the high flow impedance were convincing for the need of redrilling and provided guidance in planning a redrilling operation. Early in April 1977 a drill rig started operations over GT-2 with the goal of intercepting a fracture zone extending from EE-1. After setting a cement plug in GT-2 at 2530-m (8300-ft) depth, the sidetracking (i.e., drilling a new path from the side of an existing borehole) attempt was begun. Several attempts were necessary to achieve a sidetrack (a routine oil field operation) in the unusual environment of unbroken granite. The first trajectory (GT-2A) intersected a flow path from the pressurized EE-1 hole at a depth of 2630 m (8645 ft). However, flow impedance was still too high to be useful. A second sidetracking operation, this one from GT-2A, was accomplished and the path extended to an intersection allowing an acceptable impedance at a new total depth of 2700 m (8870 ft). This judgement was based on a 20-h pumping test performed June 3, 1977. Pumping cold water into EE-1 at 6.89 MPa (1000 psi) resulted in flow from GT-2 at 130°C with recovery of 85% of the water injected. Figure 4 shows the sequence of drill paths in plan view. The final path (GT-2B) was cased to 2613 m (8572 ft) with 19.4-cm (7-5/8-in.) casing [11].

Preliminary Energy Extraction

September 26, 1977 was the start date for a preliminary 96-h energy extraction experiment [12]. Piping and heat exchangers had been procured in advance and installed during the redrilling operations [13]. The surface loop is a piping and heat dump system enabling controlled energy extraction from the geothermal heat source. The schematic piping diagram (Fig. 5) shows the pressurized water piping from extraction wellbore through the heat exchanger (a water-to-air cooler), circulating pumps and to the injection borehole.

The borehole "christmas tree" includes valving which will allow shut-in of the underground system and installation of a pressure chamber for introduction of instrument packages into the boreholes with the system fully pressurized. The heat exchangers are horizontal finned tubes arranged in four parallel bundles and with forced air flowing vertically. The straight tubes with plugged headers may be mechanically or chemically cleaned in the event of scaling.

Heat exchanger design conditions are: 17,235 kPa (2500 psi), 12.1 MW(t)/bay of 2 bundles; and 250°C to 65°C (475°F to 150°F), 18.3 ℓ/s (290 gpm).

The circulating pumps on the cool side of the circuit are canned, multi-staged vertical turbines in a series/parallel arrangement. This provides standby equipment or the ability to run pairs in parallel to increase the flow. The pumps are each driven by 150 KW (200 hp) motors, each series combination will deliver 19 Kg/s (300 gpm), from 1725 kPa (250 psi) inlet to 10,000-kPa (1450-psi) discharge pressure.

Instrumentation will initially provide 80 channels of information for data acquisition and control signals. The incoming signals will be conditioned, and scanned as programmed by an on-line programmable calculator. All raw data is recorded on cassette magnetic tapes with periodic print-out of selected information. Primary data will also be numerically displayed from the timed scans and will be recorded on strip charts as back-up to the calculator system. Twenty channels will provide control signals from the console to operate the system.

The loop was operated with pumping service supplied by the Western Company in lieu of the pumps mentioned above which had not yet been delivered. Operation included several experiments compatible with the primary intent of loop and reservoir testing. These included the determination of the downhole water loss rates and impedance, a sodium fluorescein-dye residence time study, a water chemistry survey, seismic monitoring, and remeasurement of rock diffusion properties. The first 34.5 h were run with GT-2 effluent diverted to a catch pond rather than through the heat exchangers. The throughput of about 1125 m^3 (300,000 gal) during this period was intended to flush the borehole and reservoir and to provide baseline water chemistry data. Outlet temperature increased from static ground temperature to above boiling resulting in the steam plume shown in Fig. 6A. The loop was then closed for circulating operation for the remaining 61.5 h. Flow stabilized at about 9.6 ℓ/s (150 gpm) injection rate with gradually increasing outflow reaching a maximum of 7 ℓ/s (110 gpm). The impedance was 1300 kPa per ℓ (12 psi per gpm). Maximum thermal power and outlet temperature were 3.2 MW(t) and 132°C respectively. Injection pressure was nominally 7000 kPa (1000 psi) for 50 h and 9260 kPa (1325 psi) for 46 h. Water recovery continues long after the experiment with the hole vented to atmosphere.

Brief shutdowns were made for the dye experiment, an open drain valve and a pump changeover. The drain valve, out of sight in a pipe tunnel, had vibrated open due to piston pump pulsations. The pump changeover for service allowed loss of prime on the supercharger pump. The dye-injection studies do not show any dramatic change in residence time caused by the redrilling operation. Loop operation and the experiment were considered successful. A longer operation is planned for this winter when the vertical turbine pumps are installed. A total of 10,000-h operation is presently planned with periodic interruptions to allow instrumentation testing and diagnostic measurements or experiments for reservoir development evaluation.

DRILLING PROCEDURES SUMMARIZED

The wildcat nature of the drilling and related activities soon became apparent with such difficulties as lost circulation and swelling clays encountered in the volcanic and sedimentary overburden. Many successful uses of conventional drilling equipment and techniques did enable completion of each operation. The following paragraphs are summarization of the salient procedures reported in detail by Pettitt in references [7,8,9,10].

Drilling Generally

The overburden was drilled with steel-toothed tricone rock bits and reamers. Tricone rock bits with tungsten-carbide buttons and sealed bearings were used to drill in the crystalline basement rock (encountered at about 730 m [2400 ft]). Bits from Smith Tool Company, Security, and Hughes Tool Company were used depending on availability and apparent performance. Development or upgrading of bits by the manufacturers has shown in improved performance in subsequent operations. Records and comparisons, with qualifications due to field conditions, are reported in the cited references. Good performance rates were about 2 m/h (6.5 ft/h) and 200 to 250 m (650 to 820 ft) drilled in the granite using thin mud.

Attempts were made in early drilling of GT-2 to drill with air as the circulating fluid. Due to water entering the hole from an upper-hole source, a change was made to water circulation and performance improved so all subsequent drilling has utilized water with Aquagel added to maintain a weight of slightly less than 1 kg/liter (9 lbs/gal). Lost circulation problems necessitated many other temporary additions.

Cementing

Cement has been used for a variety of purposes throughout the drilling, including setting casing, sealing or plugging lost circulation zones, plugging the hole for sidetrack drilling, and plugging the hole to assist packer/fracture operations. The drill string was inadvertently cemented in the hole when setting under pressure and temperature was extremely accelerated. Cements for long life under high-temperature recycling applications are still in a developmental stage by various laboratories. Cooperation of the service companies to control cement used for our applications has developed as the project grew and they better understood the problems and conditions.

Packers

Inflatable packers placed in the open hole or in casing with drill pipe or in tubing to the surface are utilized for hydraulic fracturing and experiments prior to setting casing. Packers are generally reinforced elastomeric sleeves and a mechanical frame structure. Attempts to seal the open borehole were only partially successful; even in the granite the walls vary in bore and smoothness and the expansion capability is a basic design limitation. Temperature formulations are, of course, special order and limited even then. Some packers have separated, i.e., left the elastomer in the hole to be circulated out as pieces. Others have been successful in making impression records (interpreted as fracture lines) when inflated against a pressurized fracture zone.

Coring

During the planning of drilling for GT-2, it was hoped to recover granite cores on a continuous basis. This was to be done with a four-cone tungsten-carbide button bit developed for the Joint Oceanographic Institutions for Deep Earth Sampling (JOIDES), which incorporated a wire-line retrieval device operating through the drill pipe. The system was modified to use with air drilling and to enable orientation of the core. The initial attempts were not satisfactory and the several bits procured were used for normal drilling or for periodic coring only. Further modification increased core recovery rate to 70 to 90%, but the continuous recovery plan was abandoned. Diamond bits were used for periodic coring runs throughout the drilling of GT-2 and EE-1 obtaining samples from 2 to 5 ft long. Twenty-six core runs were made in GT-2 and two were made in EE-1 at the depths at which fracturing experiments were to be conducted. Several core runs were made in the redrilling operations but with less success than previously. A final core run was successfully made with a developmental bit from Smith Tool Company and adapted for retrieval by American Coldset. The bit combines roller cones and industrail diamonds (compacts). Coring in the crystalline basement has been successful, but slow.

Fracturing

The GT-1 preliminary operation included fracturing experiments. Figure 7 shows the volume-pressure relation clearly indicating the initiation of breakdown point for fracturing and the lack thereof on repumping. This and other pumping-venting sequences are reported in [14]. Typically hydraulic fracturing has been accomplished with a commercial pumping unit (truck-mounted rig by the Western Company). The breakdown pressure of 17,350 kPa experienced in GT-1 was also found effective at a depth of 2030 m (6700 ft) in GT-2. However, further experimentation in GT-2 produced apparently inconsistent results. Fracturing through perforations in the liner deep in GT-2 seemed to be straight forward, but showed no sharp breakdown pressure and repumping required higher pressures. A fracture was created at depth in GT-2 but due to an inaccuracy in pressure measurements the conditions of fracture formation were uncertain. During the EE-1 drilling fracture experiments were conducted at an intermediate depth of 1975 m (6480 ft) in a competent section of the formation. Breakdown pressure was 13,200 kPa (1900 psi) and a subsequent fracture extension pressure was measured at 9500 kPa (1375 psi). After drilling to 3100 m (10,020 ft) and not intersecting the GT-2 target fracture, another fracture in EE-1 was tried. It was hoped to intersect either the GT-2 borehole or fracture system, and to obtain better data than that from the deep GT-2 fracture experiment. A fracture was created at about 2800 m (9600 ft) after the pressure had risen to 16.6 MPa (2400 psi) at a pumping rate of 0.4 m^3/min (2.1 bbl/min). Pressure variations in EE-1 due to packer failure were then followed in the pressurized GT-2 borehole. A connection at depth had been achieved.

Directional Drilling

The Dyna-Drill was chosen as the basic tool for turning the second hole to intercept the target fracture. The Dyna-Drill is a downhole mud-driven motor which rotates the bit as the drill string remains stationary. A bent sub or adapter between motor and drill pipe causes the drilling to proceed in an arc. The everchanging bottom-hole direction must be monitored in short intervals by single-shot magnetic surveys or by an electronic transmission of such survey information to surface readout equipment. In a deep hot hole both methods have limitations. The Dyna-Drill itself depends on elastomer materials in the rotor and for seals, thus limiting equipment life and the length of each drill run. Drilling in EE-1 alternated between turning the hole -- then a straight rotary run -- then another directional change and so on. The manufacturer was able to make certain modifications to bearing seals for longer life and one unit with a modified cooling passage was held for hotter environments in the GT-2 redrill but not used.

A second aspect of the redrilling previously mentioned was sidetracking to a new direction from a point well above the bottom of a borehole. This was first attempted in the basement rock with a Dyna-Drill equipped with a diamond bit designed with cutting surfaces on the side. With light load on the bit to allow the side forces to act, a hole is started after cement plug is set in the original hole. Repeated attempts with this arrangement failed; underreaming was done to provide a shoulder or frontal cutting surface for the bit to work on. The sidetracking was accomplished with a tricone tungsten-carbide button bit on the Dyna-Drill and probably aided by the underreamed shoulder. The second sidetracking was intiated similarly, but with two changes. The new direction was from the bottom side of the sloping borehole, and the point chosen indicated a ledge or oversize bore on the caliper log.

INSTRUMENTATION AND LOGGING

A wide variety of downhole measurements have been an important part of the project at Los Alamos during drilling and for experiments. Much of the work is done by or in conjunction with commercial logging service companies. Another approach has been to develop and build instruments for particular measurements deemed necessary or not available commercially. Some commercial logging instruments have been designed or redesigned to meet our requirements. Recording has also become a cooperative affair utilizing equipment of the service company and that of the LASL in various combinations. The discussion following does not cover in detail the methods or interpretations but is intended to ennumerate and identify and does not differentiate "measuring" from "logging". While proof-of-concept will come from the operation or heat extraction process, an understanding of the application is wholly dependent on data gathered.

Several service logging companies have been employed for logs during the drilling operations and for many of the experiments between or since the drilling operations. Many of the logs were run and rerun as drilling progressed, especially in GT-2, the first hole at the developmental site and, of course, the first deep penetration of the granitic basement rock.

Surveying

During the EE-1 drilling to intercept a fracture near the bottom of GT-2, the relative positions of the two holes had to be determined. Magnetic and gyroscopic survey services were available commercially. Although gyroscopic surveys are free of effects of magnetic anomalies that might occur in the basement crystalline complex, they were then limited to a maximum hole temperature of about 150°C (300°F) or about 1830 m (6000 ft). The magnetic survey equipment was heat shielded and capable of operating at hole temperatures of about 200°C (390°F) for up to 2 h. Both surveys were used and unfortunately the results of these comparative surveys were inconclusive. The uncertainty in hole location of GT-2 after surveys in 1975 and a year later, at a depth of 1776 m (5400 ft), was about 4 m (12 ft) and probably 7 m (21 ft) at 3-km (10,000-ft) depth.

An interesting aspect of the EE-1 surveys is the consistent angular change between the magnetic and gyroscopic surveys. At depths of 1401 m (4300 ft) and 2042 m (6100 ft) -- and throughout the traces from the Precambrian surface to 2042 m (6100 ft) -- the hole coordinates obtained in the gyroscopic survey are rotated about 7.4° counter-clockwise from those obtained in the magnetic survey. This angular correction may be a function of depth (i.e., temperature) or elapsed time (i.e., precession error) or it may indicate an actual magnetic shift within the Precambrian section. While no conclusions are justified, it has been observed that there is a directionally consistent rotation of the paleomagnetic field as determined from core samples [15].

The methods utilized to determine paths for directional drilling were magnetic surveys. A magnetic survey device with surface readout and on-site computer system was used during the EE-1 operation and was also used in conjunction with magnetic single shot surveying for the GT-2 redrilling operation. The high downhole temperatures continued to create problems for the electronic guidance device.

Prior to the GT-2 redrilling, complete (to total depth) gyroscopic surveys were made of both holes with two different instruments. Two companies were employed to temperature harden their gyro equipment and resurvey both holes. The results are within limits for repeatability and accuracy and showed the boreholes to be within about 8 m (25 ft) at a depth of 2.9 km (9600 ft). The variation in surveys in one hole is ±3 m (±10 ft) at bottom. A gyro survey was made for recording the path of GT-2B (the final redrilling effort) [16,17].

Seismic-Acoustic Measurements

Acoustic and seismic measurements have been employed for a variety of purposes. Seismic monitoring related to possible earthquake activity is part of the environmental monitoring program. Diagnostic measurements are generalized as acoustic in the following discussion, and detailed data and interpretations are to be found in the references cited.

Five seismic detection stations, each consisting of triaxial geophones, have been placed in a rough circle at about 825-m (2500-ft) radius around the drill site. These contribute to environmental seismic monitoring [18] and were to test the possibility of surface detection of hydraulic fracture events and to aid in locating or mapping a fracture as it was created or extended. These stations are normally monitored continuously. Evidence of fracturing activity has not been detected on these recordings but distant earthquakes, local trucking, and sonic booms are well documented.

Other surface-to-borehole acoustic tests have been conducted with impact sources on surface and geophones at various depths in the boreholes. Surface impact tests were conducted with and without pressurization in the boreholes and fractures in efforts to locate and map the fracture extension. Explosive sources have been used similarly with charges placed in shallow surface holes -- timing of arrival times measures velocities through the volcanic sedimentary granitic combination. This provided a calibration for the surface seismic stations.

A refinement of acoustic measurements by placing a source in one borehole and the geophone package in the other has been rewarding. This technique was first utilized to determine the relationship between the boreholes. These results were approximately confirmed by the successful high-temperature gyroscopic surveys. The technique depends on measuring the transmission time and based on predetermined, in situ, transmission velocities, the distance between source and detector is calculated. Devices have been designed and constructed to provide a mechanical downhole source and to mechanically couple the detector to the borehole wall.

This system has been employed, with adaptations, for fracture mapping [19]. A series of base measurements is made with a geophone package fixed in one hole and detonator shots provided wave sources at different depths from the second hole. The base measurements are made with hydrostatic pressure in the holes. Measurements are then repeated at identical positions but with the fracture zone pressurized, or inflated. Careful analysis of the geophone traces shows a delayed arrival of the shear wave relative to the pressure wave and the base measurements. Extending the string of measurements vertically identified the upper limit of the fracture, Fig. 8, which was subsequently intersected during redrilling (Path GT-2A and GT-2B). Improvements in acoustic sources, filtering and analytical techniques (computer software) permit use of commercial acoustic logging sources in lieu of detonators.

The acoustic detection of fracture activity by geophones located downhole has also contributed to fracture mapping [20]. With geophones in place at depth, a fracture from the other hole was pressurized to the point of fracture extension. By plotting the direction from which an event approaches the detector and the pattern of arrival times (i.e., distance) a line is formed. This line is interpreted as a line in the fracture plane (Fig. 9). A series of such lines at sequenced depths provides a plot of the fracture zone.

Refinement of acoustic measurements has been possible only with developments to increase signal strength over the long cable. This has been accomplished by placing in the downhole sonde with the three orthogonal geophones thermally-isolated battery-powered operational amplifiers. Ice-filled glass dewars encase the battery-amplifier units and allow downhole operations for several hours at up to 200°C [21].

Temperature Measurements

The first need for temperature measurements came during initial drilling to ascertain rock temperatures so necessary to the project. A variety of temperature probes were designed and fabricated to suit bottom-hole conditions. Temperature measurements were originally planned for 91-m (300-ft) intervals. However, time required to approach equilibrium rock temperature discouraged this and eventually led to a technique for measuring the rate of recovery toward that equilibrium temperature and an emperical extrapolation formula which was accurate and time saving [22] The measurement technique places the sensor on bottom rock with a convection-limiting "pill" of mud surrounding and the drill pipe in the hole.

Temperature anomalies detected at certain elevations in the borehole were interpreted as indications of possible water flow zones or natural fractures. An extension of this in the completed or hydraulically fractured hole is to find flow paths to or from the boreholes in the fracture zone. Detecting the point of flow to or from the borehole requires a moving sensor in a pressurized or flowing hole. A pack-off assembly allows the movement of the logging cable with the hole pressurized. Thermistors were first utilized for such measurements by incorporation in a bridge circuit which must be staged for limited pressure ranges. The design of a thermopile sensor using conventional thermocouples and a downhole thermally-isolated reference junction has improved temperature logging capabilities. Logging rates are increased from 50 ft/min to 200 ft/min with the thermopile tool [23]. Thermal isolation for this sonde depends on ice-filled glass dewars as for the geophone package.

Geophysical Logging

The well developed and commercially available logging services are based on the sedimentary formations of interest to oil, gas, and water drilling. The physical differences encountered are more variable in such upper formations than encountered in the crystalline basement rock under Fenton Hill;

thus interpretations may not be as positive as desired. The available
logging was well utilized and correlations between the different logs
helped to confirm the interpretations.

The variations in borehole as shown by caliper logs have been correlated
with neutron and self-potential logs to identify fractures. Fractures
as detected here include those natural fractures in the granites which
were shown to be well sealed. The resistivity log suffers from the consistency and high resistance of the granites, but when used with other
indicators is confirmatory and is used as one of the guides to determining
relative permeability. The induced potential log using two holes has been
used to aid in identification of hydraulically created fractures [24],
allowing inference of total height of the fracture.

The spectral gamma log, which records total gamma activity and concentration of thorium, uranium, and potassium, leads to the lithology
definition more than any other single log [25]. In GT-2 the thorium
showed an increase from about 6.8 ppm down to 2591 m (8500 ft) to 11 ppm
at that depth; the content varied from 2 ppm to 23 ppm. The uranium
content often exhibits a high in or close to fracture zones. Because
uranium is more mobile than thorium, a depth plot of the ratio of activities
of thorium to uranium accentuates fracture zones. The uranium content
varies from almost zero to 32 ppm. The potassium content showed as
relatively low, a contributing factor to the low total gamma count. The
spectral log was again used in completion logging of GT-2B with good results
in identifying lithologic variations [26].

The full-wave sonic log shows fracture zones (unsealed) clearly, even to
seeing in some cases the later arrival of the shear wave. The sealing of
fractures inhibits the separating effect of fractures on the wave transmission. The full-wave sonic log in conjunction with the density log was
used to compute the _in situ_ elastic constants of the rock deriving values
in reasonable agreement with those measured in the laboratory from core
samples.

The neutron log, a good indicator of dry rocks when compared to the high
activity counts in zones of high water content, has been of little quantitative value and has contributed little to lithologic interpretations at
depth.

The caliper log mentioned above was also used in drilling decisions and in
planning casing-cementing operations. The cement bond log first used to
confirm adequacy of cementing was later used to detect suspected changes
such as bypassing flow past the liner in GT-2 and to explain an anomaoly
in a temperature log of EE-1 which occurred in the cased portion of the hole.

Iodine tracer logs have also been very beneficial for special questions.
During the drilling of EE-1 a tracer provided the convincing evidence and
location of drilling fluid loss through the casing (worn through by drill
pipe) high in the hole. Another tracer log confirmed the temperature log
indication that flow from EE-1 went out the bottom of the casing then

flowed upward close to the casing and finally outward to GT-2 at about 2750 m (9050 ft). This log also indicated the probable existence of more than one flow path from EE-1 to GT-2 and proportions of flow in each path.

RESULTS

Precambrian granitic rock has been drilled to a depth of 3 km (10,000 ft) and at this location and depth the temperature is 200°C. The thermal conductivity has been measured in the laboratory [27] from core samples and in situ [28]; good agreement of these efforts gives a value of 2.9 $\overline{W/m\text{-}K}$ (1.7 Btu/hr-ft-F). The temperature gradient in the Precambrian rocks, from 733 m (2400 ft) to 3 km (10,000 ft), is 60°C/km at the Fenton Hill site [22]. Determination of earth stresses at various depths have been made and are reported [29]. The relation of stresses confirms that fractures at depth are vertically oriented. Fluid losses have been studied over short terms; the results appear to be within acceptable limits. This important factor must be confirmed with long term measurements. Another condition basic to the hot dry rock concept is that of fracture extension. It is theorized that as the fracture surfaces cool stress cracks will occur due to contraction, thus extending the active (hot) surface and the useful reservoir life.

Based on an expression for change in fracture pressure under a constant water injection rate, a parameter is derived: $A\sqrt{kc}$, where A = fracture or permeating surface area, k = permeability, and c = compressibility. A historical summary of values of $A\sqrt{kc}$ is presented [28]. The history shows increases related to pressurizing experiments when the pressures have exceeded 90 to 94 bars (1300 to 1360 psi). This is then interpreted as the fracture extension pressure and from this a fracture radius (from EE-1) is calculated to be 140 m (460 ft). This approach has been in general agreement with radii predicted from the acoustic mapping efforts. It was found that after fracture extensions the hydraulic conductivity was large. This is attributed to a self-propping mechanism (i.e., the fracture faces do not seat exactly). The Na-fluorescein dye experiments showing residence time also support the concept of self-propping, as do laboratory measurements of fracture surface characteristics.

The effect of pore pressure on rock properties, such as thermal conductivity, compressibility, and permeability, has been studied. The above mentioned $A\sqrt{kc}$ parameter is a function of this changing condition, which enters into all reservoir assessments.

The flow impedance in terms of kPa-min/liter (psi-min/gal) has been determined for many conditions throughout the fracture and reservoir experimentations. It appears to be a function of absolute pressure as well as pore pressure or permeability (thus pressure and time).

The preliminary evaluation of the 96-h circulating test indicates an impedance of 31.5 kPa-min/liter (12 psi-min/gal) and a possible output of up to 5 MW(t) with the present fracture system [28].

REFERENCES

1. M. C. Smith, R. M. Potter, E. Robinson, "Method of Extracting Heat from Dry Geothermal Reservoirs," U.S. Patent No. 3,786,858, January 1974.

2. M. C. Smith, R. L. Aamodt, R. M. Potter, and D. W. Brown, "Man-Made Geothermal Reservoirs," Second United Nations Geothermal Energy Symposium, San Francisco, May 19-29, 1975.

3. D. W. Brown, "The Potential for Hot-Dry-Rock Geothermal Energy in the Western United States," published in <u>Geothermal Energy: Hearings before the Subcommittee on Energy of the Committee on Science and Astronautics</u>, September 11, 13, and 18, 1973, pp 129-141.

4. R. M. Potter, "Assessment of Some of the Geothermal Resources of the Eastern United States," published in Near Normal Geothermal Gradient Workshop, March 10-11, 1975, ERDA 76-11; Conf-750366.

5. W. D. Purtymun, "Geology of the Jemez Plateau West of the Valles Caldera," Los Alamos Scientific Laboratory report LA-5124-MS (February 1973).

6. F. G. West, "Regional Geology and Geophysics of the Jemez Mountains," Los Alamos Scientific Laboratory report LA-5362-MS (August 1973).

7. R. A. Pettitt, "Planning, Drilling, and Logging of Geothermal Test Hole GT-2, Phase I," Los Alamos Scientific Laboratory report LA-5818-PR (January 1975).

8. R. A. Pettitt, "Testing, Drilling, and Logging of Geothermal Test Hole GT-2, Phase II," Los Alamos Scientific Laboratory report LA-5897-PR (March 1975).

9. R. A. Pettitt, "Testing, Drilling, and Logging of Geothermal Test Hole GT-2, Phase III," Los Alamos Scientific Laboratory report LA-5965-PR (June 1975).

10. R. A. Pettitt, "Planning, Drilling, Logging, and Testing of Energy Extraction Hole EE-1, Phases I and II," Los Alamos Scientific Laboratory report LA-6906-MS (August 1977).

11. R. A. Pettitt, "Testing, Planning, and Redrilling of Geothermal Test Hole GT-2, Phases IV and V," Los Alamos Scientific Laboratory report (to be published).

12. H. Murphy, et al., Los Alamos Scientific Laboratory, unpublished data, October 7, 1977.

13. J. J. Mortensen, comp., "Proc. of the 2nd NATO-CCMS Information Meeting on Dry Hot Rock Geothermal Energy," June 28-30, 1977, Los Alamos Scientific Laboratory report LA-7021-C (to be published).

14. R. L. Aamodt, "Hydraulic Fracture Experiments in GT-1 and GT-2," Los Alamos Scientific Laboratory report LA-6712 (February 1977).

15. A. G. Blair, J. W. Tester, and J. J. Mortensen, comp., "LASL Hot Dry Rock, July 1, 1975 to June 30, 1976," Los Alamos Scientific Laboratory report LA-6525-PR (October 1976), p. 162.

16. D. Miles, Los Alamos Scientific Laboratory, personal communication, March 30, 1977.

17. D. Miles, Los Alamos Scientific Laboratory, personal communication, November 9, 1977.

18. R. A. Pettitt, comp., "Environmental Monitoring for the Hot Dry Rock Geothermal Energy Development Project," Los Alamos Scientific Laboratory report LA-6504-SR (September 1976).

19. J. Albright, L. Aamodt, and R. Potter, "Definition of Fluid-Filled Fractures in Basement Rocks," Proc. of the 3rd ERDA Symp. on Enhanced Oil and Gas Recovery and Improved Drilling Procedures, Tulsa, 1977 (ERDA, Washington, D.C. 1977) Vol. 2 - Gas and Drilling, pp. F-2/1 thru 2/8.

20. J. N. Albright and R. J. Hanold, "Seismic Mapping of Hydraulic Fractures Made in Basement Rocks," Proc. of the 2nd ERDA Symp. on Enhanced Oil and Gas Recovery, Tulsa, 1976 (ERDA, Washington, D.C.) Vol. 2 - Gas.

21. B. R. Dennis, J. H. Hill, E. L. Stephani, and B. E. Todd, "Development of High-Temperature Acoustic Instrumentation for Characterization of Hydraulic Fractures in Dry Hot Rock," presented at the 22nd International Instrumentation Symposium, May 1976, Paper No. ISA ASI 76222.

22. J. N. Albright, "Temperature Measurements in the Precambrian Section of Geothermal Test Hole No. 2," Los Alamos Scientific Laboratory report LA-6022-MS (July 1975).

23. B. R. Dennis, E. L. Stephani, and B. E. Todd, "A Thermopile Probe to Measure Temperature Anomalies in Geothermal Boreholes," presented at the Ninth Transducer Workshop, April 1977; also Los Alamos Scientific Laboratory report LA-UR-77-574 (April 1977).

24. op. cit., [15], p. 55.

25. F. G. West, P. R. Kintzinger, and A. W. Laughlin, "Geophysical Logging in Los Alamos Scientific Laboratory Geothermal Test Hole No. 2," Los Alamos Scientific Laboratory report LA-6112-MS (November 1975).

26. J. N. Albright, Los Alamos Scientific Laboratory, personal communication, September 8, 1977.

27. W. L. Sibbitt, "Preliminary Measurements of the Thermal Conductivity of Rocks from LASL Geothermal Test Holes GT-1 and GT-2," Los Alamos Scientific Laboratory report LA-6199-MS (January 1976).

28. H. D. Murphy, R. G. Lawton, J. W. Tester, R. M. Potter, D. W. Brown, and R. L. Aamodt, "Preliminary Assessment of a Geothermal Energy Reservoir Formed by Hydraulic Fracturing," SPEJ, Vol. 17, pp. 317-326, August 1977.

29. op. cit., [15], p. 62.

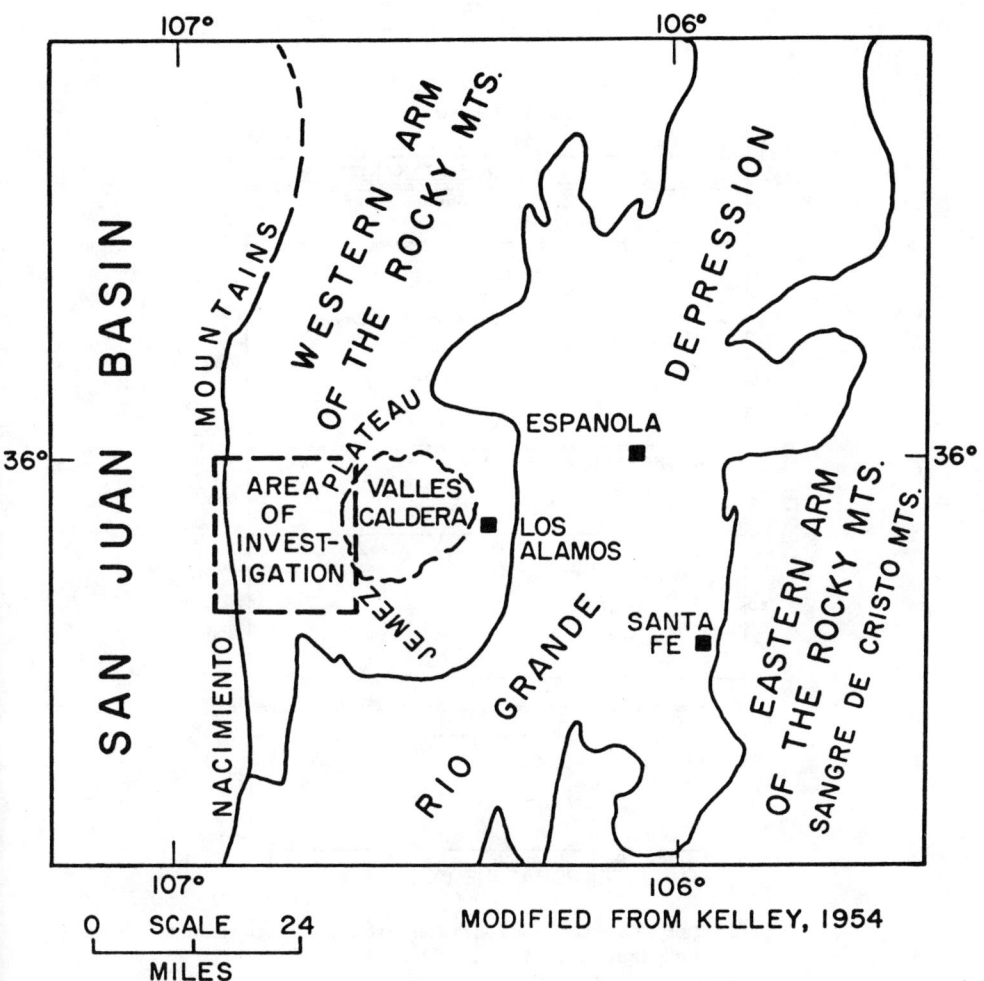

Fig. 1 Major structural features and area of investigation in north-central New Mexico.

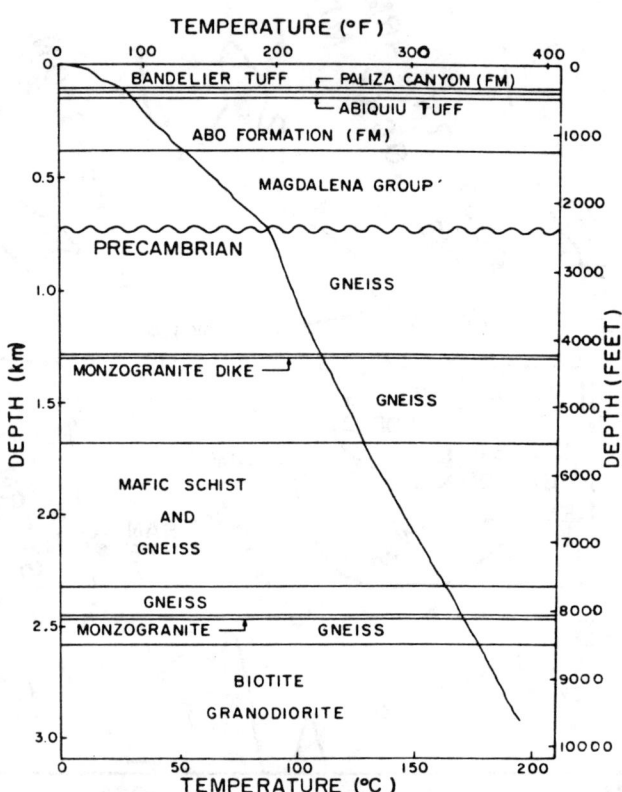

Fig. 2 Temperature gradient and geologic formations penetrated in GT-2, 1974.

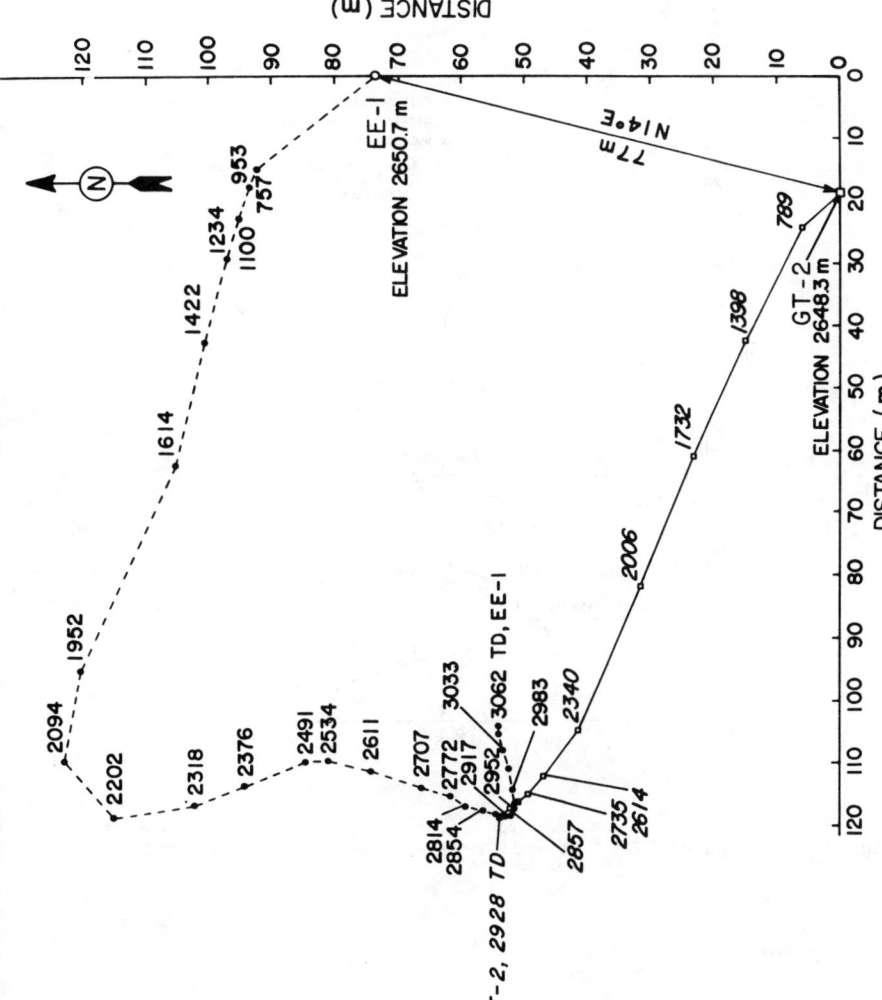

Fig. 3 Directional Drilling in GT-2 and EE-1, Plan View

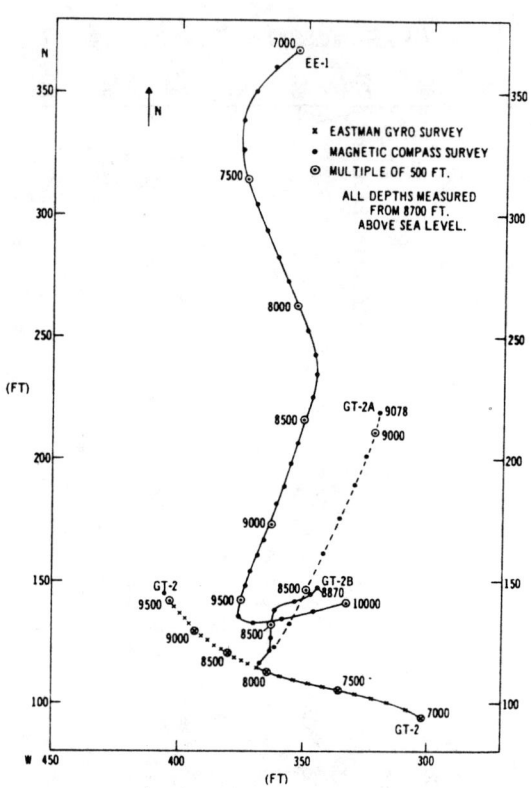

Fig. 4 Drill paths of GT-2A and GT-2B with corrected path of EE-1

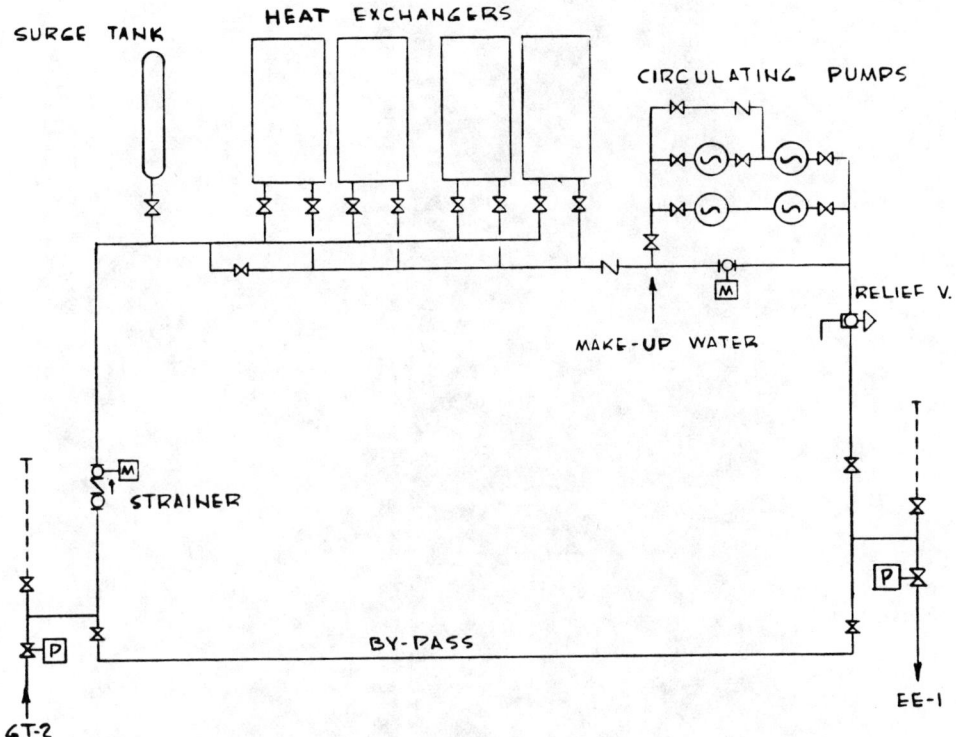

Fig. 5 Schematic diagram of energy extraction circulating test loop.

Fig. 6 Heat exchanger with pumper trucks in place for September 1977 preliminary circulation test. (77-17826)

Fig. 6A Steam plume during first phase of September 1977 preliminary circulation test. (77-17830)

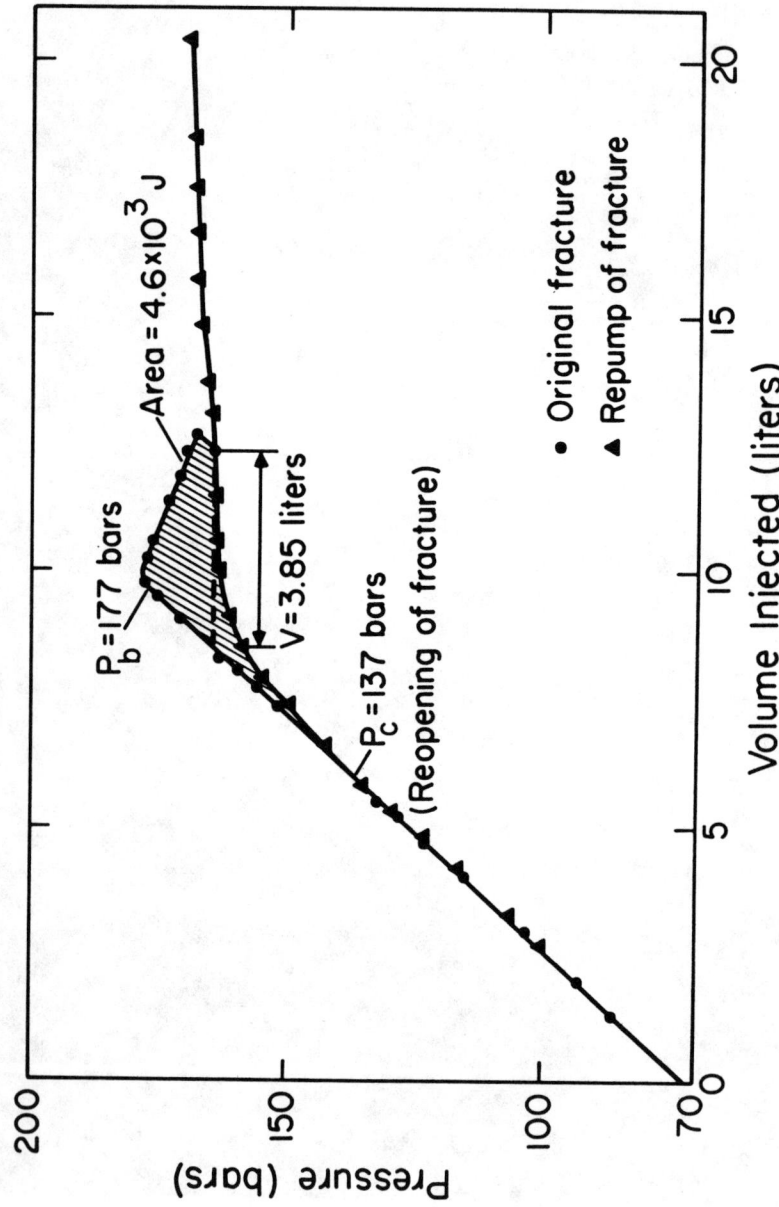

Fig. 7 Pressure-volume history for hydraulic fracturing and repumping in GT-1.

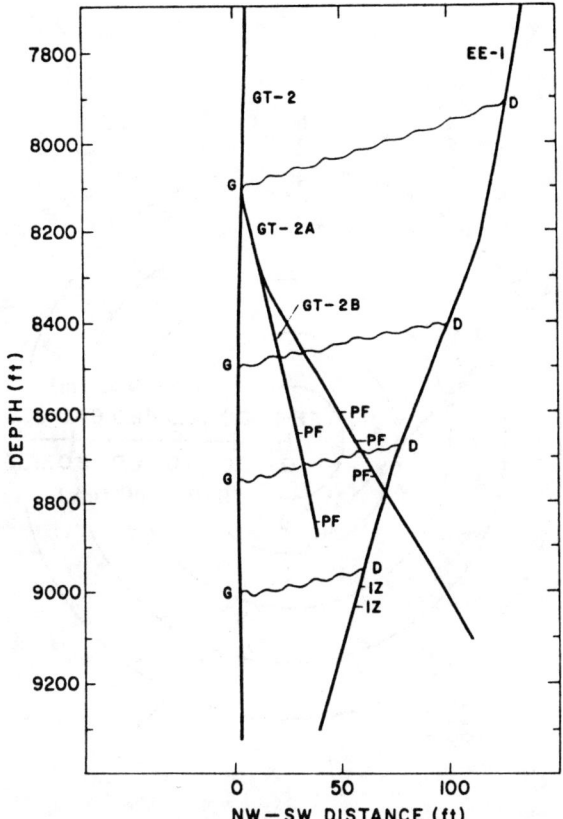

Fig. 8 Borehole elevations showing "shear-shadow" acoustic paths from EE-1 to GT-2 and subsequent redrill paths GT-2A and GT-2B. PF = pressurized fracture; IZ = injection zone; G = geophone; D = detonator

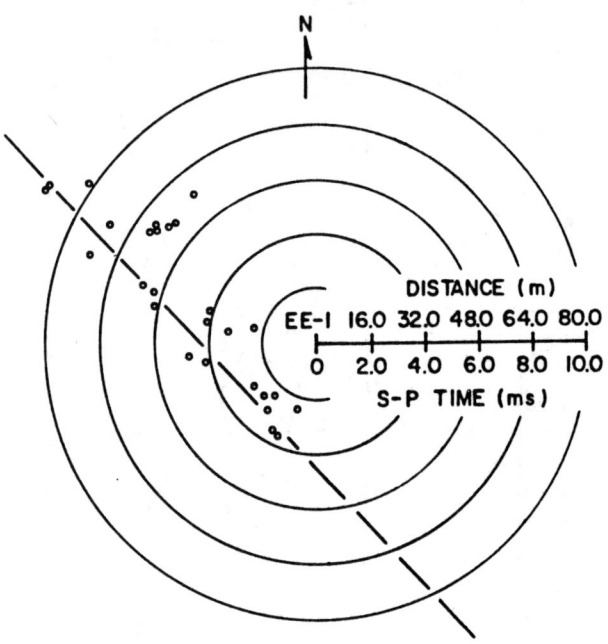

Fig. 9 Trace of GT-2 fracture relative to EE-1 based on microseismic event foci with $\phi < 20°$ for 45 m^3 inflation

GEOTHERMAL ENERGY TECHNOLOGY

Pilot Facility for the Experimental Study of Binary-cycle Conversion Systems— Thermogravimetric Loop

G. CEFARATTI
ENEL
Milan, Italy

S. AROSIO and G. SOTGIA
Politecnico di Milano
Milan, Italy

P. ALIA and G. MORANDI
CISE
Segrate, Milan, Italy

ABSTRACT

The present work is framed in the field of the research applied to geothermal energy exploitation, that ENEL (National Electricity Board) has developed for a long time. In particular ENEL took the decision to realize a pilot plant suitable for the economical exploitation of low temperature geothermal sources. The installation was designed by a study group of researchers and technicians of the three organizations to which the authors belong.

This plant is designed to test different prime movers to produce electric power. As a first step of the program, a so called thermogravimetric non-conventional system capable of supplying a 25 kW power is to be tested. The construction of the pilot plant is at present being started at ENEL power station of Castelnuovo Valcecina (Larderello field) and its completion is foreseen within the second half of 1978.

This work was carried out under contract with the Commission of the European Communities.

INTRODUCTION

Since many years ENEL started a research program in the field of low temperature geothermal sources, in view of the interest that may have as alternative sources for power production.

In fact large quantities of hot water are available particulary in Italy, that is attractive for exploitation in view of the continuous increasing in energy costs.

In this prospect, for a long time, ENEL has been interested in a non conventional system for energy conversion called "thermogravimetric"[1] [2] [3] [4] whose studies have been carried out at: Istituto di Fisica Tecnica del Politecnico di Milano.

An agreement between ENEL and Politecnico has given the opportunity for the deepening of the studies through the elaboration of mathematical models able to give all information about scaling and performances of the system both as a low enthalpy thermal energy converter [5] [6], and a system suitable for other various purposes [7] [8] [9].

Under contract with the Commission of the European Economic Community, ENEL took the decision to realize a pilot plant for use of geothermal (low enthalpy) heat to produce electric power. This utilization is obtained with different types of prime movers; as a first activity, the prime mover will be of thermogravimetric type capable of supplying a 25 kW power.

The pilot plant will allow several evaluations on the exploitation of low temperature sources. Interesting information about feasibility of plants up to 1000 kW, economic evaluations etc. can be also obtained.

The plant is suitable for testing and developing prime movers for the utilization of solar energy into power too.

The design and the construction of the plant, such as the studies of R&D in this field, have been performed in cooperation with Politecnico di Milano and CISE (Centro Informazioni Studi Esperienze, Milano).

THE THERMOGRAVIMETRIC SYSTEM

Operating principles

The operating principle of the so-called "thermogravimetric system" is based on the possibility of generating a difference in density between two branches of a closed vertical loop containing a fluid carrier (water) in liquid phase. Such a difference in density is obtained by injecting into the base of one of the two branches, an organic low-boiling fluid (working fluid) at a vapour phase and separating them at the top. In this branch, a two-phase two-component, upward flow develops while a downward flow in liquid phase with a density higher than the other one flow in the second branch. The difference in density assures the hydrodynamics of the system and creates a total head which drives a hydraulic turbine.

After separation of the two fluids at the top of the circuit, the liquid carrier fluid discharges the potential energy acquired into the hydraulic turbine and enter the circuit again. The working fluid undergoes a series of thermodynamic transformations which constitue a closed power cycle.

Referring to the key numbers in Fig. 1 and 2 and working fluid, from its injection at the base of the two phase flow duct, expands and form saturated vapour at the maximum temperature of the cycle, 4, reaches the conditions 5 of superheated vapour at the minimum pressure p_2 of the cycle established at the top of the two-phase flow duct. Because of the low ratio between the flow of the working fluid and the carrier fluid, such expansion can be considered isothermal [10].

The expansion of the working fluid absorbs an amount of heat given by $T_1 \cdot (S_5 - S_4)$; the carrier fluid evaporates partially as saturated steam at the maximum temperature of the cycle.

Owing to the high latent heat of vaporization of the carrier fluid (water), also a small quantity of evaporated fluid, absorbs a not negligible quantity of heat.

These two elements define the amount of heat Q_{HE} supplied by the source (geothermal or solar) to the carrier fluid through the heat exchanger in order to keep the working fluid at a constant temperature equal to the maximum temperature T_1 of the cycle.

When leaving the two-phase flow duct, the working fluid is subject to an isobaric cooling 5, 6 and condensation 6, 1. The evaporated carrier fluid reaches the condenser too; that's the third contribution to the condenser capacity Q_C.

Compression of the working fluid up to the maximum pressure P_1 of the cycle follows condensation and separation (by difference in density of the two liquid fluids).

The isobaric heating 2, 3 and evaporation 3, 4 are the next transformations which close the thermodynamic cycle.

The working fluid as saturated vapour is injected at the base of the two phase flow duct entering the cycle again.

The result of the above description is a system based on conventional components whose technology is well knon. The working fluids used and the temperature range in which they operate define a minimum cycle pressures above atmosphere.

The employment of direct contact exchange, especially for the 3 - 5 transformation by injection of the liquid working fluid at the base of the two phase flow duct appears to be the best solution though it rises doubts which require specific experimentation. Nevertheless one of the future aims of this research is to test such technologies. As a binary thermodynamic cycle at low temperatures, the thermogravimetric system gives interesting opportunities for the exploitation of solar energy and heat rejected from various technological processes.

Characteristics and performances

The studies on these systems have pointed out some typical characteristics of the plant which define its dimensions and its performances. The main characteristics are: T_1, T_2, G, Z/Z_{LIM}, W and working fluid (see Tab. I for nomenclature).

The power of the system can be evaluated in many ways which

emphasize various basic quantities.

Referring to the thermodynamic cycle:

$$W = G_G \cdot L_T \cdot \eta \tag{1}$$

With reference to an energetic balance concerning the two-phase flow and the carrier fluid branches one has:

$$P = g \cdot Z \cdot G_L \cdot (\rho_L - \bar{\rho})/\rho_L \tag{2}$$

L_T depends on the working fluid and the maximum and minimum temperatures of the thermodynamic cycle.

As result of previous studies the most suitable working fluids appear to be freon 11, 21, 113.

While T_2 is strongly depending on the environmental conditions (availability and kind of cold source) the maximum temperature of the cycle can assume any value. L_T increases with T_1 which, in turn increases the plan height.

This relationship can be shown considering the minimum height:

$$Z_{LIM} = \frac{\Delta p}{g\, \rho_L} + \frac{V_L^2}{2g} \left(\frac{D_M}{D_B}\right)^2 \tag{3}$$

This represents the minimum theoretically possible height when G_L tends to ∞ (or available total head tends to zero): obviously the plant height must be greater than Z_{LIM}.

Under the same conditions, if the plant height increases, its efficiency decreases because of the increased dissipations. To limit height and pressure drops, Z_{LIM} and consequently the cycle temperature T_1 will be controlled.

An increase in T_1 increases the amount of heat due to evaporation of the working fluid; consequently the efficiency will be reduced.

The mentioned effects determine the curves shown in Fig. 3. They point out how T_1 has to be contained.

Changes in plant height can be performed with a given working fluid, available power to the turbine and upper and lower temperatures of the cycle.

A reduction in diameters and a decrease in efficiency corresponds to an increased height.

The shape of these functions (not altered by a change in power and mass velocity) defines a suitable range of the ratio Z/Z_{LIM} which fall between 1.2÷1.3.

In Fig. 4 and 5 the curves refer to a 50 kW plant. The height of the plant doesn't depend on power though greater heights usually correspond to higher powers.

When even the heigth has been fixed, ther's still a degree of freedom represented by the mass velocity.

The mass velocity of the two phase flow G has an influence on diameter D_B only for the lowest values while the behaviour of the efficiency curve is quite flat through a very broad range. As a result of previous researches the most profitable G values fall between 1500÷3500 kg/m^{2}s according to the power of the plant. A qualitative analysis about the behaviour of the depending quantities of the plant at the varying of the independent parameters G, W, T_1, T_2, Z and working fluid define a profitable working range to the various powers.

In this conditions the most important quantities of the system such as Z, G, D_B are plotted in Fig. 6, 7, 8 respectively.

A mathematical model [5] [6] has been developed to predict the plant behaviour at different working fluids and operating conditions.

A wide view of the influence of the indipendent parameters on thermogravimetric system has been reached using the mathematical model and the relevant computer program. The principal assumptions presented in this model are:

1) Thermal equilibrium between the two components in each section of the two phase flow duct.
2) Isothermal expansion of the working fluid inside the two-phase flow duct.
3) No heat losses in the pipes.
4) The mixing length downstream from the injection point is negligible in comparison with the heigth of the circuit.

Furthermore the following items have been considered:

1) Pressure drops distributed and concentrated in the mono-phase pipes.
2) Pressure drops distributed in the two-pase flow duct.
3) Slip ratio in the two-phases flow.
4) Behaviour of the working fluid as imperfect gas.

Off-design performances

From a series of considerations, both thermodynamic and concerning the control of the plant [11], the best solution for the off-design behaviour is to keep constant the total head and the minimum pressure of the cycle (upper branch of the two-phase flow).

The G_G flow rate is the only independent variable in the off-design conditions.

If G_G fluctuates, the same happens to G_L and P_1.

In such conditions the thermodynamic parameters of the cycle and its efficiency appear to be constant under the various conditions. The flow pattern of the two-phase flow doesn't change; consequently the hydraulic efficiency doesn't present great alterations. Fig. 9 shows the behaviour of organic efficiency and the power of the plant at the various operating conditions and points out how the organic efficiency curve is quite flat.

Consequently the power curve is close to a straight line, and the zero power gas flow rate is around 10% of nominal value. very low for a thermal machine.

DESCRIPTION OF THE PILOT PLANT

The need to restrict as much as possible the plant heigth both for economic and technical reasons, the possibility of using an hydraulic turbine available, and the characteristics of the water for cooling and condensing have been the main elements which defined and limited the basic quantities of the plant.

The choice of the two different working fluids F_{11} and F_{21} together with the characteristics of the turbine have fixed the maximum cycle temperature, the height of the plant [12] and the power generated.

Being the basic quantities of the turbines (flowrate and total head) independent from the specific flowrate G, this paremeter has been defined in order to meet the requirements of the thermo-gravimetric system only.

Fig. 1 shows the two main loops of the plant:

1) Freon_loop

   A pump sucks liquid freon from the condenser and, after vaporisation in boiler, sends it into the mixer. Freon steam is then "recovered" by the separator and condensed again.

2) Process_water_loop

   A recirculation circuit used for maintaining the water of the loop at an imposed temperature, thus compensating heat losses, is connected with the hydraulic loop. A definite water flow is delivered by the loop, sent to and exchanger, from which - heated - it comes back into the loop.
   The turbine is Kaplan type with horizontal axis, coupled with a brake for setting the energy output.
   The working of the three above mentioned exchange-units obviously involves two more loops; the loop of geothermal steam (hot fluid in boiler and exchanger) and the loop of cooling water (cold fluid in condenser).

Fig. 10 shows the plant flow-sheet. In the figure the components of the 4 above mentioned loops are marked as:

- (100) freon loop
- (200) process water loop
- (300) geothermal steam loop

- (400) cooling water loop.

Freon loop

Its working has been described before. Nevertheless it is important to notice the existence of a tank for stocking the liquid freon and the fact that the freon steam is saturated with water steam leaving the separator: therefore it is necessary to part the two liquid phases of the condensate.

Then for the evaporation of freon, the steam-boiler and steam-drum solution is adopted, to set-up between them a thermosyphon circulation with a recirculation flowrate ten times higher than the circulation of vaporized freon.

Processing water loop

Notice also the existence of the stockage tank, indispensable for automatic or manual resetting of the level in the separator. It can be seen that this tank is maintained at the right temperature with a special branch of the geothermal steam loop. The above mentioned recirculation loop can be used to change the thermal level of the hydraulic loop for tests at different temperature.

Geothermal steam loop

The characteristics of the available geothermal steam being fixed, a deheater was used to improve the heat-rate in the exchangers.

The steam is then used in the reboiler, exchanger, and tank. The condensates are discharged through hot traps.

Cooling water loop

It supplies cold water for the deheater and the condenser. This water derives from collectors (for feeding the condensers of the Castelnuovo power plant) having a head of 13 m approximately.

Design data

- Carrier fluid	distilled or demineralized H_2O	
- Working fluid	Freon 11	Freon 114
- Upper T of the thermodynamic cycle	75°C	70
- Lower T " " "	45°C	40
- Upper p " " "	4.5 bar abs	7.5
- Lower p " " "	1.8 bar abs	3.5
- Heat exchanged in the boiler	1.3 GJ/h	
- " " " condenser	1.86 GJ/h	
- Reinstatement heat	0.63 GJ/h	
- Mass flowrate of working fluid	6625 kg/h	
- " " " water in turbine	961000 kg/h	
- Inner diameter of the two-phase duct	445 mm	

- Height of the loop　　　　　　　　40 m
- Hydraulic power of the turbine　　25 kW
- Specifications of the available
  geothermal steam　　　　　　　　172°C - 2 bar abs.

The design data have been chosen to allow the use of both freon 11 and freon 114 in the plant.

Materials

On the ground of the corrosion tests concerning endogenous steam-metal system and of the bibliography available on freon-water-metal system, we think it is possible to adopt carbon steel as the basic material for all the loop components. The same choice has been adopted for compatibility problems of freon-metal and water-metal systems. In any case a laboratory apparatus has been set up for the experimental check of literature data concerning the behaviour and compatibility of materials both referring to freon 11 and freon 114; the tests are in progress and the first results will be available approximately at the end of 1977.

The protection of the external surface against the strongly corrosive action of the atmosphere of Castelnuovo (presence of H_2S) is already guaranteed for the thermally insulated components by alluminium-sheet which covers the insulation. The protection however must be provided with painting and bitume-strips for the other components, and, for the instrumentation, using inox steel (for the parts in contact with the process fluids) and protective aluminium boxes (for the instruments).

Exchange units

All the exchange-units are conventional, with steam or freon in the shell, and are designed according to TEMA standards.

The boiler exchange units allow to disassemble the pipe-bundle, having a square pitch to clean its external surface.

Pumps

(P 101) and (P 201) are conventional, centrifugal pumps with horizontal axis. (P 101) is an hermetic type with immersed rotor, because of the physical characteristics of the freon (difficulties of seal and elimination of leakage; owing to the low NPSH (the freon being pumped quite near the vaporisation conditions) the pump is installed below the condenser.

Valves

The temperatures and pressures of circulating liquids being not excessive, ball valves are largely adopted as they offer good seal, low cost, extremely low loss of pressure, and require almost no maintenance.

Globe valves are supplied on the pump delivery to obtain a good manual regulation and identify the best working point of the pump.

Plug valves, which are the most suitable for this use, are envisaged on condensation lines.

Inox steel is used for the internal parts of the valves, seats included.

Piping and flanges

The material is carbon steel both for pipes and flanges. The choice has been made according to the ANSI B standards.

Mixers

There are two mixers assembled on the loop. Geothermal steam is deheated by injecting atomized water with a nozzle into the steam itself.

The aim of the second mixer is to achieve a uniform mixing between water and freon to minimize the mixing length along the two-phase duct and to hydraulically decouple the duct from the freon injection loop.

Instrumentation, regulation, automatisms

The instrumentation has been adopted for two requirements:

- to guarantee the control and operation of the plant by centralizing the control functions on a central panel;
- to provide all the necessary information for experimentation.

Pressure, flowrate, temperature measurements are carried out by electronic transducers.

The following parameters (important for experiments and plant operation) are controlled:

- water level and pressure in the separator;
- two-phase duct temperature;
- level in the steam-drum;
- freon flowrate in two-phase duct.

The last regulation is opered by modulating the geothermal steam feeding the boiler and then the vaporated freon; it allows to vary the energy produced by the plant.

Some interlocks and safety automatic controls are envisaged:

- to guarantee a correct operation of the system also in out-of-design conditions;
- to allow safety during any testing phase;
- to allow the shut-down of the plant;
- to put the plant in safety condition if electric supply fails.

The above mentioned presence of H_2S in the atmosphere of the installation site does not allow the use of compressed air valve actuators (on-off or modulating) and thus it is necessary to use electrical actuators. Fig. 11 shows the base lay-out planned for the plant.

Here it is clear the physical separation between the two freon and gravimetric loops.

The opportunity of this separation comes from the possibility of varying the plant once the planned experimentation will be completed: it will be then easy to use the pilot station for new experiments where freon steam is still generated: for example application of a freon turbine (or with analogous fluids).

For economical reasons a trestle for electrical lines (380 kV) is to be used as the support structure of the hydraulic loop.

The components of the freon loop are installed on an independent structure.

Realization program

The pilot plant will be located at Castelnuovo Valcecina in Larderello field (PISA).

The design of the plant with thermogravimetric loop as prime mover has been completed and the tenders for main machinary is now under evaluation by Enel.

The erection on site will start in April 1978 and the completion of the plant is foreseen within October 1978. From this date will begin the preliminary tests. Successively will begin the experimental runs.

The choice of the site of Castelnuovo is connected with the existing of the ENEL geothermal power plant, and this fact allows to use both the existing facilities and geothermal fluid (steam).

Development lines for the experimental program

A) Studies for acquiring technical and economic elements to evaluate the potential of the thermogravimetric process to produce electric energy and studies on the possible improvements of the process itself.
Starting from the experimental plant (25 kW) design the following problems will be examined:
1) evaluation of the process parameters, efficienties and costs for 100 and 1000 kW power loops generally using the same solutions and technology as the experimental one;
2) evaluation of the analogous data by introducing possible improvements into the plant, as the recovery of return

enthalpy from the separator in freon loop or direct injection of liquid freon into the two-phase duct.

B) Feasibility study for the application (to the geothermal pilot plant just builted in Castelnuovo) of a turbine directly driven by an organic vapour.

C) Experimental laboratory investigation of the metal-water-freon system chemistry to check the compatibility of the constituents (in the operating conditions) and to acquire the know-how on their behaviour. These tests will provice indications for a correct operation of the plant as well as criteria for optimizing the choice of the freon also from a chemical-technological view point.

References

1. O. Rumi, M. Silvestri, G. Sotgia:
   "Sulle possibili applicazioni pratiche di un circuito gravimetrico bifase"
   La Termotecnica, 24 n. 1 (1969), Ed. Barbieri - Milano.

2. O. Rumi, G. Sotgia:
   "Studio dei circuiti gravimetrici bifase per lo sfruttamento delle acque calde. Fluidi di lavoro, cicli termodinamici".
   La Termotecnica, 24 n. 1 (1969), Ed. Barbieri - Milano.

3. S. Pessina, O. Rumi, M. Silvestri, G. Sotgia:
   "Gravimetric loop for the generation of electrical power from low temperature water".
   United Nations Symposium on the development and utilization of geothermal sources. Pisa 1970.

4. S. Arosio, A. Muzzio, M. Silvestri, G. Sotgia:
   "A thermogravimetric pilot plant for the production of mechanical energy from low enthalpy sources".
   2. un. U.N. Symposium on the development and utilization of geothermal energy. S. Francisco, CA, 1975.

5. S. Arosio, A. Murzio, G. Sotgia:
   "Proporzionamento di un impianto termogravimento di piccola potenza, ricerca ed analisi dell'influenza unitaria fra le principali variabili geometriche e funzionali".
   Rapporto TGM/P1 Ist. Fisica Tecnica, Politecnico Milano 1977.

6. S. Arosio, A. Murzio:
   "Indagine sul comportamento fuori progetto di un impianto termogravimetrico di piccola potenza".
   Rapporto TGM/P2 Ist. Fisica Tecnica, Politecnico Milano 1977.

7. S. Pessina:
   "Applicazione di un impianto gravimetrico alla magnetoidrodinamica con metalli liquidi".
   Rapporto TRD2/MHD/1 Ist. Fisica Tecnica, Politecnico Milano 1969.

8. S. Arosio, A. Murzio, G. Sotgia:
   "Applicazione di un impianto termogravimetrico alla magnetoidrodinamica dei metalli liquidi".
   Rapporto TRD2/MHD/2 Ist. Fisica Tecnica, Politecnico Milano 1973.

9. S. Arosio, F. Ferrario:
   "Studio prelimiare nell'utilizzazione del circuito termogravimetrico come pompe di calore. Scelta del fluido di lavoro".
   Rapporto TRD2/CE/2 Ist. Fisica Tecnica, Politecnico Milano 1973.

10. P. Lupoli, A. Murzio, G. Sotgia;
    "1973 Void fraction measurement in air-water adiabatic flows into large diameter ducts by γ-rays absorption method".
    European Two-Phase Flow Group Meeting, Bruxelles.
    "1974 Indagine sulle caratteristiche del flusso bifase in condotti di grande diametro e della loro influenza sulle prestazioni di circuiti termogravimetrici".
    La Termotecnica, V 28 n. 12.

11. S. Arosio:
    "Avviamento e regolazione di un impianto termogravimetrico".
    Rapporto TGM/P3 Ist. Fisica Tecnica, Politecnico Milano 1977.

12. S. Arosio, G. Sotgia:
    "Impianto termogravimetrico dimostrativo e di ricerca. Centrale di Castelnuovo Val Cecina - ENEL".
    Rapporto TGM/DR/1 Ist. Fisica Tecnica, Politecnico Milano 1977.

TABLE I

Nomenclature

Symbol	Meaning
D_B	diameter of two phase duct
D_F	" of working fluid loop
D_M	" of carrier fluid branch
g	acceleration due to gravity
G	mass velocity of the mixture
G_G	mass flowrate of working fluid
G_L	" " of carrier fluid
h	enthalpy
H_T	total head
L_T	thermodynamic work
p	pressure
Q_I	heat supply for evaporation
Q_2	heat to be subtracted for condensation
Q_E	heat supply for expansion
Q_C	total heat to be subtracted
Q_{HE}	heat given to carrier fluid
S	entropy
T	temperature
V_L	velocity of carrier fluid
Z	height of the plant
Z_{LIM}	limiting value of Z
W	net power
Greek	
Δ	variation of a quantity
η	organic efficiency
η_T	thermal efficiency
ρ_L	density of carrier fluid
$\bar{\rho}$	average density of the two phase flow

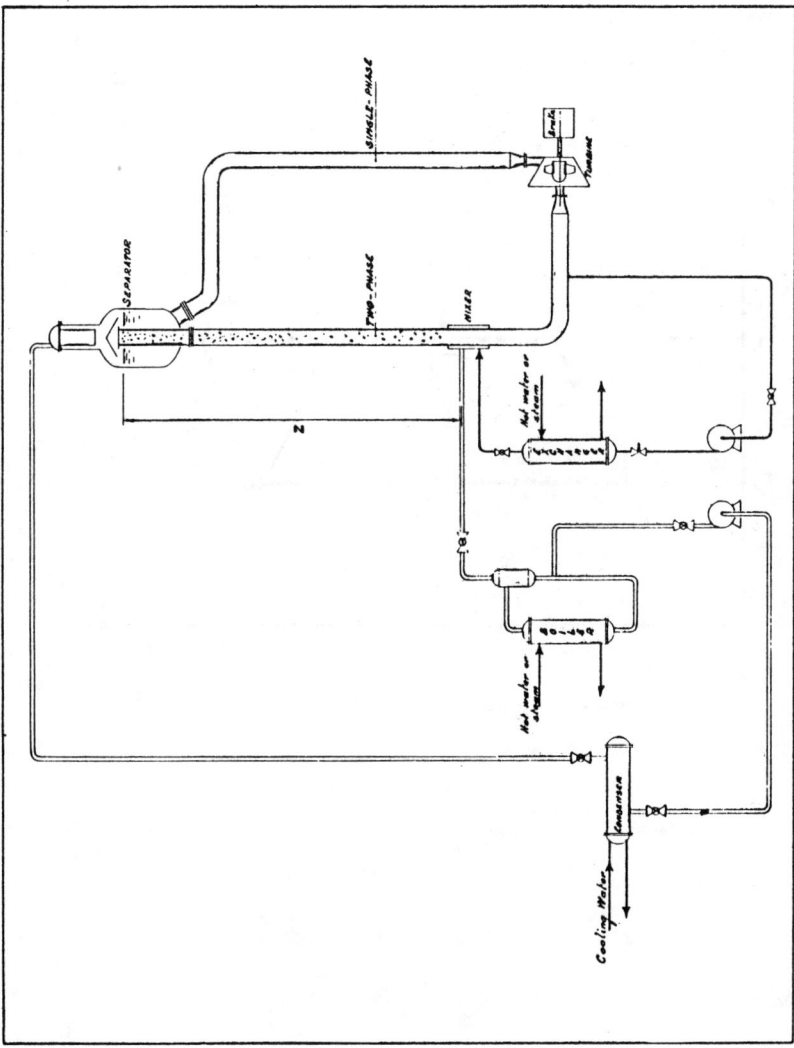

Fig. 1 Schematic diagram of the loop

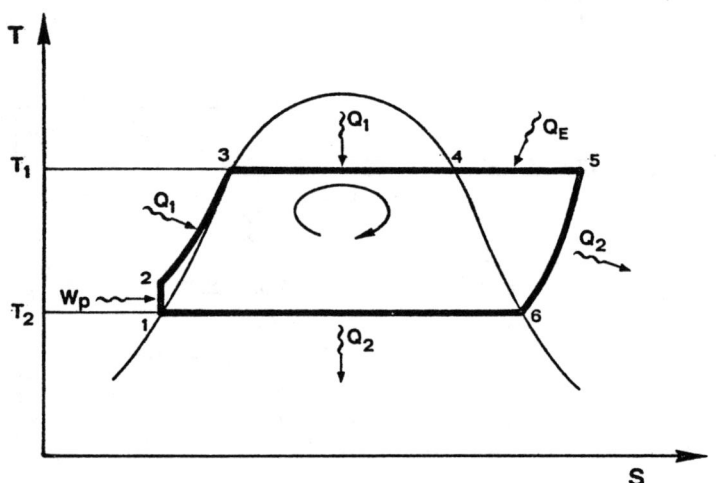

Fig. 2   Thermodynamic cycle of the circuit

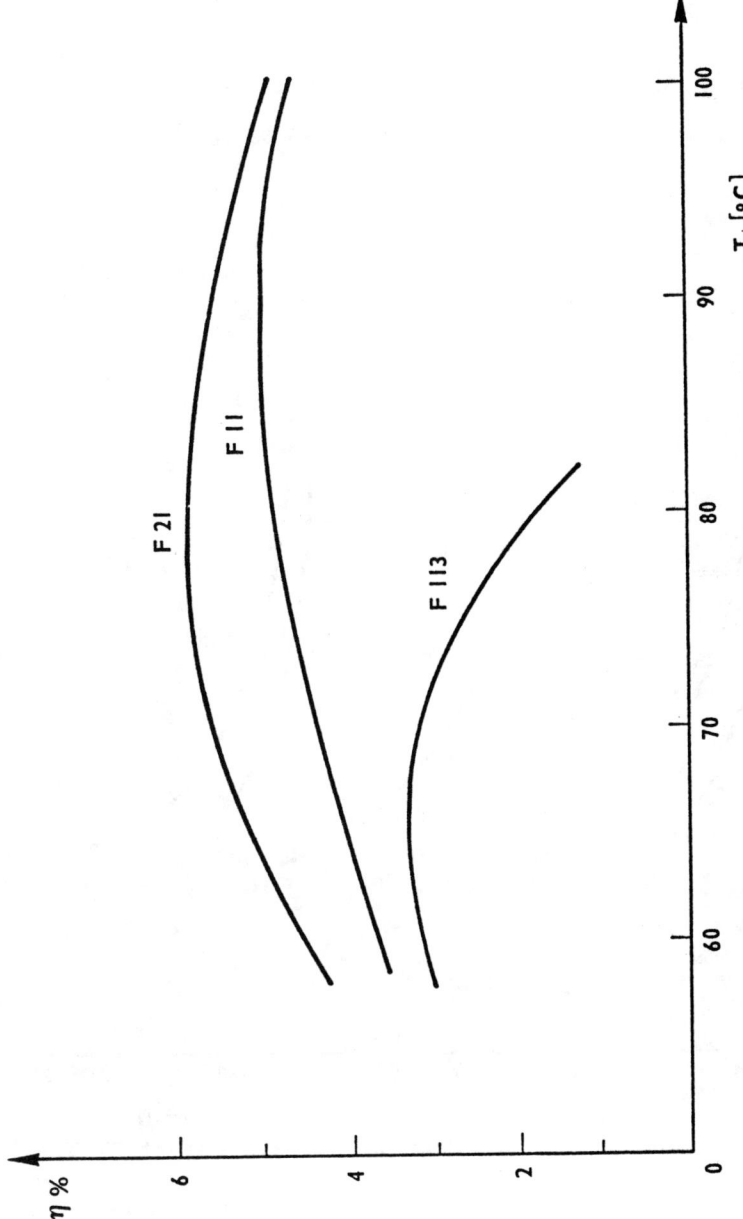

Fig. 3 Organic efficiency of plant versus T_1 (W = 50 kW)

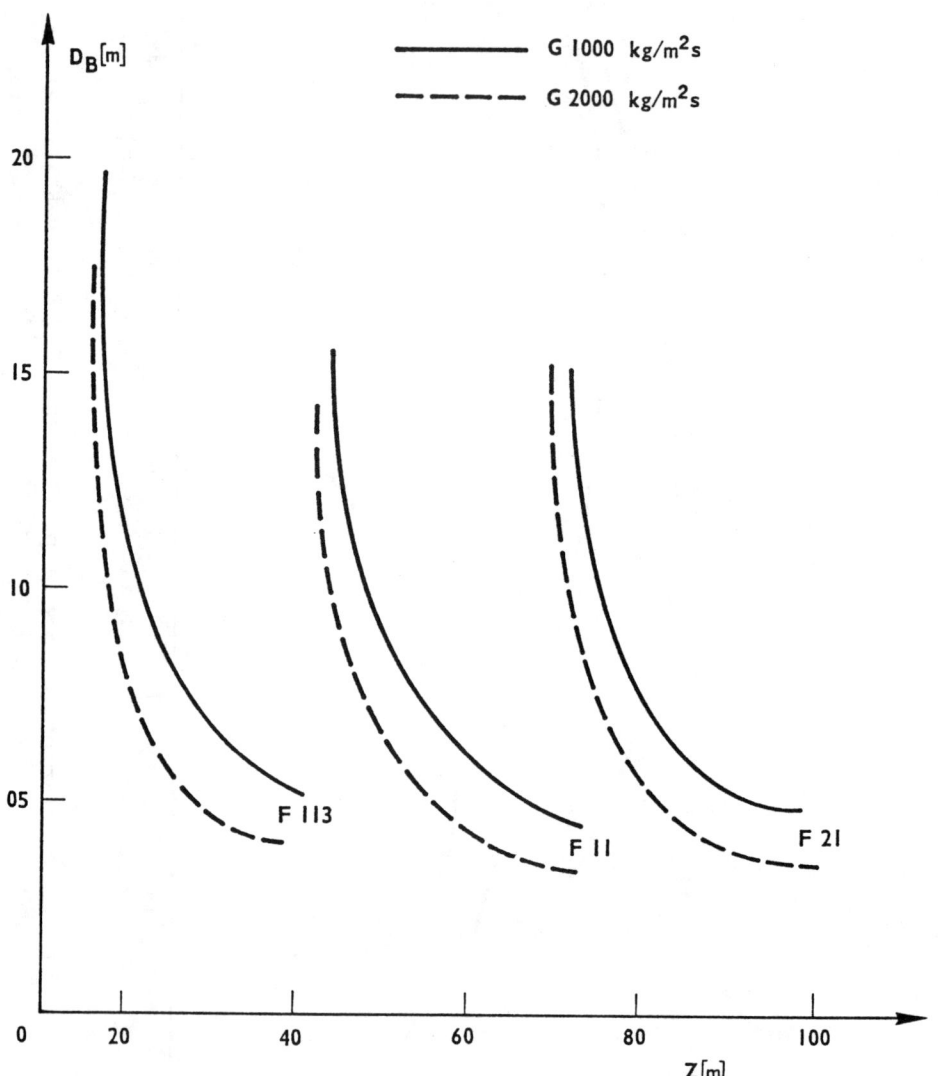

Fig. 4 - Two-phase duct diameter versus height Z (W = 50 kW)

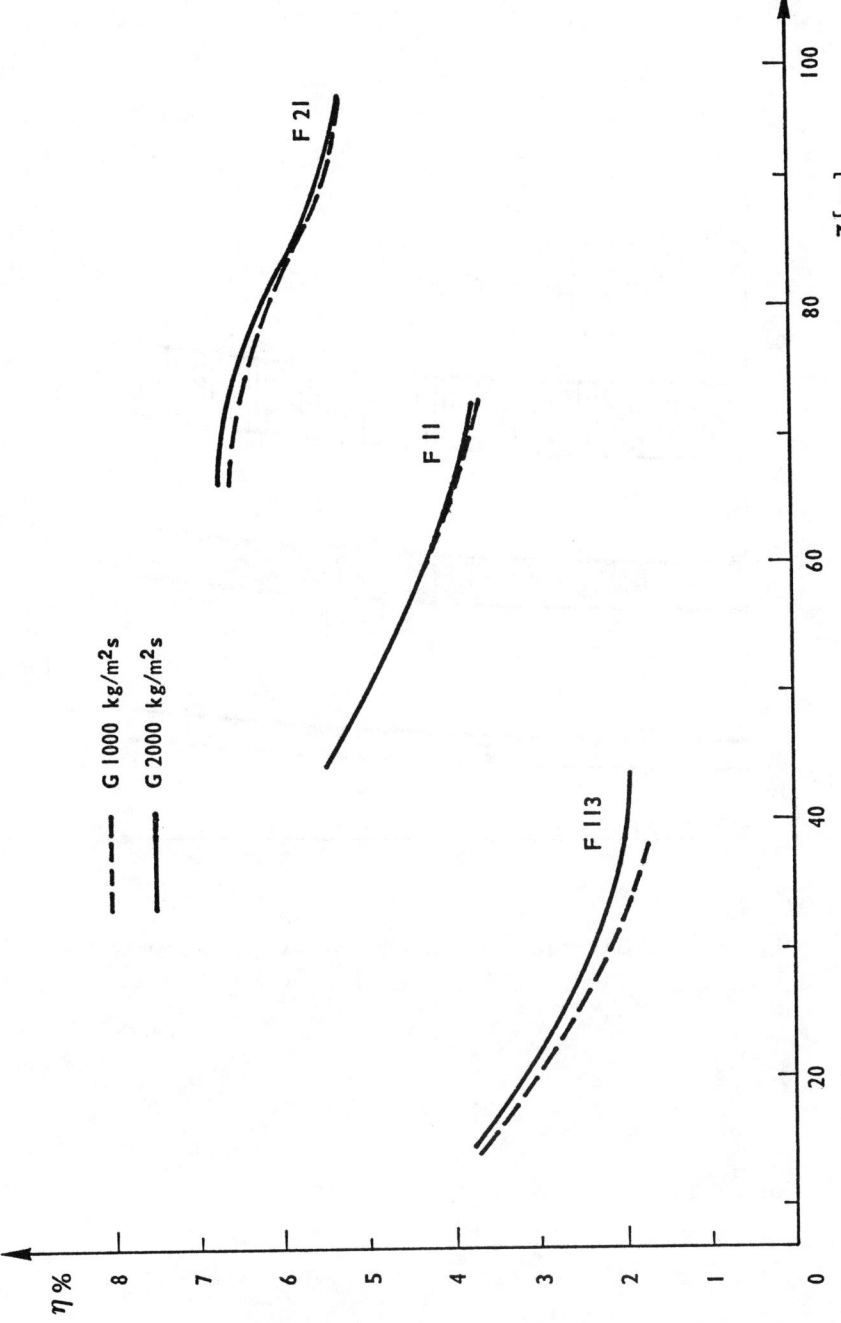

Fig. 5 - Organic efficiency versus height z (W = 50 kW).

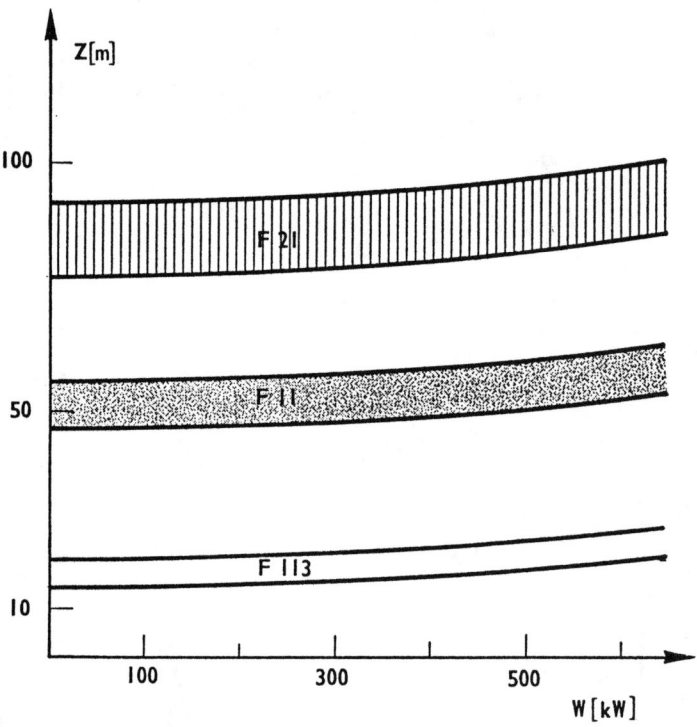

Fig. 6 Optimal range of the height versus power

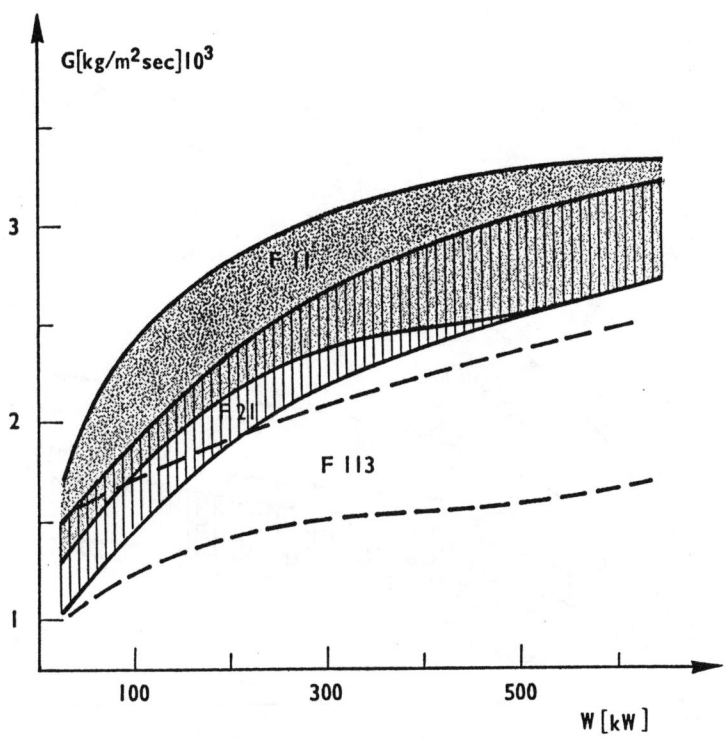

Fig. 7 Optimal range of the mass velocity versus power

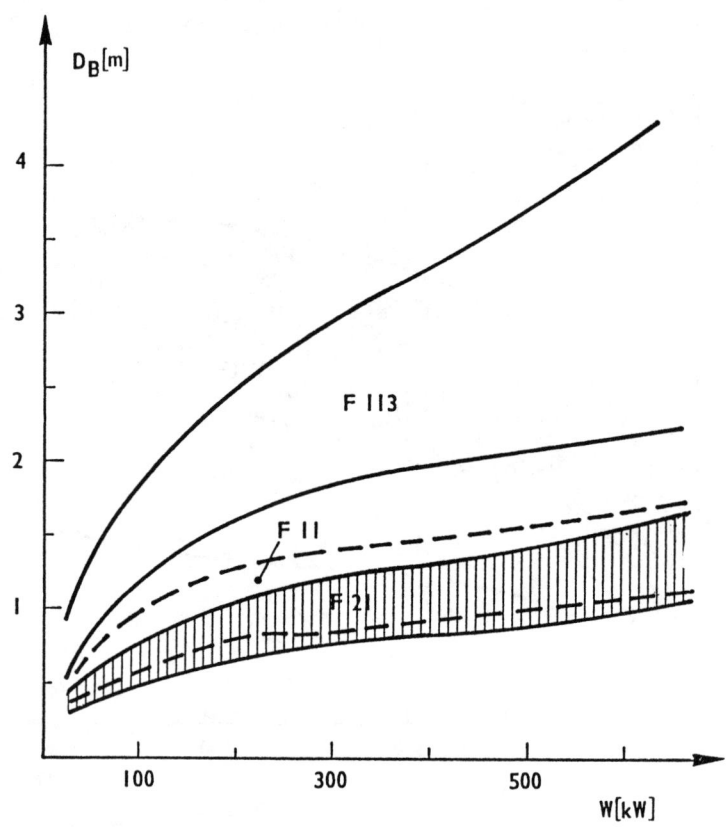

Fig. 8 Optimal range of the two-phase duct diameter versus power

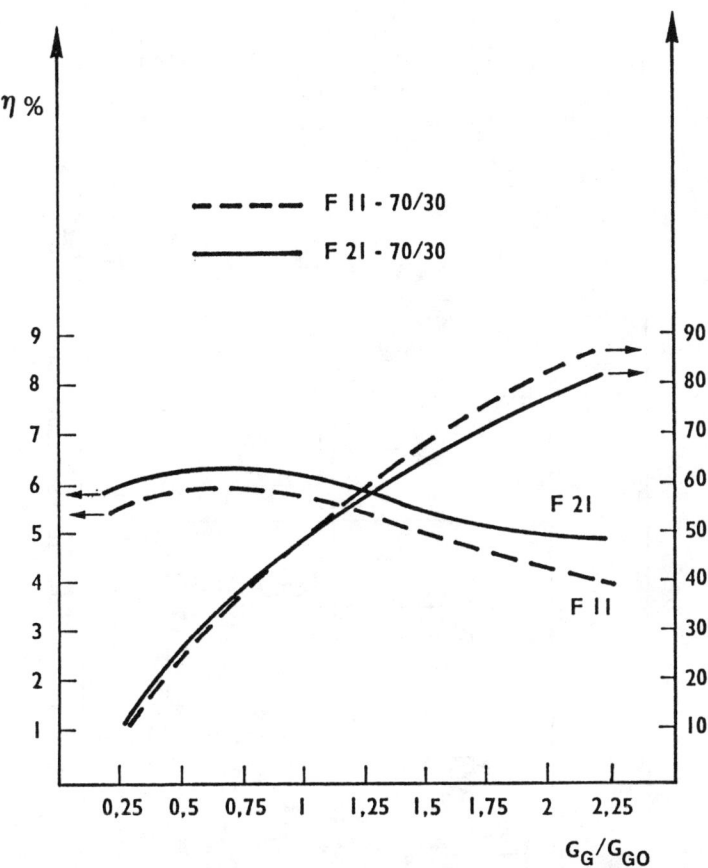

Fig. 9 Organic efficiency and power versus G_G/G_{GO} (G_G, G_{GO} are the actual and design working-fluid flow-rate respectively)

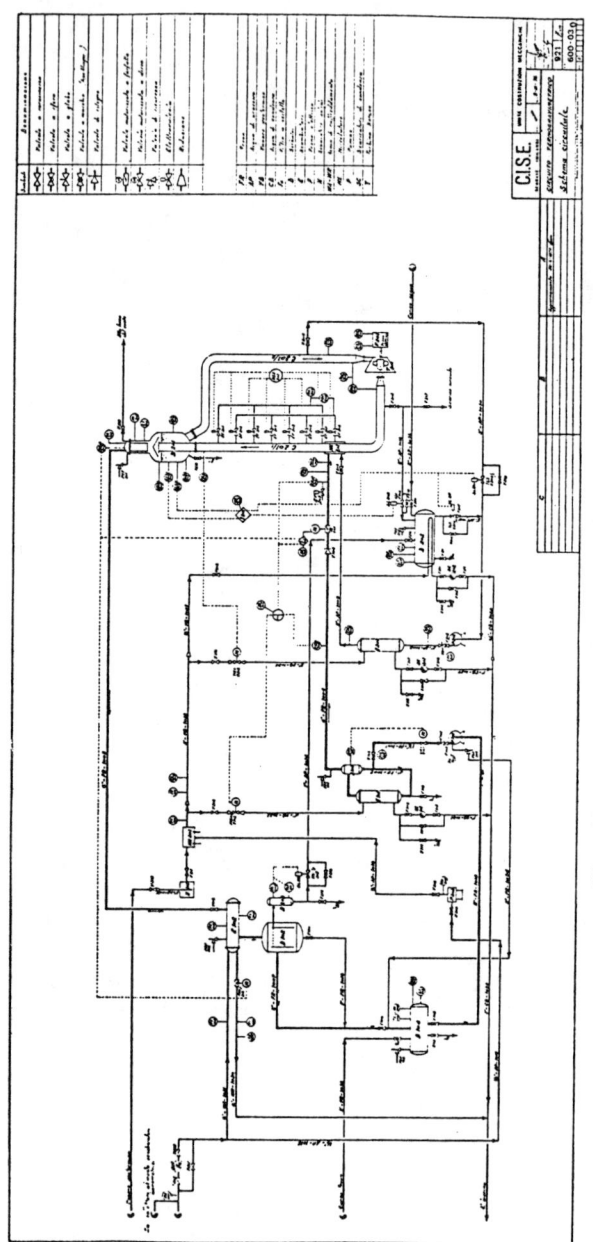

Fig. 10 - Thermogravimetric loop flow-sheet.

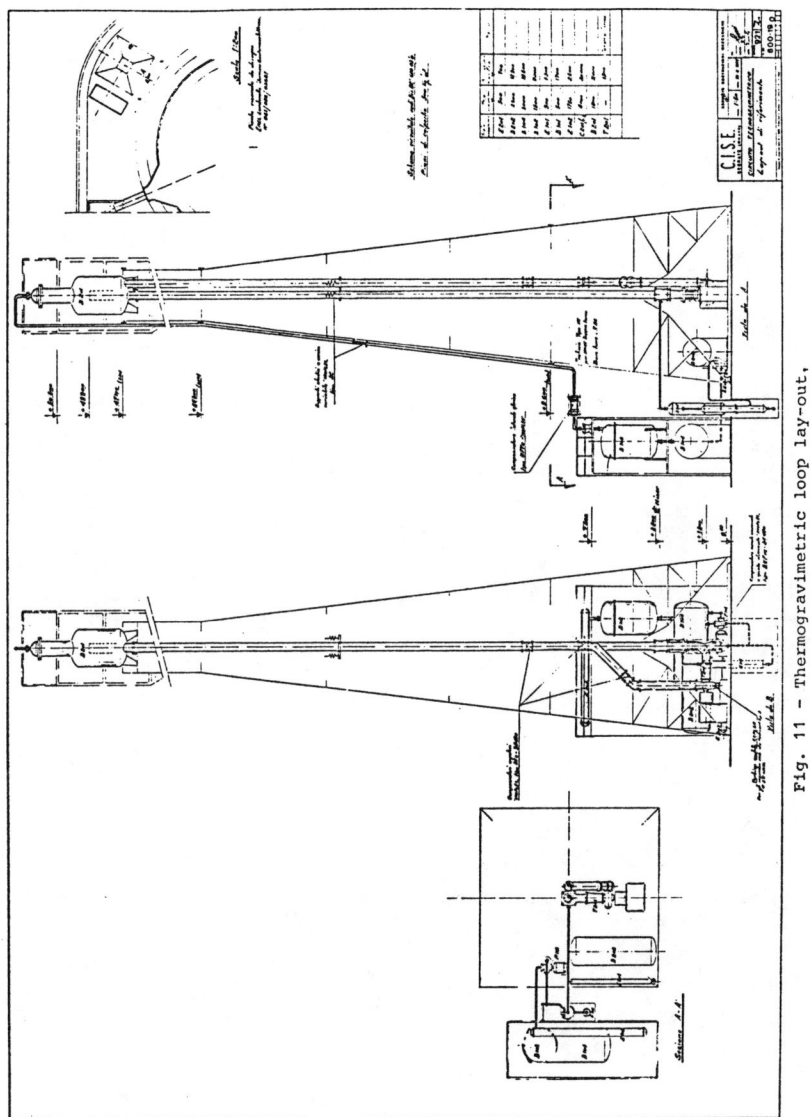

Fig. 11 - Thermogravimetric loop lay-out.

The Optimization of Alternative Energy Cycles Using Program GEOTHM*

M. A. GREEN, P. A. DOYLE, H. S. PINES, W. L. POPE, and L. F. SILVESTER
Lawrence Berkeley Laboratory
University of California
Berkeley, California 94720, USA

ABSTRACT

GEOTHM is a thermodynamic process program which will calculate a wide variety of thermodynamic cycles using a variety of working fluids. GEOTHM has a unique optimizing ability which permits a thermodynamic cycle to be optimized for minimum cost, maximum efficiency or any other user-specified parameter. GEOTHM can optimize a complicated cycle with many parts in a single step optimization. The program has been used to optimize cycles with over 20 optimizable parameters.

The optimization process is quite different for most alternative energy cycles because the energy source is often diffuse or it has a low temperature. Three examples of optimized power cycles are presented here in order to illustrate the broad capabilities of the GEOTHM code.

INTRODUCTION

Lawrence Berkeley Laboratory has been developing the GEOTHM computer program since December 1973 [1]. Program GEOTHM can be applied to a wide variety of alternative energy cycles. Until recently the GEOTHM program has been applied only to Geothermal energy cycles[2,3,4]. (The program name comes from this application.) This paper will demonstrate the use of GEOTHM in three types of power cycles. These are: 1) a Geothermal power plant which uses a binary or bi-fluid cycles, 2) a simple ocean thermal power cycle, and 3) a power cycle which is a combined gas turbine and a Rankine bottoming cycle.

GEOTHM is an extremely versatile thermodynamic cycle simulator. Its primary features are:

1. The thermodynamic processes are modularized into fundamental building blocks. The blocks can be arranged into any type of thermodynamic system.

2. The calculation of thermodynamic properties and transport properties is separated from the thermodynamic process elements.

3. The program is fast due to efficient programming in all of its iterative convergence routines(i.e., Newton's Method).

4. The thermodynamic cycle generator in GEOTHM can be used as a function generator which is steered by a mathematical optimizer. The optimization routine can design and optimize a thermodynamic cycle with respect to any user specified criterion.

5. The program is interactive. A preprocessor corrects many errors which the user makes in inputing data into GEOTHM. The program, which can be run from a remote terminal, is user oriented[5].

GEOTHM is a large computer program with over 100 FORTRAN subroutines which contain over 10,000 FORTRAN statements. Since GEOTHM is a very large program, its use has been limited to the CDC-7600 computer system.

Before proceeding to the sample cycle calculations, it is useful to review the primary features of the program. GEOTHM has subroutines which perform various thermodynamic processes. The processes that GEOTHM can model include: turbines, pumps, fans, compressors, isenthalpic expanders, flash tanks, direct contact condensers, surface heat exchangers, surface condensers, pipe, geothermal wells, and fossil fuel burners.

Fluid properties are calculated using several equations of state. The program includes equations of state for pure water and sodium chloride brines with a concentration of up to 300,000 ppm[6]. The Starling BWR equation of state is used with non-chlorinated hydrocarbons[7]. Two forms of the Martin equation of state can be used for the Freons and ammonia[8,9]. Air, noncondensible gasses and products of combustion are represented by a modified ideal gas equation of state. Thermal transport properties currently coded were obtained from the National Bureau of Standards[10].

GEOTHM can be operated in two modes, the passive design mode and the dynamic design mode. When the program operates in the passive design mode, the program designs the power plant using user specified state parameters and thermodynamic processes. When the optimizer steers the state parameters (now called optimizable parameters), the power plant is designed and optimized using the dynamic design mode. Optimization is a one step process in multidimensional space. Optimization of power cycles with up to 55 optimizable parameters is possible because the program is extremely fast. Any reasonable objective function can be optimized using the program. We have optimized geothermal power cycles for: 1) minimum plant capital cost, 2) minimum bus bar energy cost, 3) maximum energy yield per unit well flow, and 4) maximum cycle thermodynamic efficiency.

CYCLE OPTIMIZATION

The optimization process for most alternative energy cycles usually proceeds along different lines from conventional fossil fuel or nuclear cycles. Most alternative energy cycles are characterized by low energy density and or low thermodynamic availability. The former characterizes virtually all solar cycles; while the latter is characterized by the low

temperature waste heat cycles. There are large sources of energy which
have either low energy density or low temperature. The economic viability
of these cycles is questionable. Therefore, the economic optimization of
such cycles requires careful consideration of all cycle parameters.

A typical fossil fuel power cycle has relatively few parameters
which must be optimized simultaneously. The temperature and pressures
in the cycle are, in general, limited by technology. The highest temperature and pressure are therefore, nonoptimizable. In fossil fuel plants
the regeneration and reheat processes are an important part of the optimization process; a great deal of effort is expended in optimizing these
parameters. Until recently, relatively little effort has been expended
in doing a multiparameter optimization process that is needed to create
a cycle which is the most economical. The sample cycles which are given
in this report are optimized in at least seven dimensional space. They
all require careful optimization of the heat exchangers and the heat rejection system.

The three example cycles given in this report all generate 50 MW of
electric power at the bus bar. The mass flows of the *fuel stream* (the
term *fuel* here is defined in a very broad way) vary from hundreds of metric
tons per second in the ocean thermal cycle to a kew kilograms per second
for the gas turbine cycle. In each case the cost of *fuel* has a different
impact on the optimization of the cycle.

The first sample cycle, a geothermal binary cycle, shows clearly the
trade off between thermodynamic efficiency and economic viability. This
cycle illustrates that a power cycle can be optimized to maximize thermodynamic performance or to minimize energy cost. The second sample cycle,
an ocean thermal cycle, is even more extreme than the first sample cycle.
The thermodynamic availability is extremely limited by the narrow temperature difference between surface waters and bottom waters. In this cycle,
the friction loss in piping (particularly the water piping) becomes an
optimizable parameter. The third cycle, which is a fossil fuel gas turbine
cycle which tops an organic working fluid cucle, illustrates the interplay
between two different thermodynamic cycles which have different characteristics.

OPTIMIZATION OF A 50 MW GEOTHERMAL BINARY POWER PLANT

The cycle shown in Figure 1 has been optimized for minimum bus bar energy
cost. This cycle represent one of the cycles which is proposed for a geothermal power plant to be built in the Imperial Valley of California[11].
The cost coefficients for various plant components were fitted to cost
quotations obtained by a major engineering firm[12]. The heat transfer coefficients assumed for the brine heat exchanger and the condenser came from
the same source. The brine leaving the well is assumed to be pure water.
The secondary working fluid is isobutane. (See Reference 12 for other
details.)

The cycle shown in Figure 1 has six major optimizable parameters
which are: 1) the turbine inlet temperature, 2) the turbine inlet pressure

3) the condensing pressure, 4) the pinch point temperature difference for the brine heat exchanger, 5) the pinch point temperature difference in the condenser, and 6) the exit temperature of the water leaving the cooling tower. A typical passive mode cycle design of the cycle shown in Figure 1 takes about 75 milliseconds of 7600 computer time. Dynamic mode optimization takes around 20 seconds using the previously given optimizable parameters.

Figure 2 shows the optimum bus bar energy cost and the energy yield per unit well flow for the cost optimized plant as a function of the inlet temperature of the geothermal water entering the plant. Table 1 makes a comparison of various plant parameters for 50 MW net electric power plants which have been optimized for minimum bus bar energy cost and maximum yield per unit well flow. This comparison of parameters is made at geothermal resource temperatures of 175°C and 250°C. From Table 1, one can see that the optimizable parameters change considerably when a plant is optimized for a maximum theoretical yield per unit fuel flow (flow from the geothermal wells) instead of minimum bus bar energy cost. One would not build a plant which maximized the yield per unit well flow simply because it makes no economic sense.

Figure 3 shows the three dimensional surface for maximum resource utilization efficiency (yield per unit well flow over maximum yield per unit flow if the geothermal fluid were expanded isentropically from the resource temperature to the wet bulb temperature) versus the temperature and pressure of the inlet of the turbine. (All other optimizable parameters have been set at the optimum value.) This surface was generated for a 200°C resource. Figure 4 is a three dimensional plot of energy cost at the bus bar versus resource temperature and resource utilization efficiency. This surface shows that minimum cost energy is produced at a resource utilization efficiency of around 40 percent[12]. This is 60 to 70 percent of the maximum resource utilization efficiency possible for the cycle.)

Optimization of geothermal power cycles can be extended beyond plant design once the plant has been designed, the optimizer can be used to maximize electric power output at other conditions besides the plant design conditions. As an example, the GEOTHM optimizer can be used to maximize the net power output from the power cycle shown in Figure 1 when the air wet bulb temperature is reduced.

OPTIMIZATION OF A SIMPLE OCEAN THERMAL POWER CYCLE - A 50 MW PLANT

The power cycle shown in Figure 5 represents a simple ocean thermal power plant cycle. This cycle is in many ways similar to the cycle shown in Figure 1. Hot water (lukewarm ocean water) is drawn in at one end. It is used to evaporate an organic working fluid. In this example isobutane is used. (This is probably not the best working fluid to use; ammonia is probably better.) Cold water is used to cool the heat exchanger condenser. The cycle in Figure 5 was programmed to illustrate the capabilities of the GEOTHM program. It is not necessarily the best cycle to use for an ocean thermal power plant.

Since we lack expertise in ocean thermal systems, we have made a number of simple assumptions, which are given as follows:

1. The warm water inlet pipe is 500 m long extending horizontally away from the plant. The cold water inlet pipe which is 1000 m long extends to a depth of 1000 m below the surface of the water. The outlet pipes for both the warm and cold water systems are 200 m long.

2. Inlet water temperatures of 20 and 25°C are used. An inlet cold water temperature of 5°C is assumed.

3. Turbine and pump efficiencies of 85 percent are assumed.

4. Two heat exchanger U factors are assumed in this study. The heat exchangers are the major plant items, the heat transfer per dollar per degree C has an important affect on the plant optimization.

5. The total cost of the facility which includes the wessel that supports the plant is assumed to be 2.5 times the cost of the major plant conversion components. A direct plant cost factor of 1.70 is assumed. The annual maintenance cost is assumed to be 10 percent of the plant capital cost. The annual plant cost (this includes, taxes, insurance, interest on the capital expenditure and profit) is 25 percent of the plant capital cost.

6. The plant is assumed to be operating 85 percent of the time.

The pressure drop in the sea water transport pipes becomes an important optimizable parameter in an ocean thermal cycles. The cycle shown in Figure 5 was optimized with six optimizable parameters. They were: 1) warm sea water piping pressure drop; 2) turbine inlet temperature; 3) condenser temperature; 4) hot water to isobutane heat exchanger pinch point temperature difference; 5) cold water to isobutane heat exchanger pinch point temperature difference; and 6) cold sea water piping pressure drop. It is important to point out that there are other optimizable parameters such as the pressure drops in the heat exchanger which are not included. The optimized cycle shown in Figure 5 is only a *first cut* at ocean thermal cycle optimization.

Table 2 shows the effect of the inlet temperature on the optimization of a ocean thermal cycle. The assumed heat exchanger U factors were 1514 $Wm^{-2}K^{-1}$. The assumed heat exchanger cost was $18.3 m^{-2}. Table 3 shows the effect of the U factor on the cost of energy from an optimized power cycle. A factor of two increase in the heat transferred per dollar has an effect on the cost of energy and the optimizable parameters of the plant.

In Table 2 it is interesting to compare the bus bar energy cost, plant yield, and cycle efficiency with the corresponding columns on the lefthand side of Table 1. The plant yield per unit mass of water processed is equivalent in the two systems. While the resource warm water in the ocean thermal cycle costs nothing, the cost of processing the water, which is reflected in the cost of energy, is not negligible. If ocean thermal cycles are to ever be economically viable, it is clear that mulitiparameter optimization of the thermodynamic cycles will be important.

COMBINED GAS-TURBINE AND ORGANIC WORKING FLUID - 50 MW PLANT

The cycle shown in Figure 6 consists of a gas turbine cycle with an isobutane bottoming cycle. This kind of cycle, which has a high thermodynamic efficiency, could run off of small sources of gaseous or liquid fuel generated by processing agricultural or human waste (for example, sewer gas). The plant represented by this kind of cycle would have a low capital cost. The plant could be quite compact and thus suitable for use on low heating value gasses or liquid and gaseous products resulting from waste processing.

The cycle shown in Figure 6 has seven major optimizable parameters. They are: 1) the compressor exit pressure; 2) the heat exchanger pinch point temperature difference; 3) the isobutane turbine inlet pressure; 4) the isobutane turbine inlet temperature; 5) the condensing pressure in the isobutane cycle; 6) the condenser pich point temperature difference, and 7) the exit temperature of the water from the cooling tower.

Table 4 compares the parameters of a gas turbine system with and without an isobutane bottoming cycle, Two fuel costs are used in both cases. The fuel is expensive (equivalent to $13 and $78 per barrel of oil). The Table shows that the bottoming cycle is potentially worthwhile. The optimizer increases the capital cost of the plant in order to save expensive fuel. It should be noted at GEOTHM can calculate gas turbine cycles with a regenerative preheater and with cycles which use other working fluids in the bottoming loop. The optimizer will optimize the cycles for minimum cost while looking at all of the optimizable parameters simultaneously.

CONCLUSIONS

The report demonstrates how the LBL GEOTHM computer program can be used to design and optimize various types of thermodynamic power cycles. The cycles shown here only illustrate the potential of the program. The optimizer can be used to maximize the power output from power plant cycles already designed and built. It can be extended to other types of cycles such as refrigeration cycles and power cycles which derive heat from other sources. Multiparameter optimization techniques can be used to make some kinds of alternative energy sources economically viable.

REFERENCES

1. M. A. Green, H. S. Pines, "Program GEOTHM, a Thermodynamic Process Program for Geothermal Power Plant Cycles." Presented at the *CUBE Symposium* at the Lawrence Livermore Laboratory, October 23-25, 1974, LBL-3060, October 1974.

2. M. A. Green, H. S. Pines, "Calculating of Geothermal Power Plant Cycles Using GEOTHM." Presented at the Second United Nations Symposium on *The Development and Use of Geothermal Resources*, San Francisco, California, May 19-29, 1975, LBL-3238, May 1975.

3. H. S. Pines and M. A. Green, "The Use of Program GEOTHM to Design and Optimize Geothermal Power Cycles." Presented at the Eleventh *Intersociety-Energy Conversion Engineering Conference*, Lake Tahoe, Nevada, September 12-17, 1976, LBL 4454, June 1976.

4. M. A. Green, H. S. Pines, W. L. Pope, and J. D. Williams, "Thermodynamic and Cost Optimization Using Program GEOTHM." *Proceedings of the Geothermal Resources Council* Meeting, San Diego, California, May 1977, LBL-6303.

5. M. A. Green, R. N. Healey, H. S. Pines, W. L. Pope, L. F. Silvester, and J. D. Williams, "GEOTHM-PART 1, A Users Manual for GEOTHM. (Computer Design and Simulation of Geothermal Energy Cycles.) LBL Publication-202, July 1977.

6. L. F. Silvester and K. S. Pitzer, "Thermodynamics of Geothermal Brines, I., Thermodynamic Properties of Vapor Saturated NaCl (aq) Solutions from 0 to 300 °C." LBL-4456, January 1976.

7. K. E. Starling, Fluid Thermodynamic Properties for Light Petroleum Systems. Gulf Publishing Company, Houston, 1973.

8. S. L. Milora, "Application of the Martin Equation of State to the Thermodynamic Properties of Ammonia." ORNL-TM-4413, December, 1973.

9. R. C. Downing, "Refrigerant Equations." DuPont Corporation, Freon Products Division, Report No. 2313.

10. J. M. Hanley, "Prediction of the Viscosity and Thermal Conductivity Coefficients of Mixtures." *Cryogenics*, pages 643-651; November, 1976.

11. Holt/Procon, Geothermal Demonstration Plant Feasibility Study, Energy Conversion Study for EPRI. The Ben Holt Company and Procon Incorporated, Job Number 7523, April 23, 1976.

12. W. L. Pope, M. A. Green, H. S. Pines, J. D. Williams, and L. F. Silvester, "Multiparameter Optimization Studies on Geothermal Energy Cycles." Presented at the 12th *Intersociety Energy Conversion Conference*, August 28- September 2, 1977, Washington, D.C. Table 1 and Figure 4 are preliminary results presented at the 7th *Meeting of the Centers for the Analysis of Thermal-Mechanical Energy Conversion Concepts*, November 1-2, 1977 Brown University, Providence, Rhode Island.

13. J. G. McGowan, J. W. Connell, L. L. Ambs and W. P. Goss,"Conceptual Design of a Rankine Cycle Powered by the Ocean Thermal Difference." Proceedings of the *8th Intersociety Energy Conversion Engineering Conference,* 1973.

14. M. J. Zucrow, Aircraft and Missile Propulsion, Volumne 1. John Wiley and Sons, Inc., New York, 1958.

Table 1. A comparison of cycle parameters for 50 Mw power plants which have been optimized for minimum bus bar energy cost and maximum yield per unit well flow (reference 12).

PARAMETERS	MINIMUM COST OF ENERGY OPTIMIZATION		MAXIMUM YIELD PER UNIT WELL FLOW OPTIMIZATION	
	Brine Temperature		Brine Temperature	
	175°C	250°C	175°C	250°C
General Parameters				
Bus Bar Energy Cost (m$/kWh)	38.23	24.47	∞	∞
Plant Yield (kWh/ton)*	13.044	24.438	19.109	39.499
Cycle Efficiency (%)	11.30	12.40	12.88	15.88
Brine Mass Flow (kgs^{-1})	1065.	568.	727.	351.
Turbine Gross Power (MW)	67.77	66.64	65.78	72.82
Plant Capital Cost ($/kW)	849.	568.	∞	∞
Optimizable Parameters				
Turbine Inlet Pressure (bar)	43.56	61.89	62.99	129.54
Turbine Inlet Temperature (°C)	154.04	214.70	164.50	237.72
Condenser Pressure (bar)	6.26	7.28	4.29	4.19
Brine Heat Exchanger Pinch Point - ΔT (°C)	7.29	23.40	0.	0.
Condenser Pinch Point - ΔT (C°)	4.29	5.46	0.	0.
Cooling Tower Water Temperature (°C)	33.91	35.06	26.67	26.67

*kWh per metric ton of geothermal brine processed.

Net Electric Power   50.0 MW   Brine Heat Exchanger Average U Factor   1514 Wm^{-2}K^{-1}
Air Dry Bulb Temperature 48.89 °F (120 °F)   Condenser Average U Factor   ~500 Wm^{-2}K^{-1}
Air Wet Bulb Temperature 26.67 °C (80 °F)

Table 2. The Effect of Inlet Sea Water Temperature on the Optimization of an Ocean Thermal Cycle for Minimum Cost Energy.

PARAMETERS	INLET WATER TEMPERATURE (°C)	
	20	25
General Parameters		
Bus Bar Energy Los (m$/kWh)	226.7	105.1
Plant Yield (kWh/ton)*	0.212	0.0428
Cycle Efficiency (%)	0.987	1.81
Warm Sea Water Flow (kgs^{-1})	653874.	324157.
Gross Turbine Power (MW)	84.39	69.85
Heat Exchanger Area (m^2)	2669006.	1046979.
Plant Capital Cost ($/kW)	4822.	2235.
Optimizable Parameters		
Turbine Inlet Temperature (°C)	15.75	19.99
Condenser Temperature (°C)	8.61	9.70
Hot Water Heat Exchanger Pinch Point - ΔT (°C)	1.34	2.11
Condenser Pinch Point - ΔT (°C)	1.51	2.40
Hot Water Pipe Pressure Drop (bar)	0.011	0.041
Cold Water Pipe Pressure Drop (bar)	0.017	0.017

* kWh per metric ton of warm sea water processed

Plant Net Electric Power	50 MW
Cold Water Inlet Temperature	5 °C
Heat Exchanger Cost	$18.3 m^{-2}
Heat Exchanger U Factor	1514 Wm^{-2}K^{-1}
Condenser U Factor	1514 Wm^{-2}K^{-1}

Table 3. The Effect of Heat Exchanger U Factor on Cost Optimized Ocean Thermal Cycle Parameters and Energy Cost at the Bus Bar.

PARAMETERS	Case 1	Case 2
Heat Exchanger Parameters		
U Factor ($Wm^{-2}K^{-1}$)	1514.	3028.
Cost per Unit Area ($\$m^{-2}$)	18.3	18.3
Normalized Heat Transfer Cost ($W\$^{-1}K^{-1}$)	82.73	165.46
General Parameters		
Bus Bar Energy Cost (m$/kWh)	105.1	66.5
Plant Yield (kWh/ton)*	0.0428	0.0479
Cycle Efficiency (%)	1.810	1.856
Warm Sea Water Flow (kgs^{-1})	324157.	289714.
Gross Turbine Power (MW)	69.85	68.12
Heat Exchanger Area (m^2)	1046979.	532724.
Plant Capital Cost ($/kW)	2235.	1414.
Optimizable Parameters		
Turbine Inlet Temperature (°C)	19.99	19.99
Condenser Temperature (°C)	9.70	9.70
Hot Water Heat Exchanger Pinch Point ΔT (°C)	2.11	1.94
Condenser Pinch Point ΔT (°C)	2.40	2.15
Hot Water Pipe Pressure Drop (bar)	0.041	0.038
Cold Water Pipe Pressure Drop (bar)	0.017	0.016

* kWh per metric ton of warm sea water produced

Net Electric Power   50 MW
Cold Water Inlet Temperature   5 °C
Warm Water Inlet Temperature   25 °C

Table 4. The Effects of Fuel Lost on Cycle Parameters on a Cost Optimized Gas Turbine Cycle With and Without a Bottoming Cycle.

PARAMETERS	FUEL COST ($ per kg)			
	Without Bottoming Cycle		With Bottoming Cycle	
	0.10	0.60	0.10	0.60
General Plant Parameters				
Bus Bar Energy Cost (m$/kWh)	41.1	165.6	39.6	152.3
Plant Yield (kWh per ton)*	3798.	4083.	4182.	4522.
Cycle Efficiency (%)	34.195	36.747	37.641	40.702
Fuel Mass Flow (kg·s^{-1})	3.657	3.402	3.321	3.071
Gross Turbine Power (MW)	108.26	133.25	101.85	124.48
Plant Capital Cost ($/kW)	380.	478.	404.	503.
Optimizable Parameters				
Compressor Exit Pressure (bar)	10.164	17.208	9.159	15.849
Isobutane Turbine Inlet Pressure (bar)	DNA	DNA	128.245	149.829
Isobutane Turbine Inlet Pressure (°C)	DNA	DNA	383.61	326.88
Condenser Pressure (bar)	DNA	DNA	6.198	6.500
Heat Exchanger Pinch Point ΔT (°C)	DNA	DNA	6.66	1.78
Condenser Pinch Point ΔT (°C)	DNA	DNA	2.58	1.81
Cooling Tower Water Temperature (°C)	DNA	DNA	33.20	31.29

* kWh per metric Ton of Fuel Burned; DNA = Does Not Apply.

Net Electric Power	50 MW	Heat Exchanger U Factor	100 Wm^{-2}K^{-1}
Gas Turbine Inlet Temperature	1250 °C	Condenser U Factor	~400 Wm^{-2}K^{-1}
Air Dry Bulb Temperature	48.89 °C		

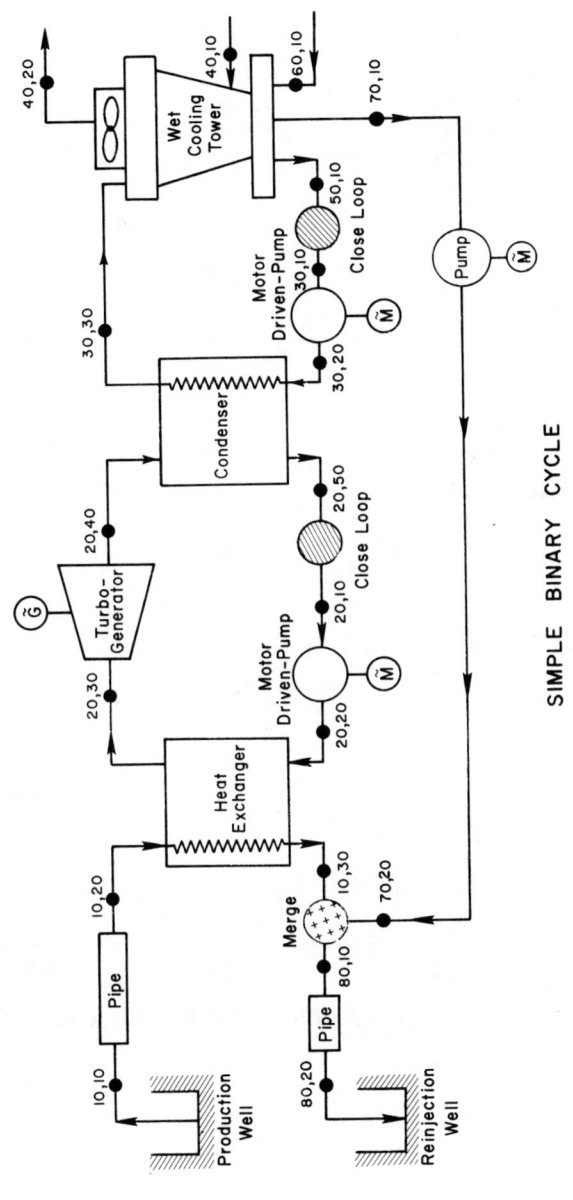

Figure 1. A Simple Binary Geothermal Cycle.

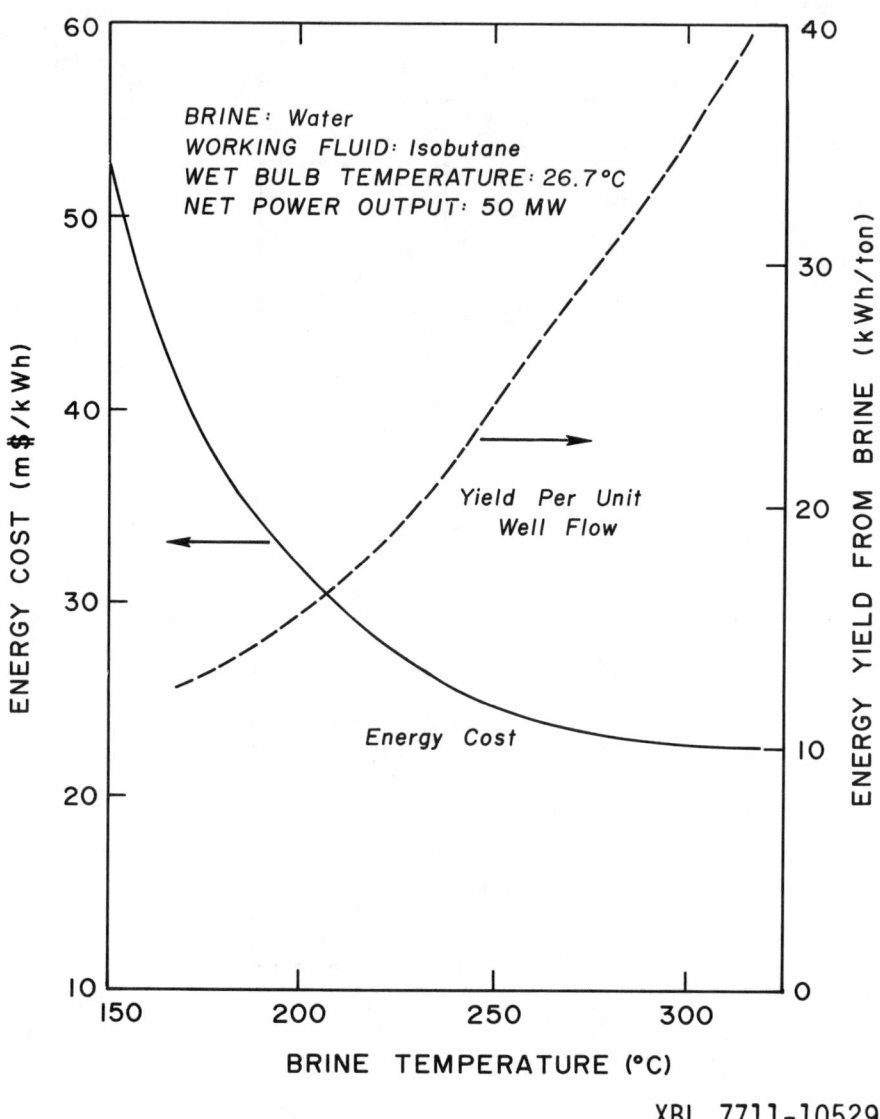

Figure 2. Electrical Energy Cost and Energy Yield from the Brine as a Function of Brine Inlet Temperature in the Cycle Shown in Figure 1.

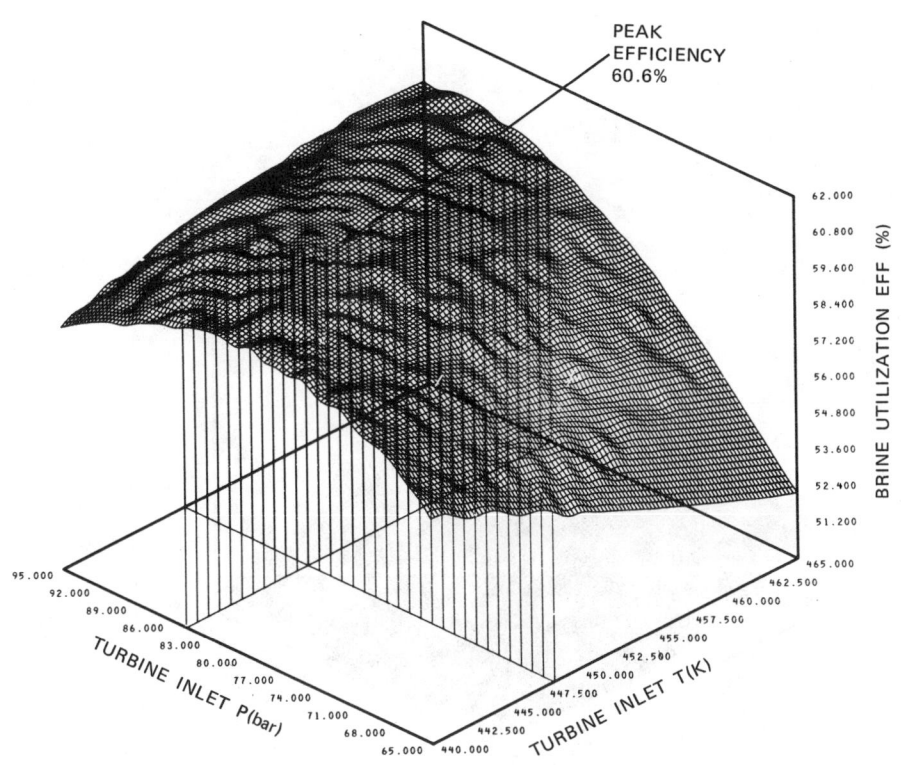

Figure 3. Brine Utilization Efficiency as a Function of Turbine Inlet Pressure and Turbine Inlet Temperature.

Resource Temperature	200 °C
Condensing Pressure	4.26 bar
Heat Exchanger Pinch Point Temperature Difference	0
Condenser Pinch Point Temperature Difference	0
Cooling Tower Water Temperature	26.67 °C

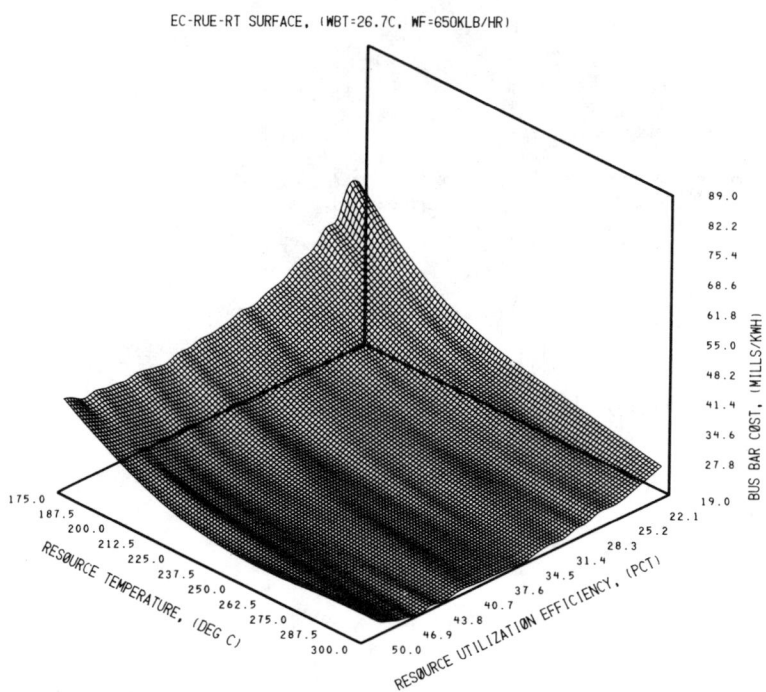

Figure 4. Bus Bar Electrical Energy Cost as a Function of Resource Temperature and Resource Utilization Efficiency (50 MWe Isobutane Geothermal Binary Cycle, Reference 12).

Air Drying Bulb Temperature	48.89 °C
Air Wet Bulb Temperature	26.67 °C
Net Electric Power	50.0 MW

XBL 7711-10528

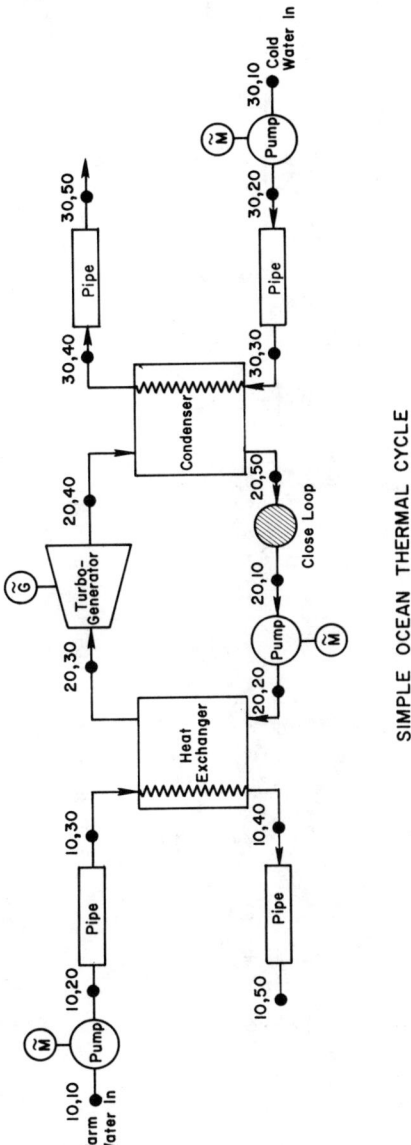

Figure 5. A Simple Ocean Thermal Cycle.

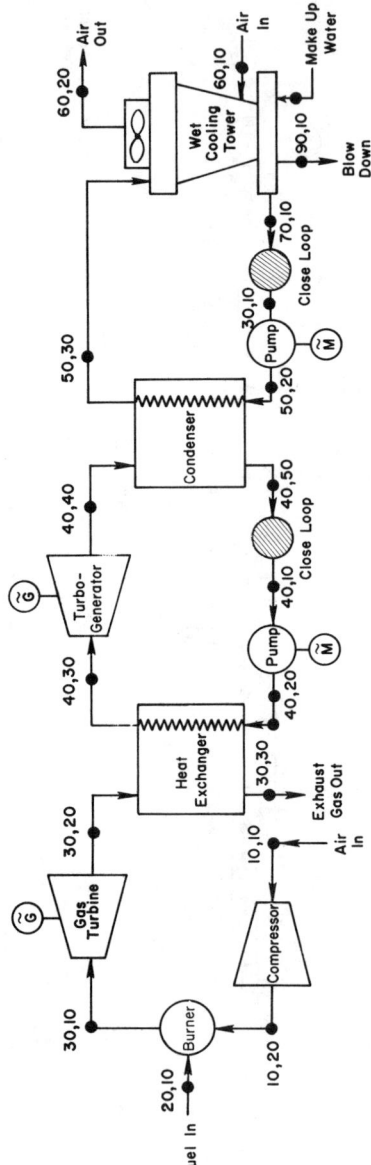

Figure 6. A Gas Turbine Cycle with a Light Hydrocarbon Rankin Bottoming Cycle.

Thermal Fracturing Patterns and Effects of an Imitation Hot Dry Rock by Impinging of Water Jets

H. KIYOHASHI, M. KYO, and W. ISHIHAMA
Tohoku University
Sendai, Japan

ABSTRACT

Effects of rock temperature and water jet speed on the thermal fracturing of imitative hot dry rock specimens have been studied with reference to the establishment of drilling and fracturing methods for natural hot dry rocks. Specimens (75 mm in diameter and 150 mm in length) were made of castable fire-resistant material. Their compressive and tensile strengths are about 22 and 2.5 kg/cm^2, respectively, in the temperature range tested. Temperatures of the specimen were room temperature, 200, 400, 600, 800, 1000 and 1140 °C. Water jet speeds used were 70, 84 and 98 m/s at the nozzle exit (1.4 mm I. D.). The water was at room temperature. The distance between the nozzle exit and the test surface of a specimen was kept constant at 40 mm. Jet operation time was fixed at 10 sec.

Experimental results indicated that fracturing patterns and effects varied characteristically with specimen temperature and jet speed. Phenomena observed in the experiments was explained by using the characteristic curve of the boiling heat transfer.

INTRODUCTION

As one of the means of new energy supply, development of geothermal energies stored in hot dry rocks without hot water circulation system has also been focused in Japan [1] . The creation of an artificial hot water circulation system in the high temperature rocks and for this purpose the establishment of drilling and fracturing methods for the hot dry rocks are necessiated to extract the heat energy from such rocks.

Kuriyagawa et al. [2] inferred the possibility of the establishment of drilling and fracturing methods for hot dry rocks from the survey of literatures. Gringarten et al. [3] theoretically investigated the heat extraction from fractured hot dry rocks, and Abe et al. [4,5] theoretically studied a method of making a vertical penny-shaped crack in hot dry rocks by injecting high pressure water. But no one reported yet how to drill holes for injecting water and extracting hot water or steam, nor how to make a penny-shaped crack in such hot dry rocks in a designated direction. As a method to designate the direction of a crack, the authors believe that it is a good idea to make a initial crack previously for leading the direction of the structural crack, but no proposal of practical methods was found yet.

High speed water jets have the enormous hydraulic fracturing power, and continuous water jets of velocities in the range of 200-800 m/s are used in coal mining and in cutting soils, relatively soft porous rocks and so on [6]. Also impinging water jets having large heat transfer power are used in cooling of high temperature bodies [7, 8]. In general, super high speed continuous or pulse water jets (up to 3000 m/s) are needed for cutting hard rocks, such as granite and basalt [9, 10]. But when temperature of their rocks is high, it thus seems possible that they might become susceptible to cutting by relatively low speed continuous jets because of unsteady internal tensile thermal stress induced near the rock surface by high cooling power of impinging water jets [11].

The final objective of our studies is to investigate the applicability of high speed water jet, whose fracturing ability is caused by the interactions between the thermal cracking and the hydraulic jet erosion, for drilling and fracturing of natural hot dry rocks. This paper deals with the experimental studies, which were carried out by using imitative hot dry rock specimens made of castable fire-resistant material instead of natural high temperature rocks, on the effects of the speed of continuous water jet, the temperature of rock specimen and the value of physical properties of rock specimen on the fracturing pattern and efficiency of them.

EXPERIMENTAL MATERIALS

Imitative rock specimens were made of castable fire-resistant material (SiO_2 : 46%, Al_2O_3 : 42%, commercial name : "ASAHI CASTER-13"). Water corresponding to 30 wt% was added to the castable fire-resistant material whose particle size was not larger than 20 mesh, then mixed and poured into cylindrical molds (75 mm I. D. and 150 mm in depth). After having been solidified in the molds for over 7 hr, the castable fire-resistant materials were kept in water for 24 hr, naturally dried in a shade for over 10 days, dried at 70 °C for 12 hr, and then dried at 110 °C for 12 hr.

TABLE 1.shows the main physical properties of imitative rock specimens at different temperatures tested. Both compressive strength and tensile strength are the test results of gradually cooled rock specimens to the room temperature. From the comparative study of compressive strength test results for cement mortars between at the cold and hot states by Hokao [12] and from the catalogue of "ASAHI CASTER", it is assumed that the compressive strength of these rock specimens in the hot state (up to 1000 °C) is almost the same as those in the cold state. Young's modulus was obtained as the secant modulus of elasticity at 0.3 % strain. Thermal conductivity was measured by the method of line heat source. In the above TABLE, *2 indicates the data from the catalogue of "ASAHI CASTER" and the values in parenthesis indicate the values obtained by extrapolation.

GEOTHERMAL ENERGY AND HYDROPOWER

EXPERIMENTAL APPARATUS AND METHOD

The main components of all the experimental apparatus consisted of a water jet generator, test chamber, measuring apparatus and rock specimen heating furnace. The outline of the first three parts is shown schematically in Fig. 1.

The water jet generator consisted of N_2 gas bomb (A) obtained commercially (pressure : 135 kg/cm^2, volume : 7 Nm^3), pressure regulator (B), cylinder made of stainless steel (F) (32.9 mm I. D. and 8000 mm in length), piston (G) (32.65 mm in diameter and 51 mm in length) equipped with two O-rings, jet on/off valve (L), nozzle (N) and some accessories: cylinder head (C), pressure gauge (D), exhaust valve (E), tail gate of the cylinder (H), thermocouple (I), insulator (J) and water feed valve (K).

Rock specimen (P) was fixed on the test stand (S) equipped inside of test chamber (O) (820 mm x 830 mm x 1140 mm) through the medium of refractory bricks (Q). Windows (U) were set on the side of test chamber to lighten inside, to discharge steam and to insert microphone for recording the noise of fracturing. The measuring apparatus consisted of pressure transducer (M) which transformed the changes of the pressure of water jets at the inlet of nozzle into electric signals, strain meter (W), trigger switch (V) which indicated the opening and closing time of valve (L), tape recorder (Z) and electromagnetic oscilloscope (X). The temperature of water jets was measured by thermocouple (I).

The spouting procedures of water jets were as follows. 1) After piston (G) was moved to the tail gate of cylinder (H), and valves (E) and (L) were opened, valve (K) was opened to supply city water. 2) After residual gas in the nozzle was almost expelled, valve (L) was closed, and (G) was moved to the cylinder head (C) through cylinder (F) by city water pressure. This state of full charge of city water can be confirmed by the stop of removing gas from valve (E) and the noise caused by the collision between (G) and (C). 3) After valve (E) and (K) were closed, and the gas was transferred from N_2 gas bomb (A), pressure regulator (B) and valve (E) were operated to adjust the pressure in (C) to reach a certain test level P_s. Then the preparation of spouting water jets was ready. 4) When valve (L) was opened, the water was discharged as a jet a certain speed from the nozzle exit. The detail drawings of the parts of nozzle (N) and test stand (S) are shown in Fig. 2 and Fig. 3.

Heating of the rock specimens was carried out in an electric furnace (electric capacity : 15 kw, inner dimension: 180 mm x 180 mm x 500 mm) equipped with a programed temperature controller.

Experimental procedure were as follows. 1) After the spouting preparation of water jets with a certain pressure P_s as above described was accomplished, a specimen heated to a certain temperature in the electric furnace at a certain heating rate was taken out by tongs, and then moved quickly to the test stand as shown in Fig. 1 (refer to Fig.3). 2) After the distance between the nozzle exit and a flat surface of the specimen reached a certain value, the rock specimen was fixed immediately. 3) After tape recorder (Z) and electromagnetic oscilloscope (X) were operated, valve Ⓛ was opened, and then the rock specimen was struck by the water jet in a certain period, to be fractured. 4) After valve Ⓛ was closed, fragments of the specimen were collected, and then one test run was accomplished.

Experimental variables and conditions were as follows. The temperature of the specimens were normal temperature, 200, 400, 600, 800, 1000 and 1140 °C. Power source pressure of water jets, P_s, was at 25, 36 and 49 kg/cm^2 corresponding to speeds of water jets at nozzle exit, U_o, 70, 84 and 98 m/s, respectively. Because normal city water was used as a jet liquid, the temperatures of water jets, t_o, were almost equal to room temperature, 10 - 20 °C. Besides, other parameters were fixed: nozzle diameter, D_o, was 1.4 mm; the distance between the nozzle exit and the tested surface of the specimen, L, was 40 mm; impinging duration of water jets was 10 sec; when rock specimens were heated, the rate of increasing temperature was about 1.7 °C/min; the time of maintaining the certain tested temperature was over 150 min; the numbers of rock specimens were five at each different condition (a combination of a certain temperature of rock specimens and a certain speed of water jets). Further, when rock specimens were moved from the electric furnace to the test stand, it took 25-40 sec before the rock specimens were struck by water jets.

The speed of water jets was calculated by following formula:

$$U_o = \sqrt{\frac{2g(P_o - P_a)}{\gamma}} \times 10^2 \qquad (1)$$

where, g = gravity acceleration (m/s^2)
P_o = static pressure of water at nozzle inlet (kg/cm^2)
P_a = atmospheric pressure (kg/cm^2)
U_o = speed of water jet at nozzle exit (m/s)
γ = specific gravity of water (kg/m^3)

The depth of fractured cavities of rock specimens, judged from the original surface of rock specimens before fractured, was measured by inserting a thin wire (0.5 mm in diameter) into the cavity and by measuring the length of the inserted parts of the wire with slide callipers. The space volume of cavities of fractured specimens was filled up with dried carborundum (60 mesh in size) whose weight could be measured. So the volume of cavities or fragments of the fractured specimens could be obtained by the known weight and the apparent specific gravity of the carborundum.

RESULTS AND DISCUSSIONS

Characteristics of Water Jets

In this water jet system, static pressure of water at nozzle inlet, P_o, changed transiently with time τ immediately after the start of shooting of water jets. That is to say, P_o increased up to the peak maximum value of about 1.2 to 2.0 times higher than the power source pressure, P_s, during 0.02 to 0.03 sec at the beginning shooting of water jets. Subsequently to the period, P_O decreased to a stationary value P_S with damping the pulsation during 0.5 to 3 sec after the start. This transient response time τ had a tendency to decrease as peak maximum pressure of P_O became high, and did not very much varied with the variation of the values of P_S.

Photographs of (a), (b), (c) and (d) in Fig. 4 show stationary water jet configurations at the water jet velocities of U_O=20, 70, 84 and 98 m/s, respectively. The jet configuration in the case of U_O=20 m/s, which was showed to compare with others, had a narrow spreading of water jet equal to nozzle diameter D_O=1.4 mm in the fairly longer distance than 200 mm. On the other hand, the jet spreadings of the water jets in cases of U_O=70, 84 and 98 m/s reached to about 2 mm at the distance L=40 mm from the nozzle exit. But no difference was found between the latter three water jet configurations.

The structures of free water jets used in our study, at the region striking the specimens, namely at the dimensionless distance L/D_O=28.6 from the nozzle exit, seemed to correspond to the initial region referring to studies by Yanaida [13]. So, the authors assumed from the data that the ratio of pressures, P_m/P_O, and the ratio of velocities, U_m/U_O at the dimensionless distance L/D_O=28.6 were both unity.

<u>Effects of Specimen Temperature and Water Jet Speed on Fracturing Pattern of Specimen</u>

Figure 5 shows photographs of the initial surface conditions of the specimens at each temperature tested. However, these photographs do not show the surface condition of the actual specimens used for the fracturing tests at each temperature tested, but show the surface condition of the specimens for the strength tests cooled slowly down to room temperature after heated up to each tested temperature. In Fig. 5, photographs of 1b to 6b and 1w to 6w show, respectively, the best and the worst surface conditions among each five pieces of specimen for each tested temperature. These photographs may give sufficient data to estimate the initial surface conditions of the specimens actually used for fracturing tests. As shown in the photographs, the initial surface condition of all specimens used for the fracturing tests at each tested temperature were sufficiently good except the specimens at 800 and 1140 °C, which had some rather large initial cracks.

Fracturing patterns of the specimens are as follows. Figures 6, 7 and 8 show the front views of the fracturing patterns of the specimens of each tested temperature struck by the water jets having speeds of U_O=70, 84 and 98 m/s, respectively. Figures 6 (b), 7 (b) and 8 (b) show the front view of the crack patterns of the restored surface of the specimens, fracturing patterns of which were shown in some photographs in Figs. 6 (a) , 7 (a) and 8 (a), respectively. The photograph number in Figs. 6 (b), 7 (b) and 8 (b) corresponds to the number in Figs. 6 (a), 7 (a) and 8 (a). Figure 9 shows typical longitudinal views of fractured specimens at the specimen temperatures 400 and 600 °C and at the jet speed U_O=98 m/s, where specimen were most seriously fractured. In this figure, photograph (a) gives the original state of the specimen, and photographs (b) and (c) correspond to photographs 71 and 77 in Fig. 8 (a), respectively. The followings are evidently drawn from above figures.

1) When the specimens were at room temperature, most part of the specimens were not damaged by the water jets having speeds less than U_0=98 m/s.
2) When specimen temperature θ was 200 °C, most of the specimens were not fractured by the water jet at a speed U_0=70 m/s. But by the water jets having speeds U_0=84 and 98 m/s, the specimens were fractured and the shape of the fractured cavities were somewhat like a crescent-shaped semicrater or like a perfect crater, respectively. 3) When θ was 400 °C, by the water jets of U_0=70 and 84 m/s most of the specimens were fractured, making cavities like the crescent-shaped semicraters and like the craters, respectively. Moreover, the crushing happened more violently in the case of U_0=98 m/s and the fragments of broken specimen head flew over to the wall of the test chamber. 4) When θ was 600 °C, the most part of the specimens was fractured, making cavities like craters by the water jets of U_0=70, 84 and 98 m/s. In the case of U_0=98 m/s the crushing occurred more remarkably. 5) When θ was over 800 °C, most of the specimens were not fractured, but a hole was made at the impinging point of each specimen regardless of the values of U_0 and in spite of the cracked initial surface conditions of the specimens. The diameter of the holes was about 4 to 5 mm. 6) The patterns of the cracks of the restored surfaces of the specimens, the shape of fractured cavities of which were like craters or which crushed severely into fragments, were formed by the radial cracks of 4 to 6 lines starting from the water jet stagnation points (in some cases the cracks had a few branches) and by the irregular polygonal cracks having their vertical angles where radial crack lines intersected or the irregular circular cracks having the stagnation points as the center points on the struck surfaces of the specimens. The areas surrounded by the irregular polygonal cracks or the irregular circular cracks tended to increase with the increase of the water jet velocity U_0.

Relations between Fracturing Effectiveness and Specimen Temperature.

Although evaluation of the fracturing effectiveness should have been considered from many points of view, the authors evaluated the effectiveness by the maximum depth and the volume of the fractured cavity of the specimens at the present study. Figure 10 shows the relations between the maximum depth from the initial surface of the specimens l [mm] and the specimen temperature θ [°C]. In this figure, open circles, open rhombuses and open squares designate the experimental points at the velocities of the water jets of 70, 84 and 98 m/s, respectively. And their closed symbols represent the arithmetic mean values of five experimental values at the same experimental condition. Also, solid, broken and chain lines were drawn as smooth curves connecting the closed symbols at each condition. If we may call the curves THERMAL FRACTURING CURVES of high temperature materials by impinging water jets, the thermal fracturing curves show the shape of N character obviously. As shown in this figure the depth l has the maximum value at the neighborhood of the specimen temperature θ=600 °C regardless of values of the water jet

GEOTHERMAL ENERGY AND HYDROPOWER

velocity U_0, and the minimum value at $\theta=900\,°C$ for $U_0=70$ m/s, $\theta=950\,°C$ for $U_0=84$ m/s and $\theta=1000\,°C$ for $U_0=98$ m/s.

Figure 11 indicates the relations between the volume of the fractured cavities, $v\,[cm^3]$, which is another index of the fracturing effectiveness, and the specimen temperature, $\theta\,[°C]$. Meanings of the symbols of the experimental point and the distinctions of the three curves are the same as those in Fig. 10. But in drawing the curves some irregular experimental points, namely, the data for the specimens shown in photograph 8 in Fig. 6 (a), photographs 35, 37, 39, 42 and 57 in Fig. 7(a) and photographs 73, 77, 81 and 95 in Fig. 8 (a), were excepted. From Fig. 11, again it is quite obvious that the relations between v and θ show thermal fracturing curves shaped like N character. The curves have the maximum points at about $\theta=600\,°C$, $500\,°C$ and $400\,°C$ in cases of $U_0=70$ m/s, 84 m/s and 98 m/s, respectively. And also, they have the minimum points at about $\theta=900\,°C$ and $\theta=1000\,°C$ in cases of $U_0=70$ m/s and $U_0=98$ m/s, respectively. When U_0 is 84 m/s, the minimum point is ambiguous.

Relations between Fracturing Effectiveness and Velocity of Water Jets.

Figures 12 (a) and 12 (b) show the relations between the mean value of the depths of the fractured cavities, $l\,[mm]$, and the water jet velocity, $U_0\,[m/s]$ (or the energy source pressure P_s) with a parameter of the specimen temperature. In this figure, l increases linearly in the case of the specimen temperature $\theta=200°C$, and exponentially and sharply in the case of $\theta=400\,°C$ and $600\,°C$ with increase of U_0. Though the value of l at $\theta=600\,°C$ are larger than that of l at $\theta=400\,°C$ in the low range of U_0, the difference between them decreases with increase of U_0. In cases of $\theta=800\,°C$ and $1140\,°C$, such tendency was found that l increased exponentially but slowly with the increment of U_0. However in the case of $\theta=1000\,°C$, we could not presume the correct functional relations between l and U_0, although l increased with increasing of U_0 as shown in the figure.

Effect of Jet Operation Time on Depth of Fractured Cavities.

Figure 13 shows the relation between the depth of fractured cavities, l [mm], and the operation time of the impinging water jet, τ [sec], in the case of the specimen temperature $\theta=800\,°C$ and the water jet speed $U_0=84$ m/s. It seems that l approaches to a certain constant value in about 10 sec as shown in this figure. This tendency is in good agreement with the experimental results shown in Reference [14] which were obtained in the case of specimen temperature and water jet temperature both being equal to the room temperature.

Relations between Thermal Fracturing Curves and Temperature Dependency of Thermal Spallability Indexes of the Specimens.

The authors pointed out in the previous chapter that the relations between the fracturing effects (depth of fractured cavities, l and volume of fractured cavities, v) and the specimen temperature, $\theta\,[°C]$ were represented

by the curves shaped like N-character having maximum values at θ=400 - 600 °C and minimum values at θ=800 - 1000 °C. As one of the means to find out why the N-shaped relations between l and θ as well as v and θ was obtained, the authors examined the temperature dependency of some thermal spallability indexes of the specimens that were usually used as indicators of thermal spallability in the case of jet piercing by high temperature flames. Geller [15] reviewed studies on the thermal spalling, and mentioned about some spallability indexes.
Acceptable indexes are as follows:

$S_1 = a\beta$ (by Mirkovich, 1968) (2)

$S_2 = a\beta d_g \tilde{\sigma_c}$ (by Calaman and Rolseth, 1961) (3)

$S_3 = E\beta/\tilde{\sigma_c} \lambda$ (by Rzhevskii, 1964) (4)

$S_4 = E\beta/\tilde{\sigma_c} a$ (by authers) (5)

where, a = thermal diffusivity (m^2/h)
  d_g = grain size (mm)
  E = Young's modulus (kg/cm^2)
  S_n = spallability index (n = 1, 2, 3, 4) (-)
  β = coefficient of linear expansion (-)
  λ = thermal conductivity (kcal/mh°C)
  $\tilde{\sigma_c}$ = compressive strength (kg/cm^2)

In order to compare the temperature dependencies of these four indexes consisting of different definitions each other, these indexes were nondimensionalized by a following equation of definition.

$S_n^* = S_{n\theta}/S_{n15}$ (n = 1, 2, 3, 4) (6)

Where, S_n^* is a dimensionless spallability index, and $S_{n\theta}$ and S_{n15} show values of S_n at the specimen temperature θ [°C]=θ and θ=15, respectively. The relations between S_n^* and θ in this experiment are shown in Fig. 14.

However, because sudden cooling was thought to induce the failure of tensile strength of material, S_2^*, S_3^* and S_4^* in Fig. 14 were calculated by using tensile strength $\tilde{\sigma_t}$ instead of compressive strength $\tilde{\sigma_c}$ in the equations (3), (4) and (5). From this figure, we can obviously recognize that the relations between S_1^* and θ do not considerably agree with the relations between l and θ nor v and θ, shown in Figs.10 and 11. The variations of S_2^*, S_3^* and S_4^* against θ do not accord with ones of l or v against θ numerically, although the tendencies of the three curves are somewhat similar to those of the thermal fracturing curves of the specimens shown in Figs. 10 and 11. As mentioned above, it is clear that the thermal fracturing curves of the specimens obtained in this experiment cannot be explained by the usual thermal spallability indexes only.

Relations between Thermal Fracturing Curve and Boiling Curve

When the temperature of a specimen is above the boiling point of water, heat transfer from the surface of the specimen to the impinging water jets should be accompanied with boiling phenomena. We know about the pool boiling heat transfer from a thin platinum wire heated by electric current to still surrounding water as an example of the basic boiling heat transfer phenomena [16]. In this pool boiling from the platinum wire, taking the heat flux from the surface of the thin hot wire, q, and temperature difference, ΔT_{sat}, between the surface temperature of the wire, θ_w, and the saturated temperature of the surrounding water, T_s, on logarithmic axes of ordinate and abscissa, respectively, we can obtained a curve like a N-shaped as shown in Fig. 15 [16]. This curve is called BOILING CURVE and used to represent the boiling heat transfer phenomena qualitatively and quantitatively. On the boiling curve, we call point C showing maximum value of the curve MAXIMUM HEAT FLUX POINT or BURN OUT POINT, point D showing minimum value MINIMUM HEAT FLUX POINT, the region of curve C-D TRANSITION BOILING REGION, the region on the right of point D FILM BOILING REGION, the region of curve A-C NUCLEATE BOILING REGION and the region on the left of point A NON-BOILING (FREE CONVECTION) REGION [17].

On the other hand, the depth of fractured cavities, l, shown in Fig. 10 may consist of the two effectivenesses, l_h and l_t, which are contributed by the hydraulic power of the water jets acting as the compressive stress and by the cooling thermal shock acting as the thermal tensile stress on specimens, respectively. Although it is very difficult to divide l into l_h and l_t exactly, the authors caluculated l_t from l and the values of l_h obtained experimentally as follows.

So as to obtain l_h, fracturing tests by the impinging water jets were carried out for the specimens at room temperature which were heated at the same conditions as previously mentioned hot fracturing test conditions and cooled down to room temperature gradually. Plotting l_h against the difference ($P_s - \tilde{\sigma_c}$) as shown in Fig. 16, the relation between l_h and ($P_s - \tilde{\sigma_c}$) may be obtained as a following experimental equation:

$$l_h = 0.28 \ (\ P_s - \tilde{\sigma_c} \ ) + 4.1 \ [mm] . \tag{7}$$

Then, l_t is calculated from a following equation:

$$l_t = l - l_h = l - 0.28 \ (P_s - \tilde{\sigma_c}) - 4.1 \quad [mm] \tag{8}$$

Figure 17 shows the relations between l_t and the specimen temperature, θ, that is, pure thermal fracturing curves of the hot specimens by impinging water jets. Consequently, as a first approximation, it is considered that l_t is in proportion to the thermal tensile stress induced in the specimen and that the induced tensile stress is in proportion to the temperature difference, $\Delta\theta$, between the inner temperature, θ, and the surface temperature, θ_w, of the specimen [18]. So, we could assume following propotional equation:

$$l_t \ \alpha \ \Delta\theta \tag{9}$$

Furthermore, as the heat flux, q, transfered from the specimen surface to the wall jet flow is proportional to $\Delta\theta$, a following equation for q is conceived:

$$q \propto \Delta\theta \qquad (10)$$

From equations (9) and (10), a following equation can be drawn:

$$l_t \propto q \qquad (11)$$

Having the ground of the above-stated matters, if we replace l_t with q and θ with ΔT_{sat} on the ordinate and the abscissa of Fig. 17, respectively, these three pure thermal fracturing curves shown in Fig. 17 might be thought to be boiling curves of a kind. Therefore, if we compare these curves with an idealized boiling curve shown in Fig. 15, and infer the signification of the pure thermal fracturing curves from the idealized boiling curve macroscopically, the followings should be pointed out.

Assumed that the maximum point and the minimum point of a pure thermal fracturing curve shown in Fig. 17 correspond to point C and point D on the boiling curve shown in Fig. 15, respectively, in the range of relatively low θ, that is, the region on the left of the maximum point on the pure thermal fracturing curve, the nucleate boiling should be thought to be the dominant mechanism of the heat transfer on the main part of the surface of the hot specimen struck by the water jet, so that q as well as l_t should increase with increment of θ. In the range of θ shown by the area between the maximum point and the minimum point on the pure thermal fracturing curve, major part of the surface of the specimen struck by the water jet should turn into the transition boiling region, where the increase of temperature causes the increase of the area on the surface of the specimen covered by the steam film having a low thermal conductivity. Consequently, q as well as l_t should decrease to the minimum value with increase of θ. Furthermore, in the range of θ beyond the minimum point on the curve, heat flux, q, from the surface of the specimen struck by the water jet to the jet flow should increase again with augmentation of θ because of increment of the radiative heat flux from the specimen surface to the wall jet flow. Therefore, l_t should increases again with increase of θ.

Moreover we can see the fact that the maximum points of the pure thermal fracturing curves move to the lower range of θ with augmentation of the power source pressure, P_s, as well as the nozzle exit velocity, U_o, of water jet as shown in Fig. 17. This tendency is well in agreement with that of the pressure characteristics of the boiling curves [17]. This seems to support the justice of above discussions.

CONCLUSION

In order to establish the drilling and the fracturing methods for hot dry rocks by high speed water jets, imitative high temperature rock specimens made of castable fire-resistant material were struck for 10 second by water jets having speeds of 70, 84 and 98 m/s at the nozzle exit (1.4 mm I. D.) in the initial region of the jet, and the fracturing patterns and effects were investigated. The results were as follows.

1). When the specimen temperature, θ, was 200 °C, specimens were hardly fractured in the case of $P_s=25$ kg/cm^2 ($U_0=70$ m/s). An incomplete crater-shaped cavity of crescent-shape was observed on a specimen surface tested in the case of $P_s=36$ kg/cm^2 ($U_0=84$ m/s), and crater-shaped cavity was observed in the case of $P_s=49$ kg/cm^2 ($U_0=98$ m/s).
2). When θ was 400 °C, by the water jets of $U_0=70$ and 84 m/s most of the specimens were fractured, making cavities like the crescent-shaped semicraters and like the craters, respectively. Moreover, the crushing happened more violently in the case of $U_0=98$ m/s and the fragments of broken specimen head flew over.
3). When θ was 600 °C, the most part of the specimens was fractured, making cavities like craters by the water jets of $U_0=70$, 84 and 98 m/s. In the case of $U_0=98$ m/s the crushing occurred more remarkably.
4). When θ was in the range of 800 °C to 1140 °C, a hole having a diameter of 4-5 mm instead of a crater-shaped cavity was observed on the surface of most specimens, independently of the speed of water jet.
5). In the case of crater-shaped cavities formed or severely fractured specimens, the fracturing patterns of surfaces restored to the original state consisted of 3-6 radial cracks centering on the striking point of water jet (sometimes with branches near the end of cracks), and of irregular polygonal cracks having their vertical angles on the radial cracks or the irregular circular cracks having the center at the stagnation point on the specimen surfaces tested.
6). The relations between the fracturing effectivenesses, whose indicators were the depth or the volume of cavities of fractured specimens, and the specimen temperature have a close connection with the boiling heat transfer phenomenon.

The phenomenon that fracturing patterns varied with the temperature of specimens and the speed of water jet could be explained by the change of the local heat transfer coefficient of impinging water jet for radial-direction. Detailed discussions on the mechanism of the phenomenon would be presented at 4th International Symposium on Jet Cutting [18].

REFERENCES

1. Ministry of International Trade and Industry, Japan: New Energy Technology Research Projects (The Sunshine Projects), Nippon sangyo gijutu shinkou kiyokai, Tokyo (1974) (In Japanese)
2. Kuriyagawa, M., Ogata, Y. and Takada, A. : Reviews of studies on fracturing of hot dry rocks, Mining and Safety, 20 (12), 13-22, (1974) (In Japanese)
3. Gringarten, A. C., Witherspoon, P. A. and Ohnishi, Y. : Theory of heat extraction from fractured hot dry rock, J. Geophys. Res., 80 (8), 1120-1124, (1975)

4. Abe, H., Mura, T. and Keer, L. M. : Growth rate of a penny-shaped crack in hydraulic fracturing of rocks, J. Geophys. Res., 81 (29), 5335-5340, (1976)
5. Abe, H., Keer, L.M. and Mura, T. : Growth rate of a penny-shaped crack in hydraulic fracturing of rocks, 2, J. Geophys. Res., 81 (35) 6292-6298, (1976)
6. Mutsnika, V. S. and Ignatova, F. I. (translated by Hokao, Z.) : Hydraulic mining and hydraulic transportation, Tokyo Univ. Press, (1961) (In Japanese)
7. Katto, Y. and Monde, M. : Study of mechanism of burn-out in a high heat-flux boiling system with an impinging jet, Heat Transfer 1974, 1V, 245-249, 5th IHTC, Tokyo, (1974)
8. Ishigaya,S., Nakanishi, S., Mizuno, M. and Imamura. T.: Heat transfer by impinging circular water jet, Trans, JSME, 42 (357), 1502-1510, (1976) (In Japanese)
9. Cooly, W. C. and Clipp, L. L. : High pressure water jets for undersea rock excavation, ASME publication, No. 69-WA/UnT-7, 1-7, (1969)
10. Harris. H. D. and Mellou, M. : Cutting rock with water jets, Int. J. Rock Mech. Min. Sci. & Geomech. Abstr. 11 (9), 343-358, (1974)
11. Pritchett, J. W. et al. : Thermohydraulic rock disintegration- theoretical analysis of rock cutting by combined thermal weakening of high-speed water jet impact, PB-244 091. (1974)
12. Hokao, Z: Thermodrilling technology for resources, Rateisu Press, Tokyo, (1976) (In Japanese)
13. Yanaida, K. and Ohashi, A. : Studies on the flow characteristics of the high speed water jets in air (1st Report), J. Mining and Metallurgical Institute of Japan, 93 (1075), 423-428, (1977) (In Japanese)
14. Yamakado. N. and Yokoyama, A. : Fracture of rock by high speed water jet. Proc. JSCE, (133), 11-19, (1966) (In Japanese)
15. Geller, L. B. : A new look at thermal rock fracturing, Inst. Mining and Metallrugy, (A), 70 (767), 133-170, (1970)
16. Nukiyama, S. : The maximum and minimum values of the heat Q transmitted from metal to boiling water under atmospheric pressure, J. JSME, 37 (206), 367-374, (1934) (In Japanese)
17. JSME : Boiling Heat Transfer, JSME, 1-14, Tokyo (1965) (In Japanese)
18. Kiyohashi, H., Kyo, M. and Ishihama, W. : Water jet Breaking of Imitation Hot Dry Rock, to be presented at 4th Int. Symp. Jet Cutting Technology, England, April 12th-14th, 1978

TABLE 1. PHYSICAL PROPERTIES OF THE SPECIMENS.

Property		Temperature [°C]						
		15	200	400	600	800	1000	1140
Apparent specific gravity	[-]	1.71	1.70	1.64	1.60	1.58	1.59	1.58
True specific gravity	[-]	2.69	2.70	2.85	2.86	2.83	2.73	2.69
Porosity	[%]	36.4	37.0	42.5	44.1	44.2	41.8	41.3
Compressive strength	[kg/cm^2]	44.0	44.2	31.8	32.9	28.7	21.7	21.6
Tensile strength by R.C.T. [*1]	[kg/cm^2]	6.76	5.98	2.56	2.59	2.72	2.60	2.49
Young's modulus	[10^3 kg/cm^2]	16.6	7.01	6.32	4.18	3.83	3.21	2.73
Specific heat	[kcal/kg°c]	(0.14)	0.160	0.213	0.236	-	-	-
Thermal conductivity	[kcal/m h°c]	0.084	0.125	0.188	0.218	-	-	-
Coefficient of linear expansion [*2]	[%]	(0.01)	0.04	0.06	0.09	0.14	0.20	-
Thermal diffusivity	[10^{-4} m/h]	(3.50)	4.58	5.41	5.79	-	-	-

※1 R.C.T. = Radial Compression Test.
※2 Data obtained from catalogue of materials of the specimen.
( ) = Extrapolated values.

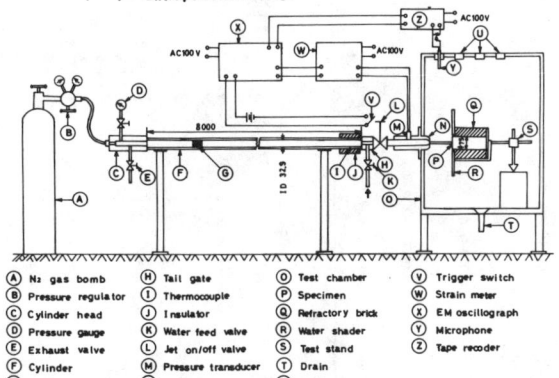

(A) N$_2$ gas bomb (H) Tail gate (O) Test chamber (V) Trigger switch
(B) Pressure regulator (I) Thermocouple (P) Specimen (W) Strain meter
(C) Cylinder head (J) Insulator (Q) Refractory brick (X) EM oscillograph
(D) Pressure gauge (K) Water feed valve (R) Water shader (Y) Microphone
(E) Exhaust valve (L) Jet on/off valve (S) Test stand (Z) Tape recorder
(F) Cylinder (M) Pressure transducer (T) Drain
(G) Piston (N) Nozzle (U) Window

Fig. 1. Schematic diagram of the experimental apparatus

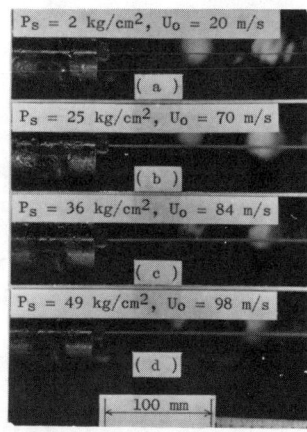

Fig. 4. Water jet configurations.

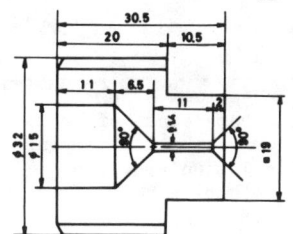

Fig. 2. Detail of nozzle design

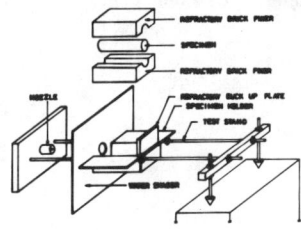

Fig. 3. A sketch of fixing of the specimen for fracturing test.

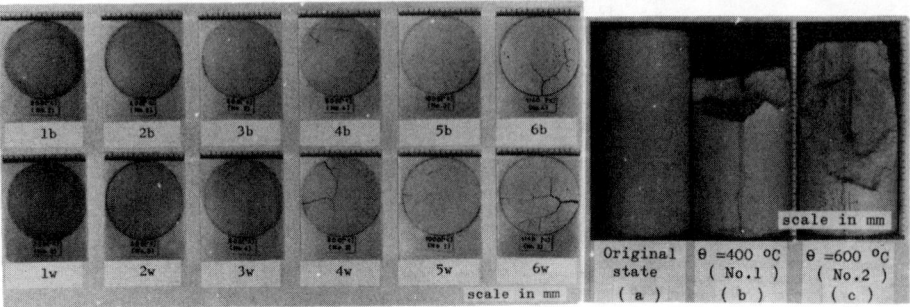

Fig. 5. Initial surface conditions of specimens. Photographs of 1b to 6b and 1w to 6w show the best and the worst conditions in each five pieces of specimen for each tested temperature, respectively.

Fig. 9. Typical longitudinal views of heavily fractured specimens at specimen temperatures 400 °C and 600°C in the case of the jet speed U_0=98 m/s.

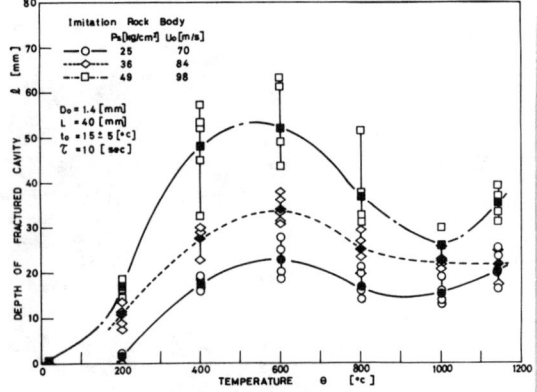

Fig. 10. Dependence of depth of cavity fractured by impinging water jet on specimen temperature. Closed figures show mean values.

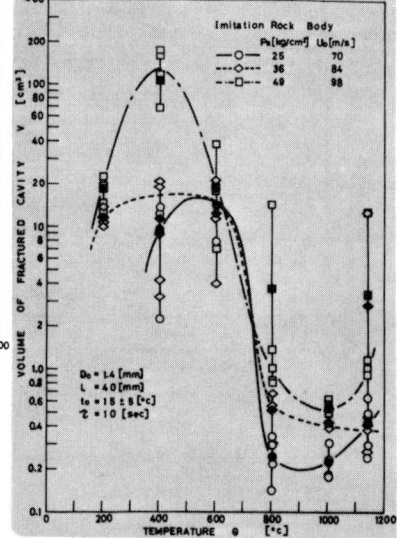

Fig. 11. Dependence of volume of cavity fractured by impinging water jet on specimen temperature.

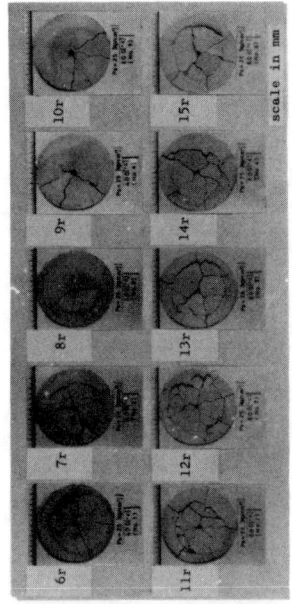

Fig. 6 (b). Front views of the cracking patterns of specimens at the states restored to the original ones in the case of the jet speed $U_0=70$ m/s. Photographs of 6r to 15r correspond to ones of 6 to 15 in Fig 6 (a), respectively.

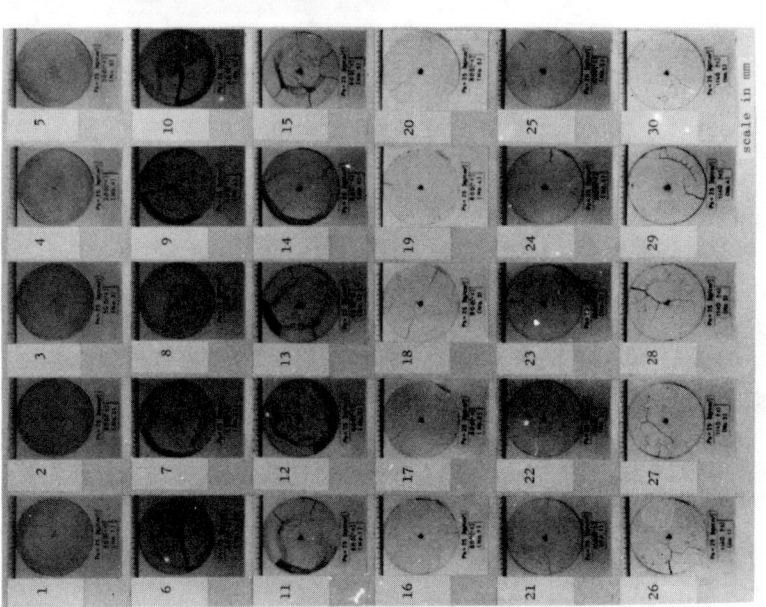

Fig. 6 (a). Front views of the fracturing patterns of specimens struck by the water jet having a speed of $U_0=70$ m/s at different specimen temperatures.

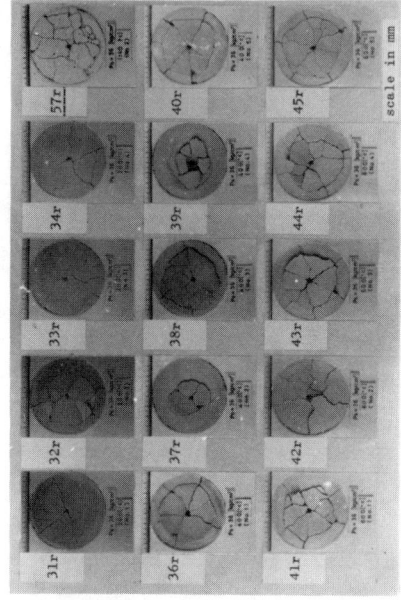

Fig. 7 (b). Front views of the cracking patterns of specimens at the states restored to the original ones in the case of the jet speed $U_0=84$ m/s. Photographs of 31r to 34r, 36r to 45r and 57r correspond to ones of 31 to 34, 36 to 45 and 57 in Fig. 7 (a), respectively.

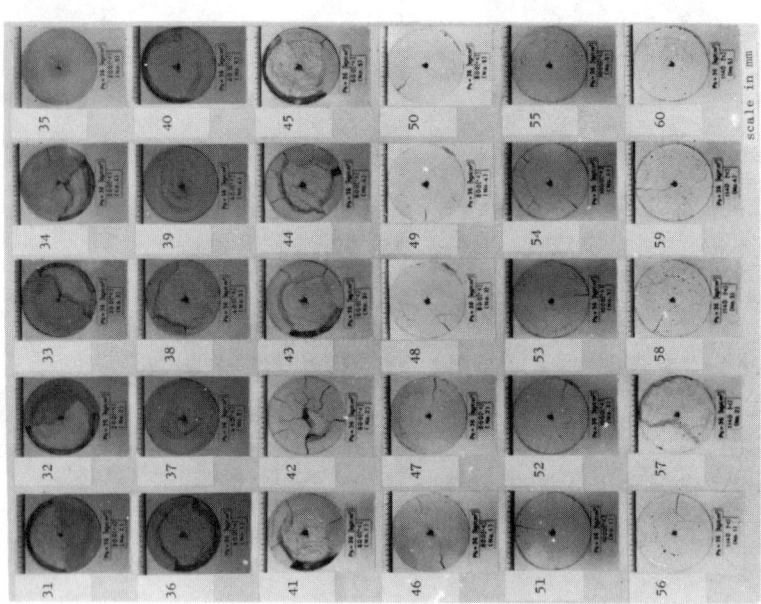

Fig. 7 (a). Front views of the fracturing patterns of specimens struck by the water jet having a speed of $U_0=84$ m/s at different specimen temperatures.

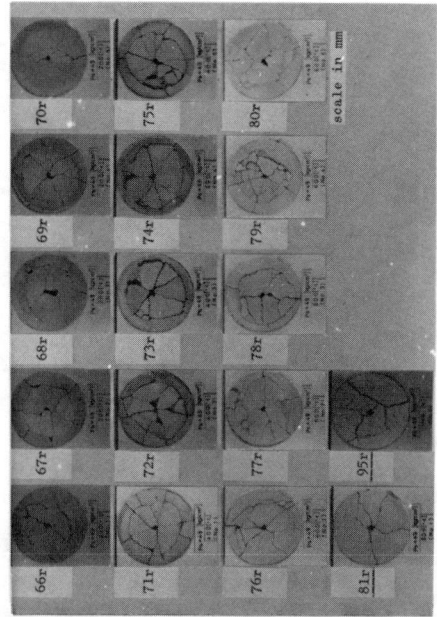

Fig. 8 (b). Front views of the cracking patterns of specimens at the states restored to the original ones in the case of the jet speed U_0=98 m/s. Photographs of 66r to 81r and 95r correspond to ones of 66 to 81 and 95 in Fig. 8 (a), respectively.

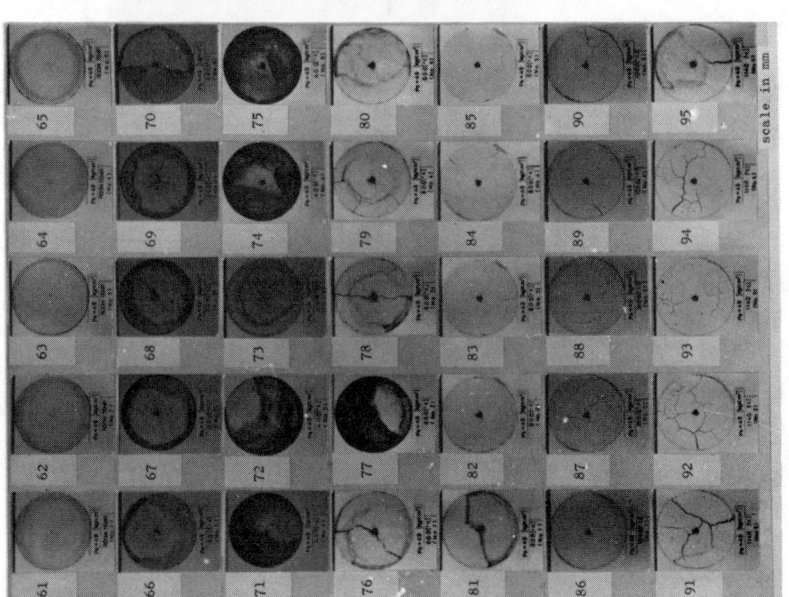

Fig. 8 (a). Front views of the fracturing patterns of specimens struck by the water jet having a speed of U_0=98 m/s at different specimen temperatures.

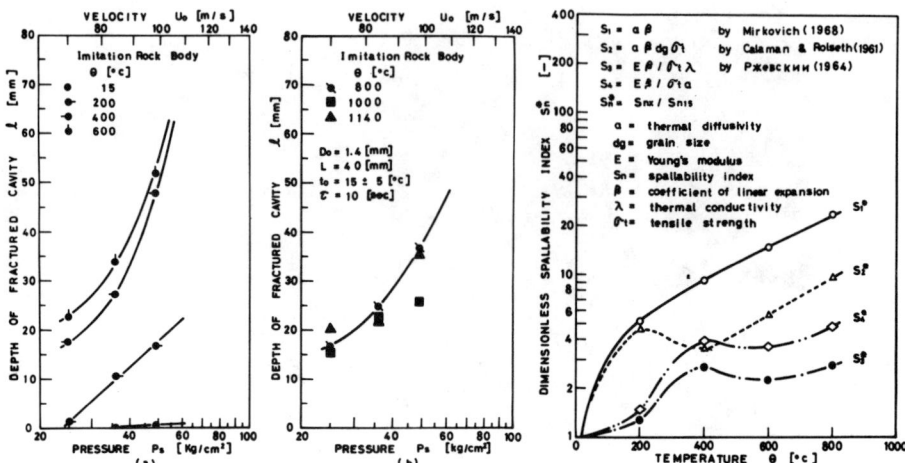

Fig. 12. Dependence of depth of cavity fractured by impinging water jet on power source pressure or on the jet velocity.

Fig. 14. Some dimensionless spallability indexes versus specimen temperature for the imitation rock body.

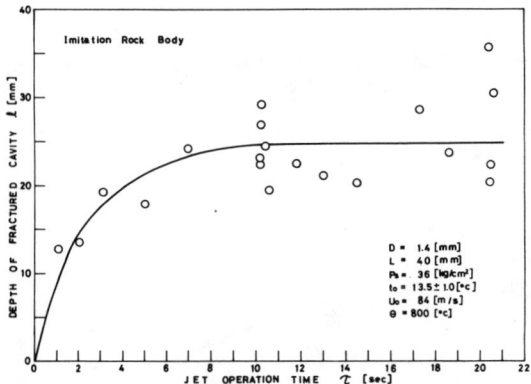

Fig. 13. Dependence of depth of cavity fractured by impinging water jet on jet operation time for specimens of temperature 800 °C.

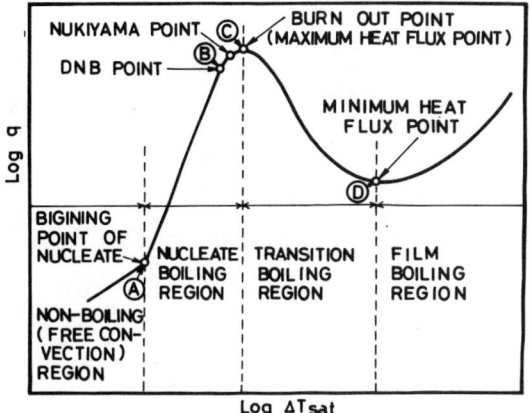

Fig. 15. An idealized boiling curve with some defined points and defined regions of boiling phenomena.

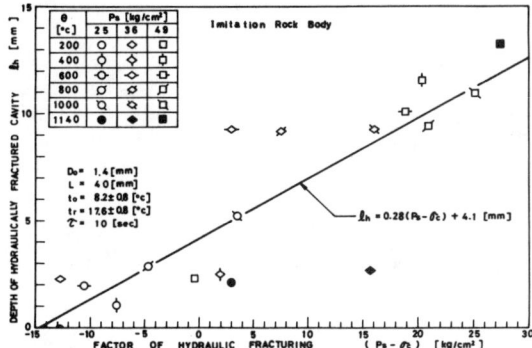

Fig. 16. Dependence of depth of hydraulically fractured cavity by impinging water jet on a factor of hydraulic fracturing ($P_s - \sigma_c$) for gradually cooled specimen of room temperature.

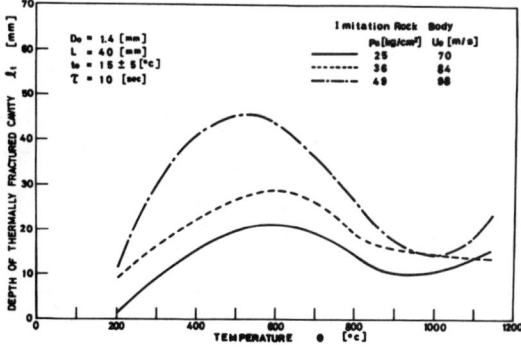

Fig. 17. Dependence of depth of thermally fractured cavity by sudden cooling effect by impinging water jet on specimen temperature.

An Exploration of the Applicability Field of Geothermal Water-fed Heat Pumps*

P. De MARCHI DESENZANI
Universita' di Pavia
Piazza Leonardo da Vinci
Pavia, Italy

G. GIGLIOLI
Istituto di Macchine
Politecnico di Milano
Piazza Leonardo da Vinci 32
Milan, Italy

ABSTRACT

Usefulness of low temperature geothermal water is still under discussion, and the examination of the present structure of energy demand could lead to the conclusion that heat available at a temperature less than $60°C$ is quite unuseful for traditional purposes.

Technological assessment of heat pump systems, however, makes indirect use of low temperature heat possible for domestic or industrial purposes.

An overall analysis has been tried in order to establish the actual field of the convenience of the practice. Energetical limitations have been found, connected with heat pump efficiency (as dependent on use and availability temperature) and with geothermal reservoir characteristics.

Economical constraints are examined; sensitivity of the heat cost to the reservoir-related costs and to the primary energy costs are outlined. An alternative two-source solution is proposed.

1) ENLARGING THE APPLICATION FIELD FOR LOW TEMPERATURE HEAT

If it is accepted that "The supply of oil will fail to meet increasing demand before the year 2000, most probably between 1985 and 1995" and that "Additional constraints on oil production will hasten this shortage, thereby reducing the time available for action on alternatives,"[1] then it must be accepted that immediate proposals are needed.

While in the long term an action could be taken in order to change the characteristics of the utilization systems and more energy conservative production processes could be developed, in the short term the large part of the characteristics of the processes are to be taken for granted. And that means, for example, that given process temperature requirements are to be fulfilled.

---

* CNR (the National Research Center of Italy) financed this research work under the "Finalized program" on geothermal energy.

It follows that when the energy source is heat (geothermal heat, in particular) there are practical constraints that limit the field of its applicability. In other words, heat is an actual energy source only with respect to those uses, whose temperature requirements do not reach the source temperature level.

As can be seen in Fig. 1, only a very small part of the global energy demand applies to heat that is available at a temperature lower than $60°C$. In spite of its large availability, low ($< 60°C$) temperature heat is then a very poor energy source.*

That is why a certain effort is made in order to find a number of new or not-well developed ways to use that kind of heat, but to which extent the operation will lead to an oil-saving result is something that is to be proved.

In fact, when we try to enlarge the users field seeking for new types of applications, we are doing a marketing operation more than finding an answer to the questions that the scarcity of conventional energy sources makes. It is of course possible that some of those applications will prove (especially in the field of agriculture) to be extremely energetically efficient, but an actual energy saving is again a matter of the future.

The simplest way to enlarge the application field of low temperature heat without an action affecting the users is then to transform it by raising its temperature level. And that is something we can do immediately, since heat pump technology is known.

2) CONGRUENCE WITH THE GLOBAL ENERGY BALANCE

Such a practice seems attractive mainly because of the jump in the demand curve in correspondence to $60°C$ (Fig. 1).

We can enlarge the users field four or five times for a difference of a few degrees in the source temperature. There are of course geographical and geoeconomical constraints that can make the use of geothermal heat impractical or impossible in a number of cases, but it must be considered that those constraints are perhaps softened, if the tie between the geothermal water temperature and the use temperature is softened.

There are examples of geo-heat application for space heating purposes,[4] but no aprioristic statement can be made about the actual advantages of the practice.

---

* When we speak of low temperature geothermal water, we exclude the possibility of electric energy production. Electric energy could be economically produced, with relatively new techniques, even when the source temperature is $80 \div 90°C$,[2,3] so that heat use is a problem only for water available at a temperature less than $80°C$.

In principle, we want to substitute, to fulfill a given duty, the oil supply with geo-heat: but we actually supply the user a secondary heat, that is fabricated using geo-heat <u>and mechanical power</u>. A certain amount of mechanical power might also be necessary for geo-heat extraction.

As is shown in Fig. 2, we meet the demand of a given amount of heat at temperature $\bar{T}$, by the means of the consumption (in terms of primary energy) of an amount of heat at a temperature $\bar{T} - \Delta T_1$ and of an amount of heat at a temperature $\bar{T} + \Delta T_2$ (that is the one at which the mechanical power is generated).

The demand of heat at the higher temperature is actually the feeble point of the practice: That the demand for oil (the energy source that we want to save) will decrease, is still something to be demonstrated. It is true that the heat pump application has always the advantage of freeing the user from oil, because he can drive the compressors with electric motors, the electricity having been produced with any energy source, in the place, the way and the scale that are the most convenient, but that would not mean that geo-heat acts as a real energy source.

To prove that primary energy demand (other than geothermal) will decrease, we have to prove that the ratio between the energy we need for a geothermal heat pump and the heat we use,

$$R = \frac{(P_P + P_{HP})/\eta_{MPP}}{Q_U/\eta_Q} \qquad (1)$$

is less than unity. In other words, we have to find the conditions under which the geo-heat pump application would be energetically advantageous.

3) EXPLAINING THE CONDITION FOR ENERGETICAL CONVENIENCE

We prefer the energetical approach to the economical approach, even if, in the case of the heat pump, economical constraints will be more restrictive than energetical constraints, because what is said about energy is to a certain extent true always and everywhere, while economical variables strongly depend on particular situations.

In any case, geo-heat pump economical behaviour will also be outlined. For the moment, we have to consider the condition for energetical convenience in an explicit form.

User and reservoir characteristics are correlated: expressing the power requested for compression, when the user's requirements are known, as

$$P_{HP} = Q_U/\varepsilon \qquad (2)$$

Expressing the power requested for the extraction as

$$P_P = \frac{g\Delta Z/\eta_P}{c(T_g-T_f)} (1-1/\varepsilon) Q_U \qquad (3)$$

as results by the introduction of the energy equation at the evaporator

$$\dot{m}_g c(T_g-T_f) = (1-1/\varepsilon) Q_U \qquad (4)$$

we define a parameter

$$\psi = \frac{c(T_g-T_a)}{g\Delta Z} \qquad (5)$$

which expresses the energetical characteristics of a geothermal reservoir,* and the parameter

$$\lambda = \psi \, \eta_P \qquad (6)$$

which is another function of the geothermal reservoir characteristics, as the pump efficiency can be considered a datum. We also define the parameter

$$\rho = \frac{T_g-T_f}{T_g-T_a} \qquad (7)$$

that expresses the intensity of use of the available geothermal energy, so that the energetical convenience condition was finally written as

$$\sigma = (1-1/\varepsilon)/\lambda\rho + 1/\varepsilon < \frac{\eta_{MPP}}{\eta_Q} \qquad (8)$$

with the specification that ε and ρ are evaluated at

$$T_f = T_{fopt} \qquad (9)$$

---

* ΔZ is to be conceived as a <u>dynamical</u> variable, therefore dependent on the well flow rate. However, we assumed that ΔZ was an independent variable with respect to the heat pump flow rate, and therefore known when the reservoir characteristics are known. The fact is that we want to define a field where the geo-heat pumps can be used profitably, and a field where geo-heat pump application <u>is not to be taken</u> into consideration, and that is a general approach, that is to be maintained free of particular occurrences. With that hypothesis, the amount of water used for heat pump purposes and that extracted from a (single) well are not necessarily the same.

It must be noticed that we maintained the same hypothesis for economical considerations, thus being allowed to define a cost of the well per length unit and per flow rate unit.

The field of energetical convenience, as a function of user and reservoir characteristics, could then be found, provided that the expression for ε was known.

4) HEAT PUMP COEFFICIENT OF PERFORMANCE

The value of the heat pump coefficient of performance ε is a function of
- the thermal levels at which the thermodynamic cycle works,
- the cycle organization (single or double stage compression with economizer, superheating, subcooling, regeneration, etc.),
- the working fluid nature (critical temperature and pressure, normal boiling temperature, latent heat of condensation and evaporation, viscosity, conductivity, etc.)
- the thermodynamic behavior of different types of machines in the various power ranges.

The difference between the coefficient of performance ε of a real heat pump and ε_i of an ideal heat pump is related to the non-reversible entropy production (non-isentropic compression, non-reversible adiabatic expansion, heat transfer with finite temperature difference, heat exchangers pressure drops, etc.) as is evident from

$$1/\varepsilon = 1/\varepsilon_i + \frac{T_{min} \Sigma \Delta S_{irr}}{Q_K} \qquad (10)$$

In the geothermal heat pump temperature range, we consider numerous working fluids, like refrigerants, R11, R12, R113, R114, R21 and other fluorinated hydrocarbons with higher molecular weight and critical temperature.

Fig. 3 shows the dimensionless performances $\varepsilon/\varepsilon_i$ of a few fluids. The values of heat exchangers pressure drops are usual and the centrifugal compressor adiabatic efficiency is $\eta_c = 0.8$ (which is acceptable for a high enough power level). The thermodynamic advantages of R11 are evident; anyway, we have to emphasize that the choice of the working fluid does not depend only on thermodynamic criteria but also on other technical and economical considerations (volumetric flow rates, heat transfer coefficients, mixability with lubricating oils, non-flammability, chemical stability, availability and cost).

To really develop heat pumps for high temperature use (i.e., with condensation above 75 ÷ 80°C, research on new working fluids has to be carried out.

For the energetic and economical evaluation of geo-heat pumps, we assume the ratio $\varepsilon/\varepsilon_i$ as constant in the temperature range we consider. In fact, if we change the thermodynamic cycle organization (from single- to-double compression stage with economizer, regeneration and so on), we can achieve similar values of $\varepsilon/\varepsilon_i$, also at the higher temperature of condensation T_k.

5) ENERGETICAL APPROACH: CALCULATION RESULTS

No relevant constraints result from the above condition when the case of a geothermal system with no reinjection of the extracted water is examined.

The upper limit of ΔZ cannot act as a means of selection of the favorable cases except when geothermal water temperature is very low (up to 30°C) and use temperature is quite high. As the interesting range for use temperature is 60 ÷ 80°C, a slight energetical gain could be obtained even in unexceptional geothermal situations.

On the contrary, the energetical criterion might provide significant restrictions on the use of a system with reinjection, or, if reinjection is mandatory, it would lead to the exclusion of geo-heat pump adoption in a number of cases.

Orientative indications about the applicability of the system (heat pump and reinjection) can be obtained by the means of a simple change of the scale of the diagram, provided that a part of the losses could be considered as a constant and a part of the losses could be considered proportional to ΔZ, with a given coefficient. The lower scale in Fig. 4 is an example of such an operation, for a hypothesized situation.

Reservoir modelling would, however, be necessary to obtain significant indications, so that the problem is left open to further investigation here.

6) ECONOMICAL CONSIDERATIONS

Although the economical approach is always time and place dependent (neither the absolute cost values, nor the cost ratios can be supposed independent of the particular moment and of the particular country to which evaluations are referred), nevertheless it is necessary to have a sense of the limits that economical considerations will provide a hypothesized technical solution.

Assumptions were made, in order that the cost of the heat produced by the geo-heat pump system could be compared to the cost of the heat produced by a conventional system. As for the energetical evaluations, cost variations due to the consideration of actual mass flow rate were not taken into consideration: Maintaining the hypothesis of no dependence of well dimensions or well flow rate on heat pump system water flow rate, we supposed that the cost of the well could be expressed as a cost per length unit and per flow rate unit.

The cost of the geo-heat pump system (with no reinjection) was then expressed as:

$$K = a(1-1/\varepsilon)/\lambda\rho + b/\varepsilon + (C/\Delta T_{ml})(1-1/\varepsilon) \tag{11}$$

where:

$$a = \frac{K_w}{\tau} \eta_P \frac{h_w}{g\Delta Z} + \frac{K_P}{\tau} + K_{ME}$$

$$b = \frac{K_c}{\tau} + K_{ME}$$

$$C = \frac{K_{ev}}{\tau} \frac{1}{U_{ev}}$$

while the comparison criterion was

$$K < K_{conv} \qquad (12)$$

where:

$$K_{conv} = K_Q^* \qquad (13)$$

The cost of the heat pump system was evaluated assuming for the characteristic parameters (ε, ρ, ΔT_{ml}) their energetically optimum values: Also, if an optimization based on economical values produces a reduction of the overall cost, no significant variations will occur for what concerns general indications.

Sensitivity of the heat cost to the main reservoir and water characteristics (temperature gradient and water composition, in particular) was tested assuming a set of different values for K_w and for K_{ev}. It is, however, to be remembered that the reference situation (Italian economical system) has particular connotations.

In fact, we think that a given economical-industrial system (a country) can be represented, to a certain extent, by the means of a given set of cost ratios and a given cost value (the cost of electrical kWh, for example), but that such a set (apart from its time-related variations) is, in general, different from place to place.

That last statement will be confirmed in Fig. 5: Even for energy, the costs as well as the cost ratios show relevant differences.

It must be added that from Fig. 5, and with the aid of a few other data, we arrived at two other basic assumptions. The first is that the kind of use (industrial versus domestic, in particular) of the produced heat is not indifferent to determine the conditions under which geo-heat

---

* The cost of the heat release device (condenser or boiler) was not taken into consideration: That means that it is considered to be not much different in the two cases. Such an assumption will generally play in favour of the conventional plant, but not to a relevant extent.

pump application is convenient;* the second is that (always in Italy) the cost of mechanical energy is roughly the same whether it is produced by electricity or produced directly from fuel.

7) ECONOMICAL APPROACH: CALCULATION RESULTS

For explanation purposes, it would be better to write Eq. 11 in the form

$$K = b \cdot \sigma + (a-b)(\sigma - 1/\varepsilon) + (c/\Delta T_{ml})(1 - 1/\varepsilon) \qquad (14)$$

Calculations show (Fig. 6) that in the cases when heat pump application is economically feasible, the large part of the cost (nearly 2/3 of the total cost is due to that term even if the heaviest conditions are applied to well and evaporator cost) is due to the term $b\sigma$.

That does not mean (Fig. 7) that the influence of the geo-system thermal gradient or of water composition is neglibible: On the contrary, there is so little place for costs other than those that appear in the term $b\sigma$, that only in very favourable cases does the geo-heat system result economically convenient.

But if we remember what b represents, we arrive at the conclusion that it is the cost of mechanical energy, together with its ratio to the cost of fuel, which is the most limitative factor for geo-heat pump application: Even in the limited case of zero cost for well and evaporator, the maximum economically acceptable energy consumption is, for industrial users, only 60% of the energetically acceptable one. The situation looks better if heat is used for domestic purposes; an enlargement of the field of economical convenience can also be forecast in the near future, if the cost of fuel increases more than the cost of electricity; that will happen if satisfactory replacement energy sources can be found for electricity production.

No optimistic provisions, however, can be made for the very short term. That is not a comforting conclusion, but the important fact is that it can be modified if more energy-conservative technical solutions are adopted.

8) AN ALTERNATIVE GEO-HEAT PUMP SYSTEM

We said that the ratio of fuel cost to mechanical energy cost might increase in the future; with this, we refer to substitutive ways for electricity production: We then suppose that pumps and compressors are driven by electric motors.

---

* Industrial fuel costs less than domestic fuel, as well as industrial electricity costs less than domestic electricity, but the definition of "domestic" and "industrial" is different: actually based on quality of fuel, based on quantity of electricity.

Mechanical energy can, however, be produced in on-site plants, so that the conversion into electricity is not necessary, except for the part that is given to the pumps (if the geothermal water level is so low* that the pump cannot be coupled with a ground-level motor).

We have said that (in the Italian case) there is not a significant economical advantage to produce the same quantity of mechanical energy using thermal motors instead of electric motors, but it should be noticed that in the first case we can act on the quantity of the needed mechanical energy.

It is clear that we refer to total-energy devices: If the heat is discharged from the power cycle at a sufficient temperature level, it can be used for heating purposes (Fig. 8), so that the geothermal heat demand will be reduced. Sufficient temperatures can be obtained using organic fluid power cycle$^{(6,7)}$ engines, as well as normal Diesel engines.

A secondary effect can also be produced on the heat pump efficiency, if the waste heat exchanger is in series with the heat pump condenser. Even if we take into account the primary effect only, we can see that the cost of heat is reduced by nearly 40%, with decisive effects on the field of the possible economical applications.

---

* But there are cases (Iceland) of a 60 meter shaft.

REFERENCES

1. Wilson, CL, <u>Report of the Workshop on Alternative Energy Strategies</u>, McGraw-Hill Book Company, New York, 1977, pp. 3-4.
2. Moskvicheva, VN, "Utilization of Heat of Geothermal Springs and Waste Hot Waters in Freon Operated Power Plants," International Seminar on "Future Energy Production - Heat and Mass Transfer Problems," Dubrovnik, August 1975.
3. Gaia, M, Macchi, E, Angelino,G, "Design and Performance of a Low Temperature Heat Engine," to be published.
4. Aureille, R, Lamethe-Parneix,D, "Possible Applications of Geothermal Energy in France," International Seminar on "Future Energy Production - Heat and Mass Transfer Problems," Dubrovnik, August 1975.
5. See Ref. (1), p. 87.
6. Angelino,G, et al., "A Proposal for a Low Pollution Central Heating System for the City of Venice," 6th International Congress of Climatistics," Milan, March 1975.
7. Angelino,G, Ferrari,P, Giglioli,G, Macchi,E, "Combined Thermal Engine - Heat Pump Systems for Low-Temperature Heat Generation," The Institution of Mechanical Engineers, proceedings 1976, Vol. 190 27/76.

LIST OF SYMBOLS

a,b,C	see Eq. (11)
c	specific heat
g	gravity
h	height
K	cost
$\dot{m}$	mass flow rate
P	mechanical power
Q	heat
R	see Eq. (1)
S	entropy
T	temperature
U	overall heat transfer coefficient
Z	pump total head
Δ	difference
ε	heat pump coefficient of performance
η	efficiency
λ	see Eq. (6)
ρ	see Eq. (7)
σ	see Eq. (8)
Σ	sum
τ	utilization time
ψ	see Eq. (5)

Subscripts

a	ambient
c	compressor
conv	conventional
ev	evaporator
f	final
g	geothermal
i	ideal
irr	irreversible
HP	heat pump
k	condensation
min	minimum
ME	mechanical energy
ml	mean logarythmic
MPP	mechanical power production
o	evaporation
opt	at optimum conditions
P	pump
Q	heat production
U	user
w	well

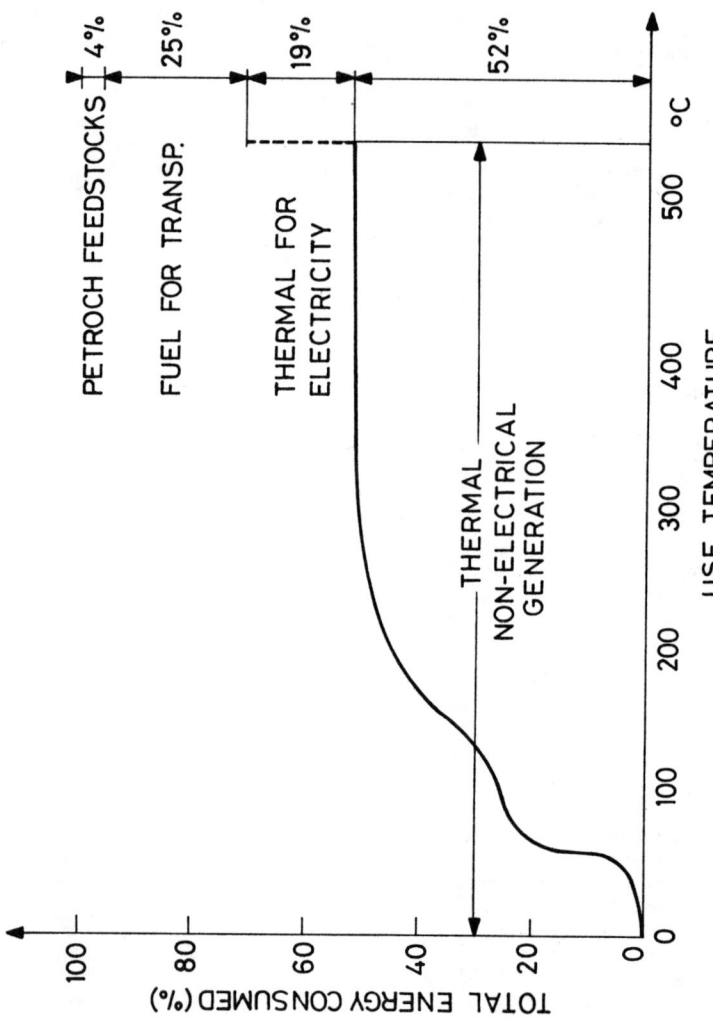

Fig. 1 - Use temperature of thermal Energy - D.E. Anderson "Solar Thermal System Requirements"

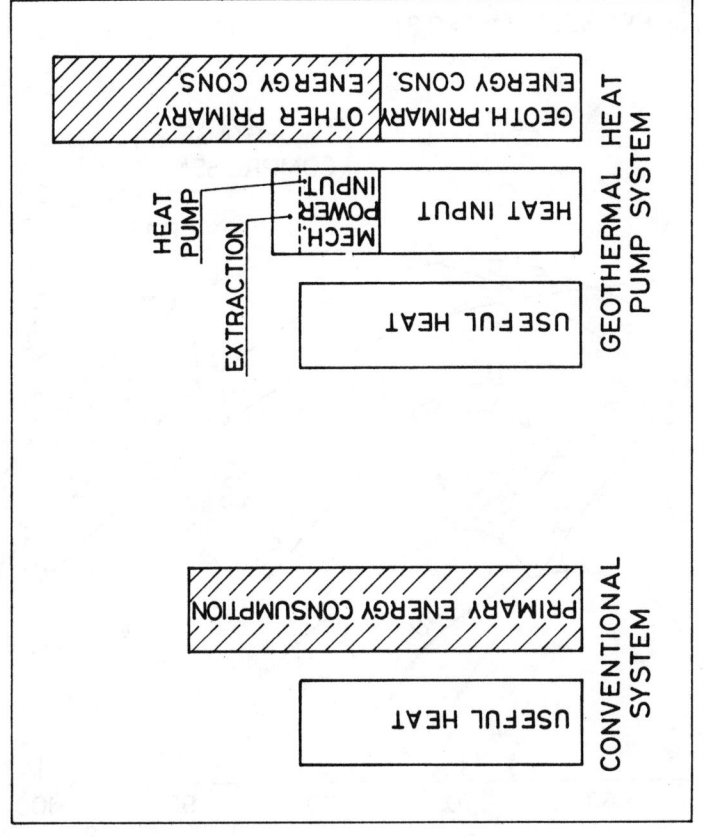

Fig. 2 - Geo-heat pump qualitative behaviour related

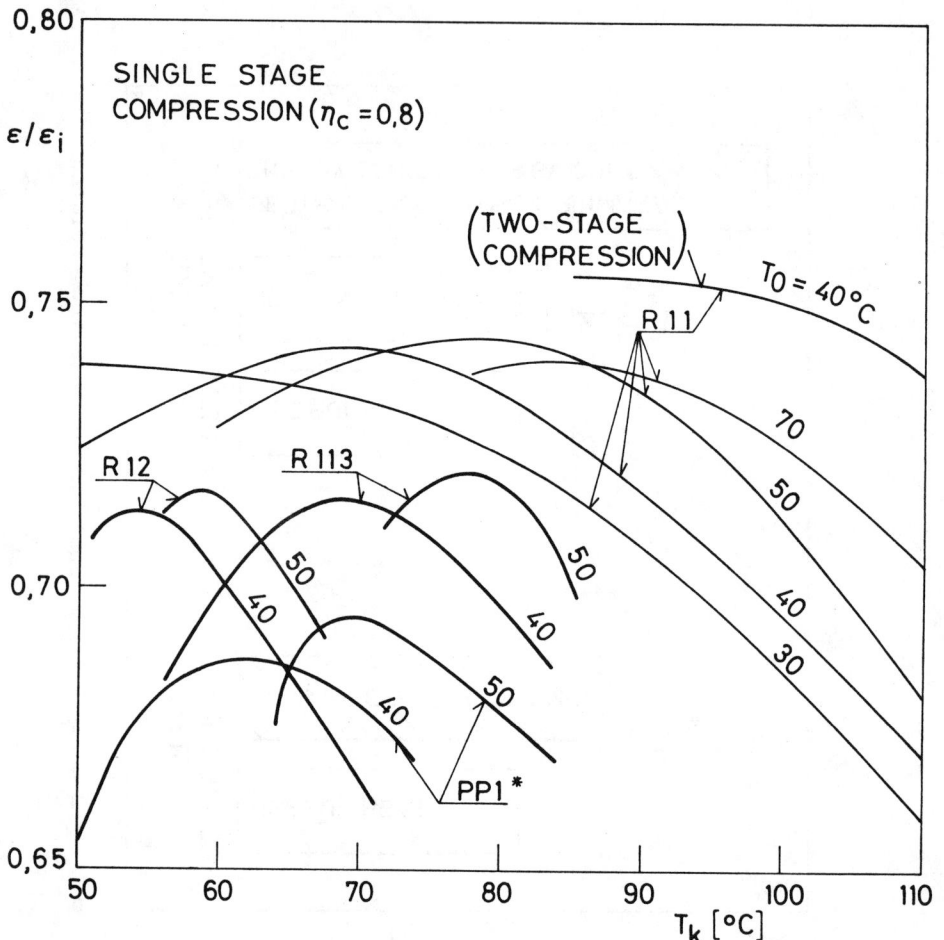

Fig. 3 Heat pump real/ideal performances ratio for several fluids and several cycle temperatures.

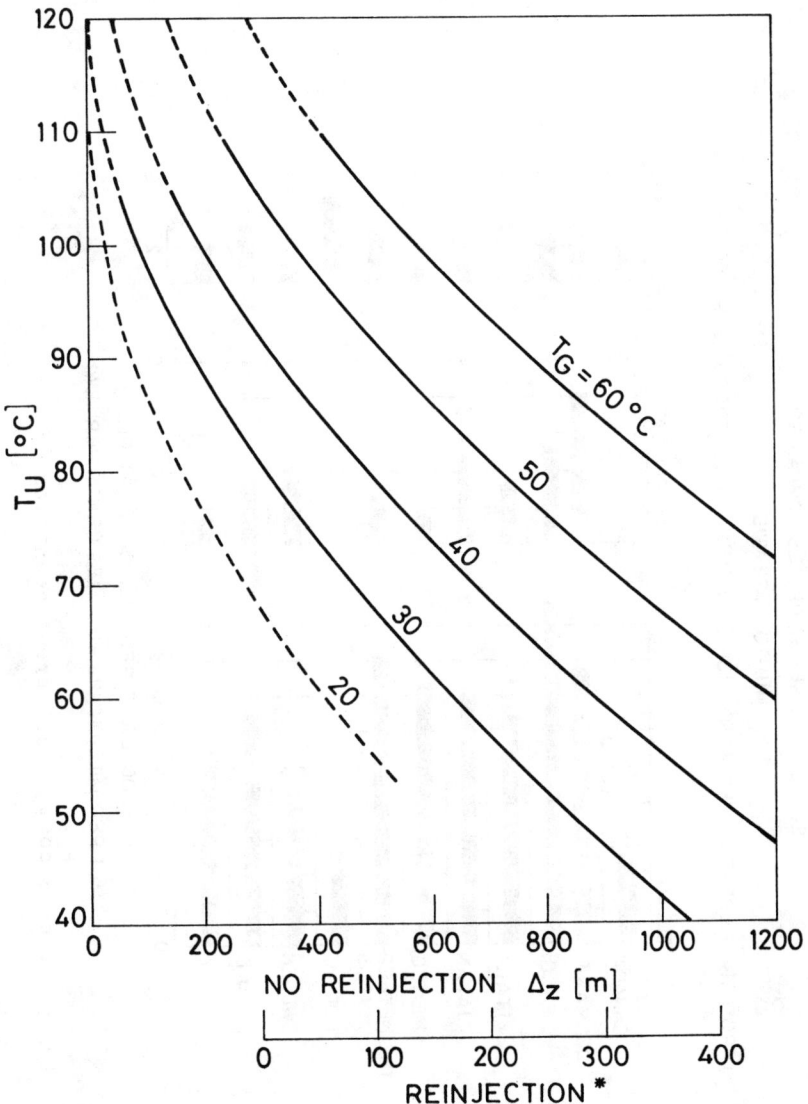

Fig. 4 - Upper limit for use temperature as a function of extraction pump total head (energetical criterion)

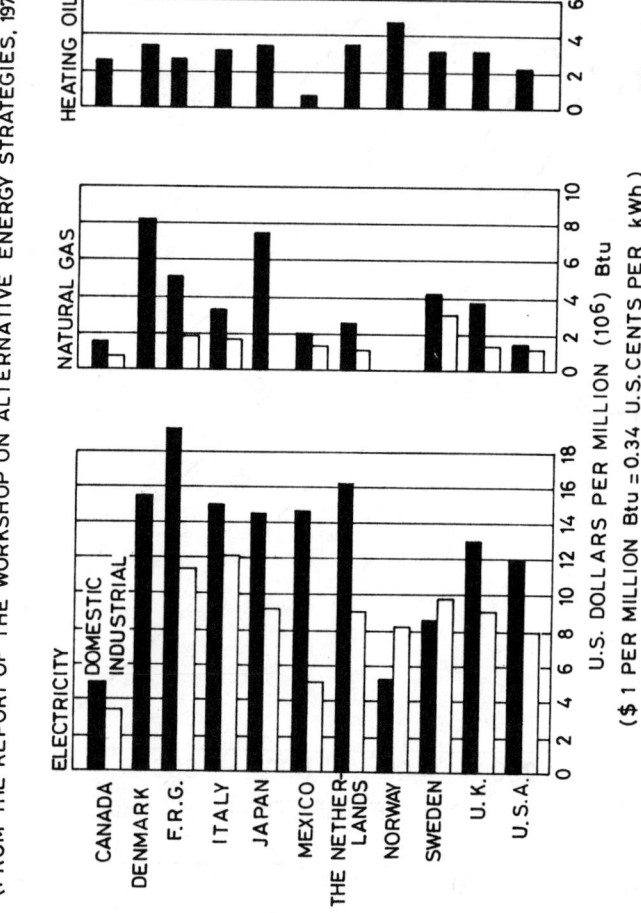

Fig. 5 - Energy cost in different countries [5]

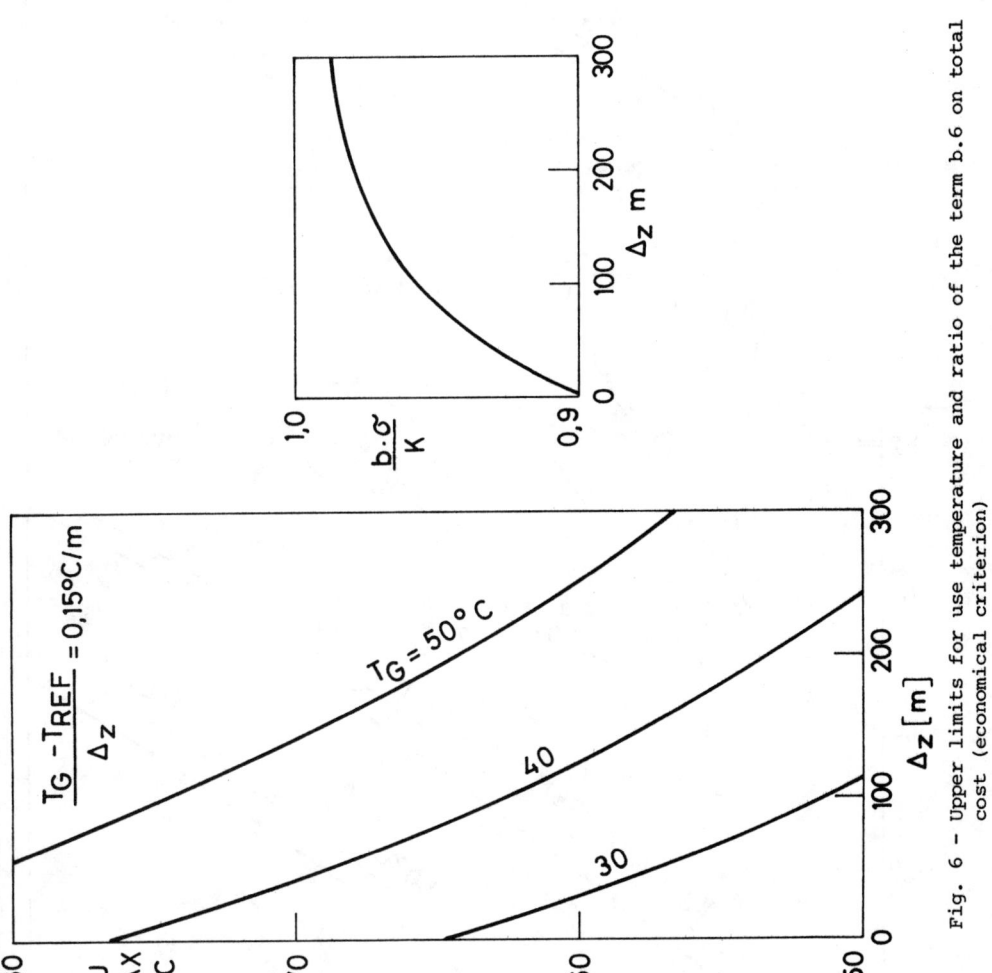

Fig. 6 - Upper limits for use temperature and ratio of the term b.6 on total cost (economical criterion)

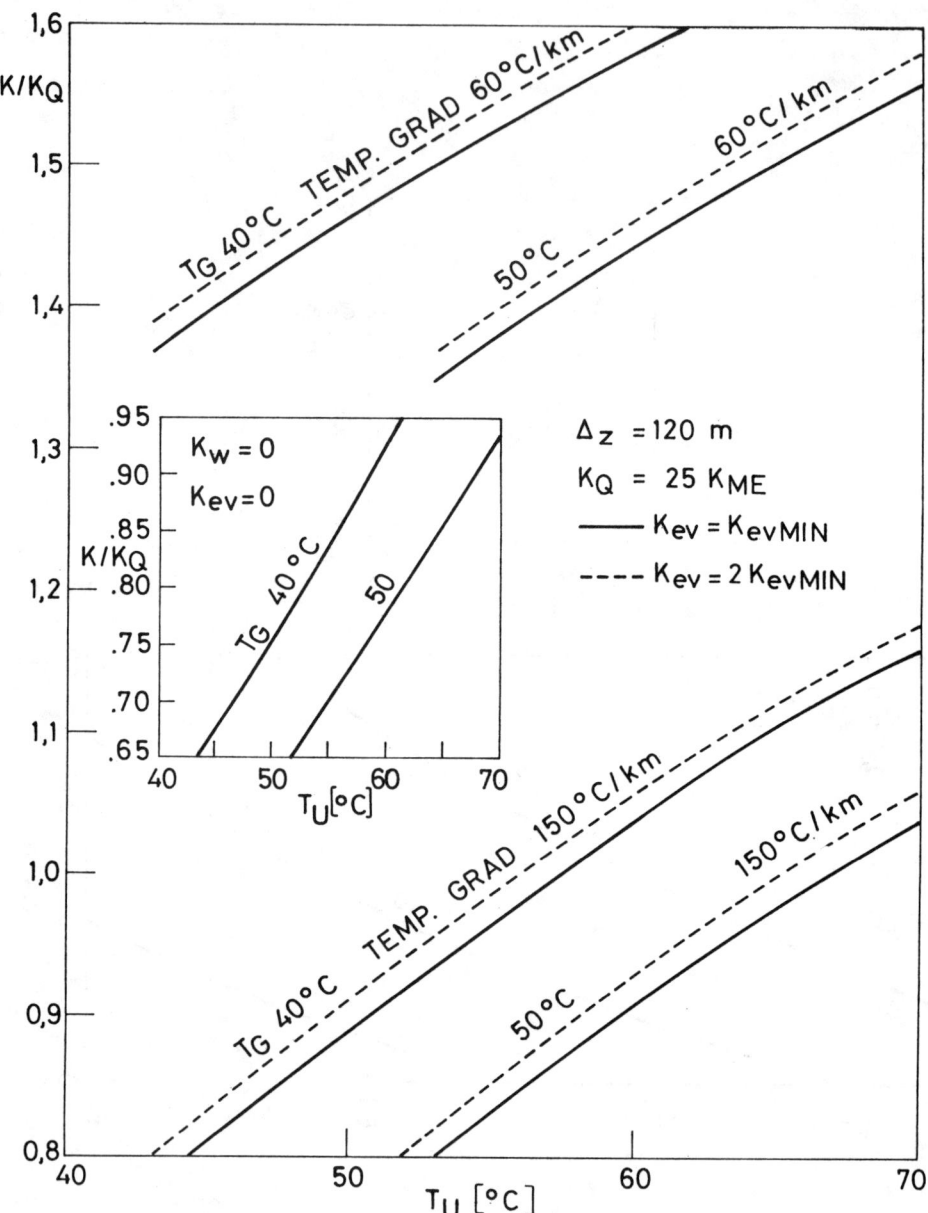

Fig. 7 Ratio of the heat costs for heat pump and conventional system as a function of use temperature.

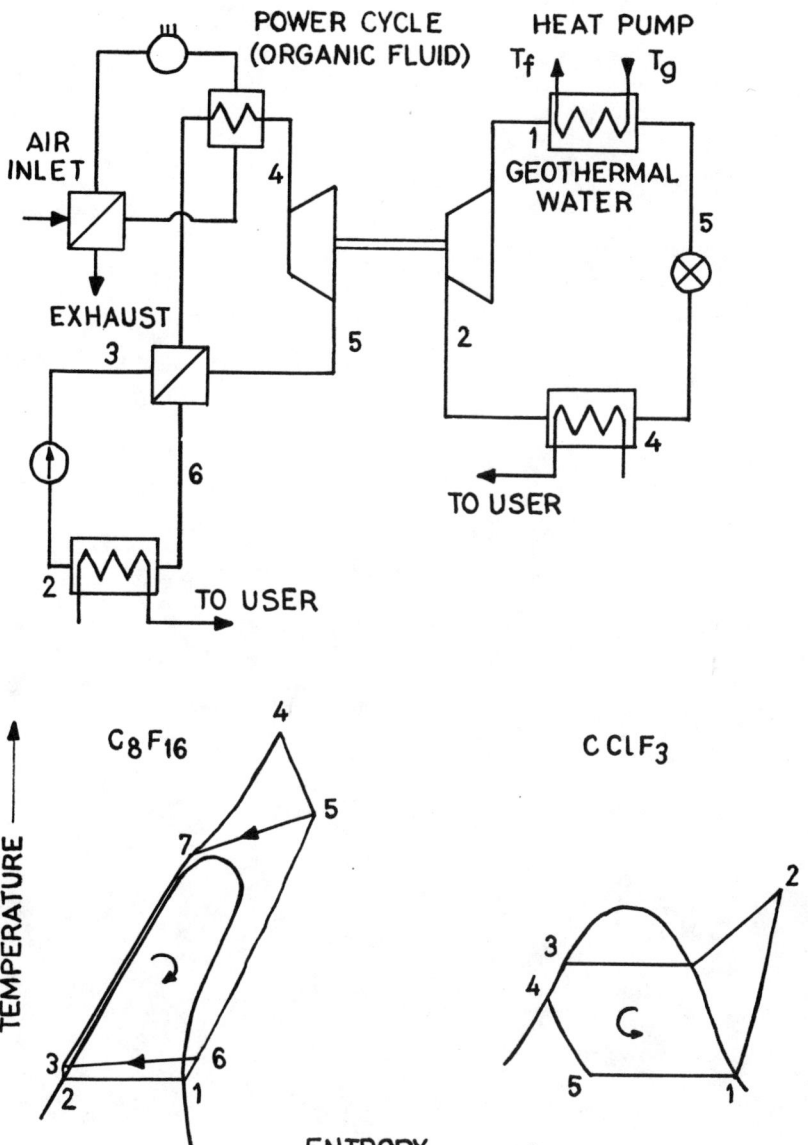

Fig. 8 - Cycles configuration and component organization of an organic fluid engine - geothermal heat pump system

Geothermal Community Heating Systems

J. F. KUNZE, R. C. STOKER, and L. E. DONOVAN
EG&G Idaho, Incorporated
Idaho National Engineering Laboratory
Idaho Falls, Idaho 83401, USA

ABSTRACT

The use of geothermal energy for central heating systems of towns and cities appears to be the practical means by which geothermal energy can make a very significant contribution to the nation's energy budget. It is estimated that possibly 4.22×10^{18} Joules (4 Quads) of our present 14.7×10^{18} Joules (14 Quad) annual use for space heating could be economically supplied by geothermal waters. The reason that geothermal space heating is likely to be more significant than geothermal electricity is that the latter is economically attractive today only for high temperature (>200°C [390°F]) resources. Although it will eventually be competitive down to 140°C to 150°C (285°F to 300°F), the resource is expensive to tap with deep wells and piping the water several km is expensive (but technically quite feasible). These lower temperature resources are much more abundant, and likely to be found in or near many of the large cities of the West. Even so, the capital investment is high and one or two intermediate depth geothermal wells can generally supply the needs of several thousand homes. Therefore, one is inevitably led to the conclusion so common in other energy fields; the economics of size determines the feasibility of the system.

TWO CASE STUDIES

Two geothermal case studies have been made in cities of the Intermountain West, including exploration for and tapping of the resource. One is the city of Boise,[1] capital of Idaho, population of 100,000. The other is the small town of Sugar City [2] devastated by the Teton Dam collapse of June, 1976. It is now being rebuilt, and central-district heating is a viable candidate.

Boise Example

Boise, Idaho has the oldest major geothermal district space heating system in the nation. Operation began in 1892, and at one time 400 homes were served. Each of these homes typically had peak heating needs exceeding 105.5×10^6 Joules/hr (100,000 Btu/hr), and the geothermal fluid reaching the radiators in the homes was 74°C to 77°C (165°F to 170°F). Then, in the 1940's, natural gas was introduced to Boise and the load eventually dropped to only 150 homes. However, within the last year, additional homes have been added and today the system serves approximately 200 homes.

The system's gross income is approximately $50,000 per year. Despite the obvious advantages and desirability of using geothermal energy, the system has been on the verge of being disbanded for economic reasons, and three years ago was taken over by a cooperative organization of the property owners.

The geothermal resource in Boise appears to be quite abundant (Fig 1), a fact confirmed in part by the drilling of two successful geothermal wells in the summer of 1976 by the Idaho National Engineering Laboratory. The wells are within the city limits, only about 1.6 km (1 mi) from the state capitol building complex and Boise State University. *The buildings and their position with respect to the resource are shown in Figure 2. More complete details are in Reference 3.

The well drilling has confirmed the presence of the geothermal resource within fractured and faulted zones that are located in the foothills on the north side of Boise. It appears the hot water circulates along fractured and faulted zones oriented N-S and NE-SW. The hot water moves out of the mountains from the north and northeast until it encounters the NW-SE oriented Front Fault. The water then moves up and across the Fault. The hot water is subjected to cold water dilution as it moves on out into the valley. See Figure 1 for further system detail.

Another hypothesis recognizes the source of hot water as a normal western temperature gradient under the Snake River Plain. Geochemistry indicates that the reservoir temperature is in the neighborhood of 110°C (230°F) and this can be expected at a depth of only 2 km (6600 ft) with the appropriate conductivities for the overlying saturated alluvium, basalt, and sediment. This theory implies a more general, wide-area resource, with the Boise Front Fault merely serving as a conduit to bring the hot water to the surface. Cold water dilution also occurs as the water moves upward and outward along the Fault.

CURRENT COSTS AND FUTURE PROJECTIONS

The ten state buildings selected have a combined yearly heat requirement of a net 94.9×10^{12} Joules per year (90,000 million Btu/year) and a design basis heat load of 52.7×10^9 Joules per hour (50 million Btu/hr). This makes the system far from being trivial in size, and it would seem to be adequate in size to be attractive for geothermal development. The natural gas fuel costs for the 1976-77 heating season were approximately $300,000.

---

* Only ten of the total buildings, all having hot air or hot water systems, were considered as being the type most compatible for conversion to geothermal heat.

During the summer of 1976, two exploratory test wells were drilled in an attempt to confirm if an adequate resource could be found. These wells are small bore wells 16 cm (6-1/2 in.) and were not drilled for production purposes. Both wells proved successful, tapping a 77°F (170°F) resource at about the 274 m (900 ft) depth. The wells are 305 m (1000 ft) apart on the surface, are presently fully cased (9 cm [3-1/2 In.] and 11 cm [4-1/2 in.] liners) and are now being tested. Preliminary indications are that the two wells, when pumped, should deliver a combined total of approximately 31.50 liters/second (500 gallons/minute) with drawdown of 91 m (300 ft) over several years of operation. Since the wellhead temperature has been measured at 74°C (165°F), the inlet temperature to the heat exchangers should be about 71°C (160°F), and the discharge 49°C (120°F). With ρ = 977 kg/m^3 (61 lb/ft^3) or 982 kg/m^3 (8.2 lb/gal), the required flow rate to supply the ten building heat demand is 161 liters/minute (2550 gal/min).

The estimated capital cost for supplying this water to the buildings, disposing of the cooled water by injection wells, and making the necessary additions of pipes and heat exchangers to the various buildings is approximately $3,000,000, as summarized in Table I.

About 13% of these costs are in the building heating and ventilating system modifications. But nearly 60% of the capital investment is in the pipelines to carry the fluid to the buildings and away from them to the disposal wells.

For the Boise area, with 3222 degree C-days per year (5800 degree F-days per year), and a winter design temperature of -23°C (-10°F), the annual utilization factor is only 21%.

It is no wonder a geothermal system finds it difficult to compete. Who would dream of operating a capital intensive nuclear power plant, for instance, on only a 21% load factor? On the other hand, utilities routinely use gas turbine peaking units, low capital cost, high fuel cost, at load factors within the 10% range. A similar solution must be found to get better utilization from a high capital cost geothermal heating system.

Figure 3, is a histogram plot of annual temperature frequency distribution in the Boise area. From this figure, it is apparent that designing to -23°C (-10°F) design with the basis heat load, the distribution system is poorly utilized. Whereas, if the design basis were in the range of -6.7°C to -1.1°F (20°F to 30°F), the system would be effectively utilized during most of the heating season. Below the -6.7°C to -1.1°C (20°F to 30°F) design temperature, the deficiency would be taken up with a fossil heating system. Presently, existing installed heating systems can be utilized in this capacity.

By adding buildings to the basic distribution system, each with fossil peaking capabilities, a system can be installed with minimal additional cost that can effectively triple the number of similar type buildings served. Table II shows a relative comparison of the different type systems illustrating the favorable effect of peaking on the delivered cost of geothermal heat.

Sugar City Example

Sugar City, a small town located in Madison County, Idaho, between the forks of the Teton River, at approximately the center of the flood plain, was the first town directly in the path of the Teton Dam flood crest. Reference Figure 4.

The city limits of Sugar City are a rectangle 805 m(1/2 mi) by 1609 m (1 mi). About 1/2 of this area is laid out in blocks 140 m (460 ft) square and much of the remainder is planned for development. Figure 5 shows the layout of the town with the projected additions on the east side. Prior to the flood there were 218 homes within the city limits with approximately 20 homes within 1 block of the city limits. Plans for four additions within the city were underway which would have added nominally 100 more residences. It was estimated, then, that existing plus near-term development, excluding the flood effects, would have resulted in about 300 living units in the city.

There are a few (approximately eight) commercial buildings in town but most of these are vacant or have been converted to apartments. These are either single or two story buildings averaging about 465 m^2 (5000 ft^2) each. Most are of brick or stone construction. Plans for rebuilding the commercial area are not firm at this time. Church and public buildings total almost 9290 m^2 (100,000 ft^2).

For the purposes of this case study, average values for homes and buildings were used in heat loss calculations for the winter assuming that all new construction will be well insulated (to what is often referred to in this area as "electric heat standards"), and that older buildings will be upgraded with better insulation. The following, therefore, was assumed:

1. The average distribution of homes will be seven per city block. This distribution allows for 224 homes and apartments in the presently plotted city.

2. Homes will be rebuilt with floor areas averaging 130 m^2 (1400 ft^2) per hours.

3. Houses will be well insulated and in general, be of good quality construction, with heat loss values of nominally 322×10^3 Joules/hr°C (550 Btu/hr-°F).

4. Non-residential buildings would average 2601 m^2 (28,000 ft^2) of floor area per city block, and cover ten blocks. Heat loss values for these buildings are assumed to be 369 Joules/hr°C (0.63 Btu/hr°F) per foot of floor space.

On the basis of weather data compiled over the last five years (Figure 6), a winter outdoor design minimum temperature of -32°C (-25°F) was selected.* This represents, then, a residential heating demand figure of 11.6×10^9 Joules/hr (11×10^6 Btu/hr) and a commercial heating demand figure of 16.9×10^9 Joules/hr (16×10^6 Btu/hr) for a total city demand of 28.5×10^9 Joules/hr (27×10^6 Btu/hr). Applying the average yearly temperature distribution data, the yearly heat load for the city becomes 68.6×10^{12} Joules/hr (65×10^9 Btu/hr).

CURRENT COSTS AND FUTURE PROJECTIONS

Natural gas is the principal space heating fuel presently used in the homes and businesses in Sugar City. Some fuel oil and some electric resistance heating is employed. Table III lists typical costs of space heating fuels today for the Sugar City area.

Table III shows the obvious fuel cost advantage of geothermal, if the geothermal water supply (the wells) are considered as a capital, not a fuel cost. The only geothermal energy costs would be the electric pumping costs and operation of the compressor motor in the case of heat pump boost. It should be cautioned, however, that many commercially proposed geothermal projects would have the water supplied by a "resource company" to the user's heating system. In such a case, the user pays a fee based on the quantity of geothermal water he uses. Separate accounting of this type would significantly alter the costs as shown, since profit necessary to operate on risk capital would be included.

The relatively attractive present costs of oil and gas (compared to electricity) are likely to become far less attractive in the future. For instance, if natural gas prices continue to rise at 8% per year, then in 1985, the average homeowner in the Sugar City area would be spending 1/10 of his annual income just to keep warm.

An approximate minimum estimate for the geothermal system would be $1,000,000 including some exploration costs. Most of this cost would need to be expended in the initial phases. Only some of the well drilling cost could be deferred until the system expands in later years.

A system cost of $1,000,000 amortized over 25 years at 7% interest represents an annual cost of $86,000. Operating and maintenance costs are estimated as $50,000/year. This cost added to the "fuel" costs shown in Table III make the geothermal system competitive with current natural gas

---

* Figure 6 represents average temperatures for the last five years and even though used for preliminary design and scoping purposes it does not reflect the extreme lowest winter temperatures. For example, the coldest temperature for the five year data was -39.4°C (-39°F).

costs. Obviously, the attractiveness depends on the loan terms that can be negotiated. Furthermore, an estimated $150,000 minimum "front-end" cost must be arranged for, a cost that could be lost without any useful resource being encountered if the first wells are unsuccessful. If they prove successful, however, this "front-end" cost would become part of the total project costs indicated. The remaining difference between the $1,000,000 investment plus the operating costs and the revenues from the system lead to the return on investment. Obviously, this is unattractive compared to today's gas costs unless much of the $1,000,000 investment comes from low cost municipal bonds or other low interest non-equity arrangements. "Tomorrow's" gas costs should give a more favorable prospect, however. Compared with electric heating costs, the ROI on a full $1,000,000 investment is nominally acceptable in today's commercial market place.

Figure 7 shows a typical layout for a district heating distribution system. The layout is valid for geothermal water distribution, direct use, as well as secondary circulation, heat pump applications or a central boiler plant.

CONCLUSIONS

Fossil peaking is a virtual necessity to make district geothermal heating competitive with conventional fuels in a moderate climate condition, as indicative of the contiguous 48 states. Furthermore, it makes economic sense to use fossil peaking, at low capital cost, in order to get the best utilization factor out of the high capital cost geothermal system. Such considerations do, however, impose a burden of size, making the most attractive systems those that gross over a million dollars in annual revenues, based on today's (1977) competing fossil fuel prices.

By such minimal use of fossil peaking on the coldest of days, the annual energy use of fossil fuel for peaking is only 5 to 10% of the current annual use. This means a 10 to 20 fold reduction in fossil fuel consumption and air pollution. Of course, such a geothermal system will become even more attractive if (as expected) fossil fuel prices rise substantially more than the cost of other goods and services in the future. Thus, the temperature at which fossil peaking should be used will most likely depend on the relative price of fossil fuels compared to the cost of living in the future.

REFERENCES

1. Kunze, J. F.; Stoker, R. C.; Donovan, L. E.; "A Geothermal Space Heating System for Boise, Idaho," TREE-1167, November 1977.

2. Kunze, J. F.; Lofthouse, J. H.; Stoker, R. C.; "The Potential for Utilizing Geothermal Energy for Space Heating in Re-Constructed Sugar City, Idaho," TREE-1016, January 1977.

3. Kunze, J. F.; "Geothermal Space Heating - The Symbiosis with Fossil Fuel," Presented at the 12th Intersociety Energy Conversion Engineering Conference, Washington D.C., August 28, through September 2, 1977.

TABLE I

Summary of Capital Costs for 161 lpm (2500 gpm) Geothermal Heating System for Boise

---

A. **PRODUCTION SITE**

    Five wells (one spare for contingency) $180,000
    305 to 457 m (1000 to 1500 ft) deep

    Pumps, surge tanks, controls, weather   120,000
    protection $300,000

B. **DISTRIBUTION SYSTEM TO BUILDINGS**

    1609 m (1 mi) large 30.5 cm (12 in.) dia main
    2414 m (1½ mi) of 7.6 cm (3 in.) dia main

        all concrete asbestos, insulated with urethane and buried 1.8 m (6 ft) below grade, at $213/m ($65/ft) in streets

        plus grade crossings, river crossings $850,000

C. **DISCHARGE SYSTEM**

    5 wells, 457 m (1500 ft) deep $200,000
    pumps, controls   90,000
    Pipelines to wells   300,000
        $590,000

D. **BUILDING MODIFICATIONS**

    Piping and heat exchangers $280,000

        $2,020,000

Engineering services at 25% plus contingency at 25%   1,136,000

    Total cost of 10-building system $3,156,000

TABLE II

Project Size	Capital Cost	Operating Cost	Fossil Fuel	Delivered Cost of Geothermal*
10 buildings	$3,000,000	$61,000 or $0.66/billion Joules ($.07/therm)	--	$281,000 or $3.03/billion Joules ($.320/therm)
20 buildings	$3,500,000	$94,000 or $0.51/billion Joules ($.054/therm)	$ 22,000	$370,000 or $1.99/billion Joules ($.210/therm)
38 buildings	$5,000,000	$190,000 or $0.47/billion Joules ($.050/therm)	$107,000	$660,000 or $1.62/billion Joules ($.171/therm)

* Includes operating and amortized capital costs. Interest rate used was 6%.

TABLE III

HEATING SEASON ENERGY COSTS

(Does not include costs of capitalization)

	Home	Commercial Block	Total City
Natural Gas ($2.46/billion Joules [26¢/therm])	$390	$12,500	$211,000
Fuel Oil ($105/m^3 [40¢/gal])	460	14,700	248,000
Electric Resistance (2.2¢/kW-hr)	775	25,000	420,000
Coal-fired Boiler, central community heating ($53.6/thousand kg [$40/ton] delivered from Wyoming)	290	9,300	157,000
Geothermal * Direct Use Pumping Costs	70	1,000	25,000
Low Temperature Geothermal (32°C [90°F]) Central Heat Pump COP = 4 and pumping costs	280	9,000	152,000
Individual Heat Pumps COP = 3.5 and pumping costs	310	9,900	168,000
Cool Water Heat Pump COP = 3.0 and pumping costs	380	12,200	206,000

* Does not include royalty payments to the geothermal rights owner, but does include both production well and reinjection well pumps as well as distribution pumping.

** COP = coefficient of performance = $\frac{\text{Heat output}}{\text{Work input}}$

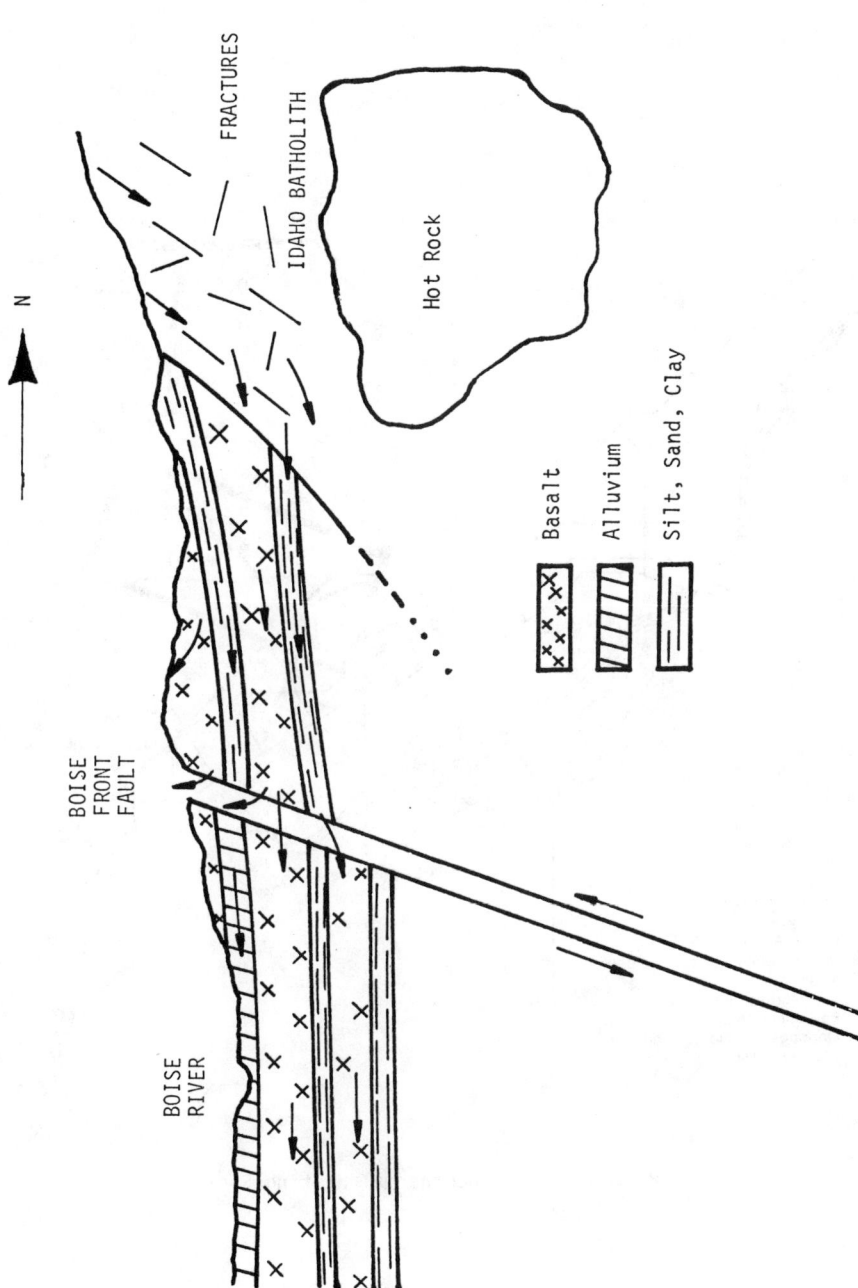

Fig. 1 Boise Front Fault Cross Section and Water Flow Schematic

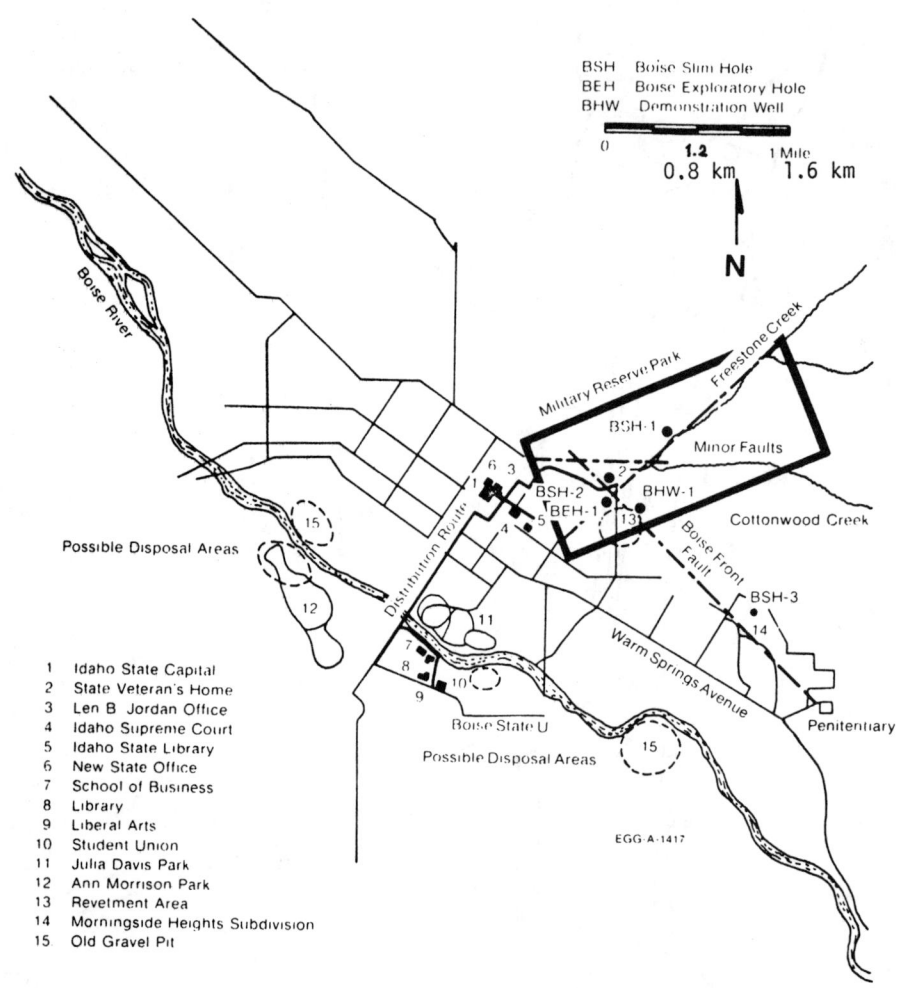

Fig. 2 Proposed Boise Geothermal Heating System

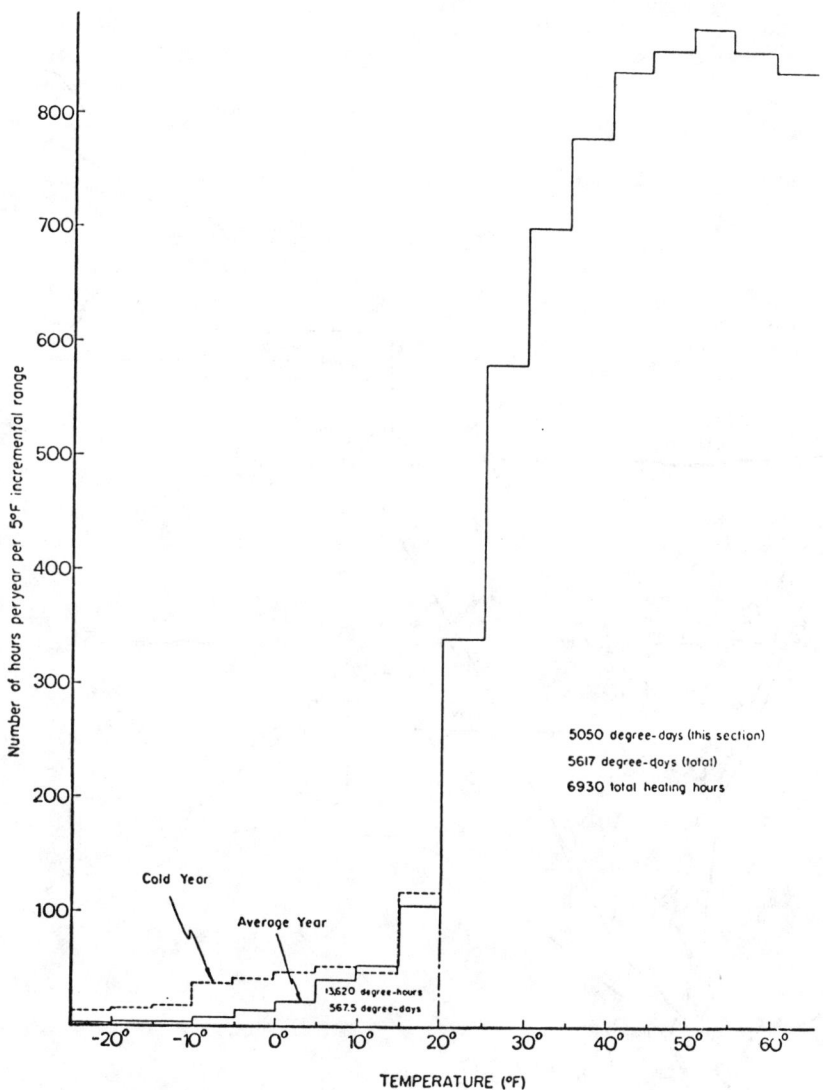

Fig. 3 Boise, Idaho - Average Annual Temperature Histogram for Heating

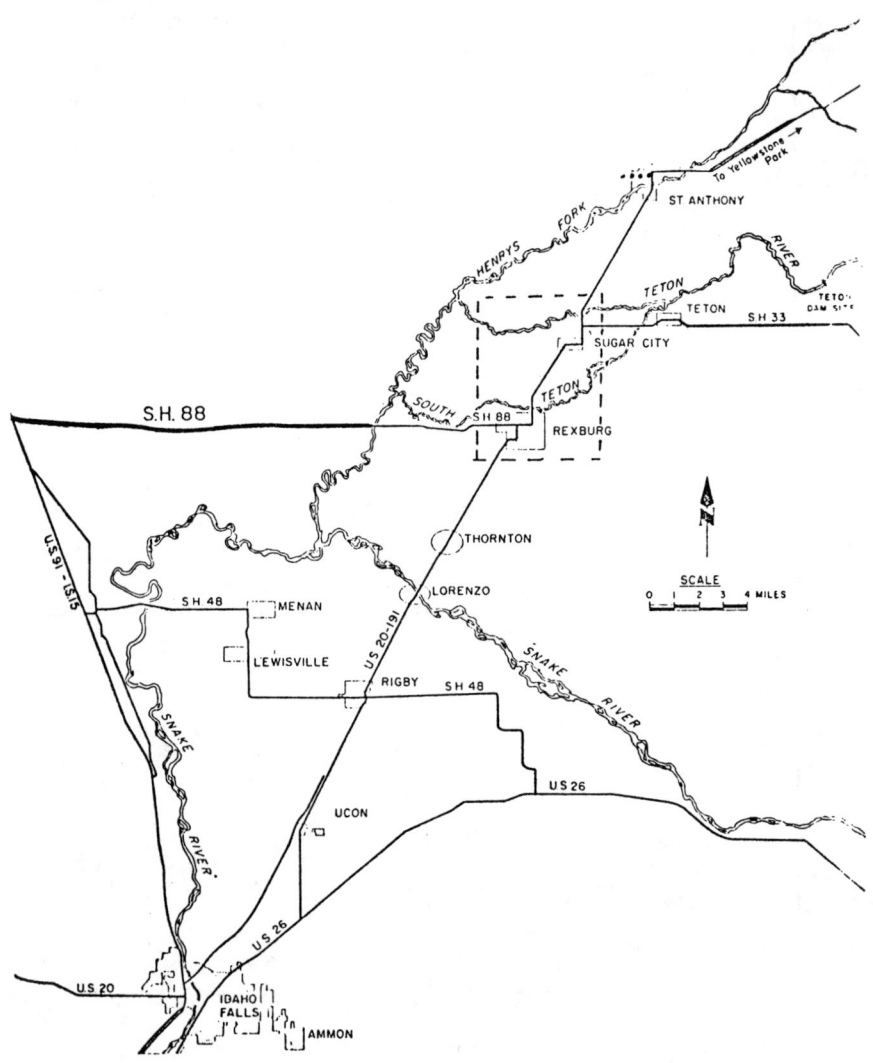

Fig. 4 Map showing Teton Dam site, Sugar City and surrounding communities

Fig. 5 Pilot plan for Sugar City including undeveloped additions on the east side

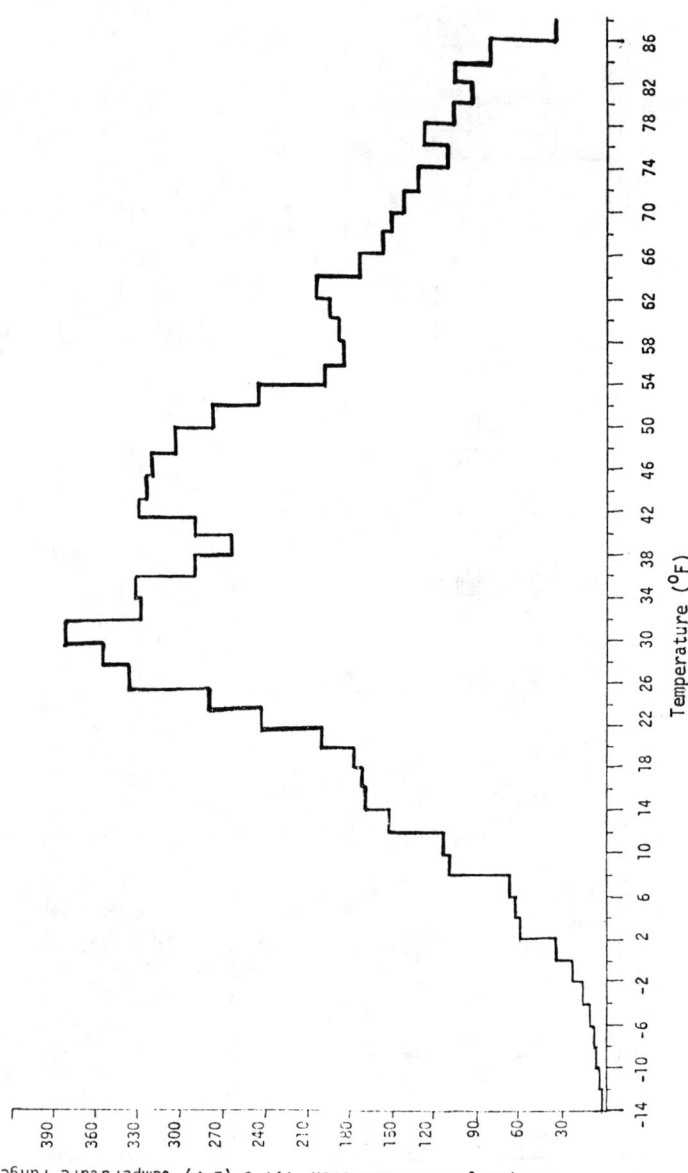

Fig. 6 Average temperature data for Sugar City

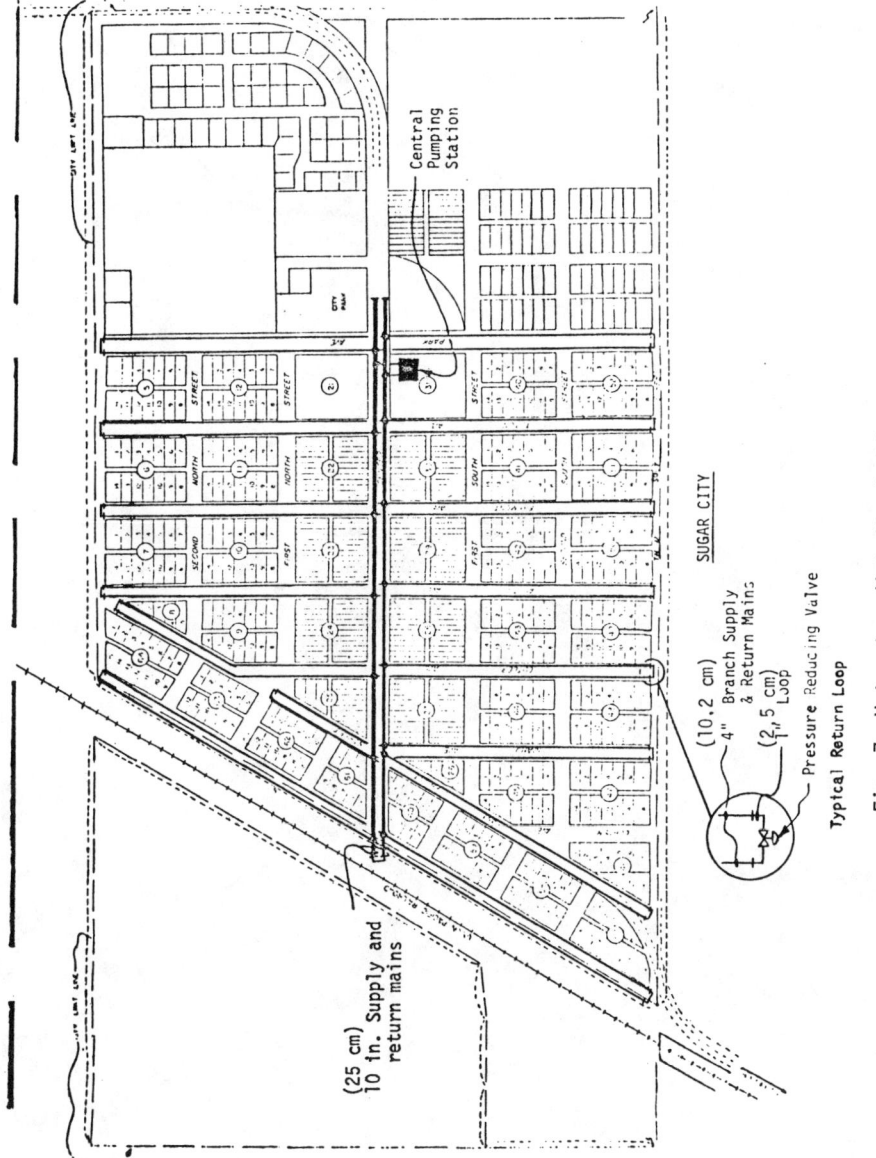

Fig. 7 Hot water distribution layout

Conserving Electric Power by Geothermal Refrigeration—Cooling and Freezing

EDWARD F. WEHLAGE
ISGE, P. O. Drawer 4743
Whittier, California 90607, USA

ABSTRACT

True energy conservation is possible through substitution of geothermal heat for electric motor drive compressors in the refrigeration industry that is typified by the *Great American Ice Cream Cone*. Perhaps 5% of the U.S.A. annual energy budget goes into refrigeration services. Consumers may face priorities with power for frozen foods and industrial processes. People will be reluctant in accepting the changes.

Alternatives include complete elimination of cooling. Substantial blocks of power can be replaced with geothermal refrigeration service at +5° C and as low as minus 50° C. The author believes this is practical. Total energy systems may work with geothermal but direct applications for cooling appear more acceptable. Several types of geothermal refrigeration systems are now apparently available with present technology. Capital, cash flow and incentives have bearing. If energy conservation is to be war, then some drastic changes are outlined.

INTRODUCTION

Electrical energy may be saved when the combination of thermal energy from terrestrial heat can be applied for the production of useful refrigeration effect in process work, cooling and freezing as well as comfort cooling services.

All modern cooling and freezing services might be typified by *The Great American Ice Cream Cone*. It demands tremendous respect as a luxury and it becomes an important food. Many industrial processes are supported through such use. People consider it an essential service, but it is also a huge consumer of electrical energy to drive the compressors creating the *state of energy deficit* we recognize as refrigeration.

Since an energy consuming popular service is at stake, thoughts about alternatives assume signifigant importance and stature.

Estimates suggest that some 5% of the annual U.S.A. energy budget is spent for mechanical refrigeration. Once the World's Fair *CORNUCOPIA* (St. Louis, 1904), our great ice cream cone has a common bond of *REFRIGERATION* with all frozen foods, space conditioning and vast industrial processes - all using *POWER*.

Few products may be cooled unless there is power for refrigeration compressors. This energy dependence is going to force consumers around the world to establish some priorities for the consumption and pricing of future energy sources.

The double scooped delectable splendor of toasted almond fudge with a golden crown of lemon and lime required relatively few watts to prepare, freeze, store and transport, yet in the aggregate it is *BIG BUSINESS FOR ENERGY*. While its cost has already been increased, its future faces restructured utility rates with 20% to 50% increases, and governments seek "energy de-centives" to discourage large consumers, and add energy taxes to force conservation by increasing production costs.

Conservation realities may soon force a new appraisal of the World War I and II declaration that *our cone was the edible symbol of morale*. Some alternatives are needed and one of them will be geothermal heat for true conservation in producing refrigeration effect at $\pm 40°$ F ($+5°$ C) and as low as minus $60°$ F (minus $50°$ C) without electric motors driving compressors.

A LITTLE BACKGROUND

Refrigeration has become so well established over the last 125 years that today a crude ice maker becomes a boon for primitive people in the jungle or desert. Only a total dislocation in energy sources will quickly loosen the connections between people and cooling.

A few centuries ago, Hippocrates (460 - 377? B.C.) observed:

> "..........and most men would rather run the hazards of their lives or health than be deprived of the pleasure of drinking out of ice."

Natural ice lost its market to the more modern artificial variety due to the vicissitudes of the weather, sanitation, and its inability to satisfy a mass market. Around 1850 the growth of practical steam power brought the development of mechanical refrigeration that was followed by electrical motors and compressors which today govern refrigeration in the food industry, chemical processes, and comfort cooling of all types. In 1850 ice cream sold for 65 cents a quart. What price for 1999 A.D.?

In the U.S.A., 750 million frozen Eskimo Pies are sold annually and seven ice cream plants are said to be operating in Moscow. The entire frozen food industry is scarcely 40 years old, but Iceland ships 90% of its fish in frozen form. Like the men of Hippocrates, a lot of people will resist any curtailment in food and freezing operations. They have come to expect these for survival in our present social and industrial orders.

ALTERNATIVES - LIKE GEOTHERMAL COOLING

Clearly, we shall need alternatives. We ought to be preparing for more of them in meeting our energy dilemma, especially in the case of applied experience for refrigeration services.

1. Consider efficiency improvements in process & equipment

Conservation will result with improvements in equipment and operation efficiency. One step is the requirements set up for improved design of apparatus in the refrigerated air conditioning industry with better heat transfer and lower power inputs.

2. Minimizing waste - another old energy story

Energy inputs can be reduced by minimizing waste in processes and distribution. Many large operators already use regenerative cycles to transfer heat to and from entering raw material and finished products. Unfortunately, most efforts to improve efficiency and reduce waste will recover only a small percentage of the total required energy input.

3. Pay higher - and higher - prices, or simply just stop

People may elect to simply pay more and more up to the limit of their resources. This may curtail demand and use, but the human factor described by Hippocrates still applies. He cautioned that men would hazard their lives for ice, and surely that includes their purses.

It may be possible to simply eliminate things like refrigerated products in the face of economic dislocation and discomfort. In banishing our ice cream cone we might follow the Japanese warlords at the onset of World War II. They simply eliminated ice cream at the market place.

A question remains as to how far most people will or can go in dropping frozen foods, critical chemical and pharmaceutical items, climate control for computers, plastic fabrication, and the photographic industry - and so on ad infinitum. There some some interesting essential services in alternative energies.

4. Use an *indigenous - previously unused* alternate energy

Geothermal energy - THE INDIGENOUS INDUSTRIAL TYPE HEAT OF THE EARTH. - stands ready at many locations to replace enormous blocks of motorized compression power with geothermally energized refrigeration. In those areas of the world where it is readily available this heat can offer continued energy services with true conservation by replacing fossil fuel consumption.

TOTAL ENERGY VERSUS DIRECT HEAT EXCHANGE SERVICE

Temperatures around +40° F (+5° C) are required for most comfort cooling and many industrial services. In frozen food operations temperature requirements are typically minus 20° to minus 35° F (minus 29° to minus 37° C). Some industrial processes might require minus 60° F (minus 50° C). The author's work confirms a belief that these levels are practical with known technology modifed for geothermal duty.

Some industrial plants use compressors with thousands of horsepower from motors and steam turbines. Others operate with heat that might otherwise be wasted or lost. Not all of these plants are compatible with known geothermal locations - but some are.

Plastic forming plants frequently use several hundred horsepower to cool molds and product. Mass production mushroom growers may require about 1,000 horsepower to cool the production area. The author is currently working on a geothermal cooling proposal with a single owner in food freezing with about 5,000 horsepower serving a single growing area.

Probably the least efficient enthalpy use for geothermal energy is the production of electrical power. It has long been recognized that factory power plants requiring low pressure process steam may produce a kilowatt of electricity at a far lower heat rate than a conventional utility unit. This in-house *total energy* has lagged because of high investment costs and low utility power prices. Energy priorities and high costs are giving new emphasis to this type of operating in *CO-GENERATION* proposals. What is essential is a satisfactory capital structure, a cash flow adequate to cover the costs, and external incentives creating an enthusiastic management environment.

Co-generation applications can be energy productive for some of the known geothermal installations. Only a few fortunate areas deliver geothermal heat hot enough for large scale power generation. Low pressure gives high steam consumption per kilowatt. Where industrial total energy matches the required industrial steam demand the result is efficient enthalpy utilization much like the Tasman Paper Company's mill in New Zealand.

Current outlook for geothermal total energy systems in the U.S.A. is bleak although fuel availability, increased costs and new incentives for conservation may ultimately break the barriers.

Sources at $\pm 300°$ F, too low for electric production, will utilize direct transfer systems and depend on utility power lines.

A BALANCED CONCEPT - INDIGENOUS HEAT/UTILITY ELECTRIC

An ideal plant would combine generation and process services to match electrical needs with process heat requirements.

Geothermally energized refrigeration units without compressor motors still require electrical power to operate internal pumps and controls, circulating pumps, cooling towers, etc.

Replacement of compressors leaves the essential rotating power requirements to a utility system. These loads are smaller, and relatively constant, so they provide a good utility system load factor. Another opportunity in genuine electrical power utility conservation is created.

This balancing concept presumes the application of some conventional technologies currently at hand, while simultaneously expanding installations with geothermal heat at various temperatures less than the level of $\pm 300°$ F $(150°$ C$)$.

WHAT IS THERE TO WORK WITH?

Several established technologies may be fully adapted to this unique source for cooling energy. Probably two cooling systems in the world currently use geothermal heat sources. None is believed to operate with a chilled medium (brine) lower than the $\pm 40°$ F $(+5°$ C$)$ range. One is located in New Zealand.

1. Conventional heat pumps

Heat pumps use refrigeration to raise heat levels of water and air. Heat may be taken from the earth with well water or by using coils buried in the earth. Electric power acts as a lever, lifting heat from the earth for useful applications.

2. High temperature heat pumps

Geothermal fluids at $+90°$ F $(+32°$ C$)$ provide a heat source to add electric power in a recently developed high temperature unit for boosting water temperature to $\pm 230°$ F for processes.

These are high efficiency, special construction machines.

3. Mechanical drive refrigeration units

Electrical power generated with geothermal heat may be used to operate conventional electric compression systems. Prime movers like steam turbines or other rotary expanders may also operate rotating compression machinery of many types. Proposals have been made for combining such units with other types of heat utilization and secondary refrigeration equipment along lines familiar to power plant engineers for absorbtion unit schemes.

Vapor turbine drives have been considered for both electrical generation and mechanical drive systems. None of these seem to have been actually tried in refrigeration duty.

4. Lithium bromide and water absorbtion systems

Heat actuated, closed system, refrigeration units, similar to old time flame units and modern gas fired refrigeration systems are described as *absorbtion refrigeration units*.

Water contained in a sealed high vacuum chamber boils at a low temperature and carries away heat. With an absorbent it may recycle again and again into the absorbent. Nearly 200 pairs of such affinties are known. One of the practical ones is lithium bromide as the absorbent and water as the refrigerant. The refrigerant must then operate above water's freezing point - generally about $+40°$ F ($+5°$ C) in the circulating fluid, known also as "brine".

This type of system is used in the two known installations for geothermal heat use. These are commercial air conditioning unit water chillers judging by available information. The machine in the U.S.S.R. may have been a special design.

5. Water and ammonia absorbtion machines

Systems planned for low temperature operation use ammonia as the refrigerant and water as the absorbent. There are also absorbtion units operating with direct firing of combustibles.

At the turn of the 20th Century many of the meat packing plants over the world used this type of system. Some are still used in chemical plants and oil refineries. Many are currently in use with oil fuel.

Modifications are required for geothermal service. None are now designed for such service. Inevitably there will be some. These units are high investment items so the cheaper energy is offset by high fixed charges. Experimental installations will inevitably be expensive. However, these units should prove out very well with geo-heat and justify regular production.

6. Steam jet vacuum cooling units

Steam jets may be used to produce vacuum (negative pressure) in closed systems to reduce the boiling point of water. Where geothermal steam exists with sufficient pressure for jet operation such a unit might be applied. At least one commercial unit has been marketed for industrial service with waste steam. The jet is not recognized as an efficient device, but it is practical.

PROBLEMS - MOSTLY UNRESOLVED

Geothermal was long touted as a pure, clean and environmentally safe source of heat. Unfortunately experience has been showing some negative ecological factors for consideration.

Wells are often so expensive as to make installations economically questionable. One the positive side, wells unable to meet the original requirements for electric power production may be written down and made acceptable for alternate use in industrial service for processes including refrigeration.

Utility supplied electrical energy is still one of the better bargains in today's energy field and it offers stiff opposition for budding technologies outside the utility system domain.

All costs for untried and experimental geothermal refrigeration systems will be high and designs are not yet ready. In the U.S.A. we are woefully short of hands-on experience for the design construction and operations. Our country has the best drillers and a low use record.

Industry does not follow ideas on new energy. Social problems are involved. Geothermal heat is not found in large cities. New communities for workers are needed. Economics are complex and generally our laws do not allow much progress in untried energy options - let alone freedom in the older ones.

In the U.S.A. we have been short of money for research; short of a true demand for conservation; short of interest; short of experience; and short of will to experiment; and possibly to face some failures.

WHERE CAN WE GO?

Our President has been quoted as saying we should approach our energy dilemma as if we were going to *WAR*. In that case we need to start some mighty drastic moves, and pretty soon, too.

If we are at war, then we must move toward practical research, empirical economics and pragmatic politics. Problems for geothermal developments are energy related problems, and are therefore inextricably tied to local, state, regional, and national politics - in the U.S.A., and every other country. All efforts must be directed toward the maximum replacement of fossil fuels, pound by pound, barrel by barrel. Refrigeration with geo-heat, *with jobs for people, must share the same approach.*

Among others, we might begin in 10 areas of concern:

1. Create better understanding of geothermal energy

The people who promote, explore and drill for geothermal deposits speak a different language from those who actually put it to use. The general public is not really aware of its value and the working tradesmen do not understand how it functions.

When solar energy techniques started to break down into smaller and understandable terms more people began to see its new merit while awareness has boomed - and it sells.

Lack of popular acceptance and the ability to comprehend some reality for geo-heat in everyday life accomplishes little to accelerate government support. Geo-heat must be made actionable for industry. Geothermal refrigeration will follow any growth.

2. Improve practical heat transfer methods

Geothermal refrigeration is totally dependent upon good heat transfer from the fluid source to the final product. We know a lot, but not yet enough.

Direct contact heat exchangers may be developed for getting the heat into a refrigeration cycle, but there may be too much loss of the vapor medium. If carbon steel successfully serves and no fouling or corrosion results, perhaps stainless steel or Admiralty type metals will no longer be thought mandatory.

3. Build up reserves of operation and maintenance experience

Encourage hands-on experience with actual operation and spread this working knowledge throughout industry. Learn about maintenance practices and get word around about solving operation experiences so that failures create progress in technology.

More users, more factories, more applications, *mean more jobs.*

4. Recognize that social problems go with new services

People are an essential part of the development beyond simple technology considerations. In due course, housing and jobs will be a problem in geothermal communities fully as critical as any of today's financial, legal and institutional problems. A nation's energy dilemma can only be resolved through the combined efforts of *business, governments - and people.*

5. Changes in financial philosophies

New functional financial practices for *working geo-heat* can be adopted which differ from those made available for prospecting and drilling, promoters and landowners.

Fast write-off, or accelerated depreciation should be both legislated and practiced for new process applications where geothermal heat replaces fossil fuels. Special 1, 3, 5 and 10 year maximum write-off periods must be available for the *industrial end consumer* and not for explorers, promoters, developers, oil companies, and/or energy holding companies - except as users.

Loan guarantee programs should be limited *to the application of geo-heat,* rather than to serve as an incentive for financing more well drilling programs across the nation.

6. Direct subsidy for *each kilowatt* (thermal/electric) used

If we are at war, our entire society owes something to the energy battle at home. A direct subsidy may be a desirable incentive if paid for each kilowatt (thermal or electric) consumed in a process industry that has been relocated or constructed *especially* for the use of geothermal heat to replace oil or gas and electrical power otherwise consumed in the factory.

7. Establish geothermal industrial energy parks

Locate new industrial energy parks at prime geothermal areas to gather and serve industry and commerce with central heat, refrigeration and power supplemented with judicious boosting from auxiliary fossil fuels.

This wider base provides more flexibility for financing and resolving technical problems arising from geothermal field operation and too large for an individual factory operation.

State, regional and federal participation and subsidy will be clearly necessary while encouraging drastic changes in operational procedures with geothermal heat in food processing, canneries, growers, chemical plants, preservers, plant services and plastic processors, or clothing makers and wood workers.

Tax free warehouses, cold storage facilities, and commerical - wholesale distribution centers can be utilized as incentives to attract businesses to a geothermal service business community.

8. Government assumption of underground energy resources

Countries outside the U.S.A., where geothermal energy has been successfully applied claim generally that every underground resource, like geothermal heat and energy, are government monopolies - "property of the crown" in New Zealand.

Iceland, Greece - and New Zealand - make these resources available by permit or through the offices of national producing authorities or energy corporations.

It sometimes appears that groups and individuals in governing circles of the U.S.A. may be totally dedicated to the idea of a national ownership for all energy producing and distribution or ownership services, and that geothermal energy may be an entering wedge for reaching such a goal.

Government clearly intends to stay in geothermal energy activities and manage its growth with controls on every aspect of its use.

Already there are so many legal and agency problems in the management and promotional areas of geothermal services, and seemingly they will never be resolved, that this *wartime* approach might just make the right circumstance for nationalization and government purchase of every underground energy resource - thus allow the government to settle every problem.

9. Exile energy industries capable of going geothermal

Few industries will move voluntarily to geothermal use. Incentives are needed. Lack of energy could be one of them. If the State Of California wanted to curtail certain segments in gas or oil use, it could easily mandate such action.

For example, the State could simply declare that no natural gas or fuel oil could be used in greenhouses over 1,000 square feet after 1982. Remember, we are talking about *war*. There is enough geothermal energy in California to run every existing greenhouse, and throw in some extra water and electricity.

The State probably has the authority already, or it might legislate it, to void property taxes for 10 or 20 years for an industry persuaded to drop all rights for gas or oil and convert to 100% geothermal heat for low grade energy needs.

10. Put alternative energies *on the firing line*

There must be no reason for business or industry to avoid use of new heat sources like geothermal heat, but no business manager can keep his job if he accepts an unreliable replacement for conventional fuels. After all, fuel and power may constitute a small portion of his manufacturing budget, even when the utility bills reach substantial dollar amounts each month.

To serve industry, alternative energies have to FIT on the firing line!

CONCLUSION - 300 MILLION KILOWATTS EACH YEAR

Refrigeration in industry - freezing and cooling - with geothermal heat taken from the Earth will become another building block for the structure of *conservation in electric power.*

One installation, replacing equipment with 1,000 horsepower of electric motor driven compressors will release about 1,000 kVA in power plant generating capacity and more than 6 million kilowatts in energy consumed annually - at 75% load factor.

Reasonable goals for the Western U.S.A. installations alone may eliminate 50,000 horsepower in compressor electric motors. Less than 5% of the original power would drive the new machines. Over *300 millions of kilowatts annually* would be available that utility systems could deliver elsewhere for critical needs.

This offers *true electric conservation with geothermal* while it helps the freezing and cooling industry keep on delivering lots of fine ice cream cones for America!

E N D

REFERENCES

Dickson, Paul; *THE GREAT AMERICAN ICE CREAM BOOK,* 1972, Athenum

Ruppright, Siegfried; *THE ABSORBTION MACHINE COMES BACK,* 1937, Refrigerating Engineering, August, 1937.

Wehlage, Edward F., P.E.; *NON-ELECTRIC COOLING AT +4° C,* ISGE, Transactions, Paper No. 7405, Vol. 1, No.1, September 1974.

Wehlage, Edward F., P.E.; *THE BASICS OF APPLIED GEOTHERMAL ENGINEERING, CHAPTER 12, REFRIGERATION,* Geothermal Information Services, West Covina, CA.

Wehlage, Edward F., P.E.; *GEOTHERMAL CIVIC INDUSTRIAL PARK MONOGRAPHS, PART NO. 2 of 4, TAX FREE WAREHOUSES WITH GEOTHERMAL HEAT SERVICES,* ISGE Transactions.

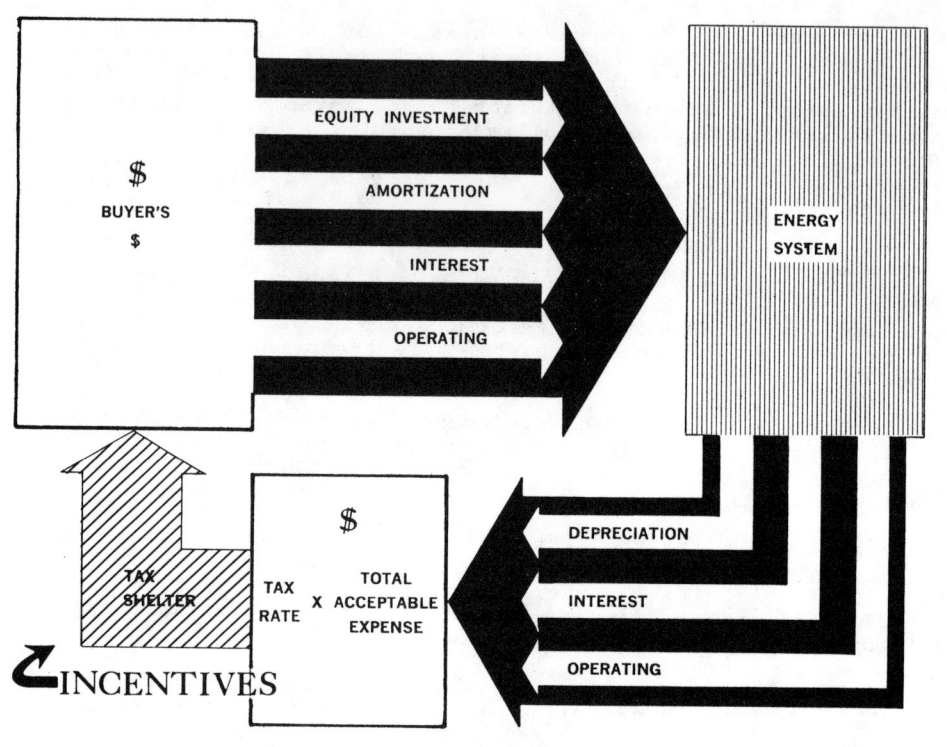

FIGURE 1. INDUSTRIAL PURCHASERS OF ALTERNATE ENERGY SYSTEMS DO NOT PURCHASE ALTERNATE ENERGY, AND TOTAL POWER SYSTEMS, WITHOUT FINANCIAL INCENTIVES WHICH MAY BE FOUND INTERNALLY FOR AN ORDINARY FINANCIAL PROCESS, BUT PARTICULARLY AS THE RESULT OF EXTERNAL INCENTIVES, ESPECIALLY FROM THE TAX STRUCTURE. USUALLY THE BUSINESS MANAGER IS NOT A POWER EXPERT, BUT A CONSUMER.

GEOTHERMAL ENERGY AND HYDROPOWER

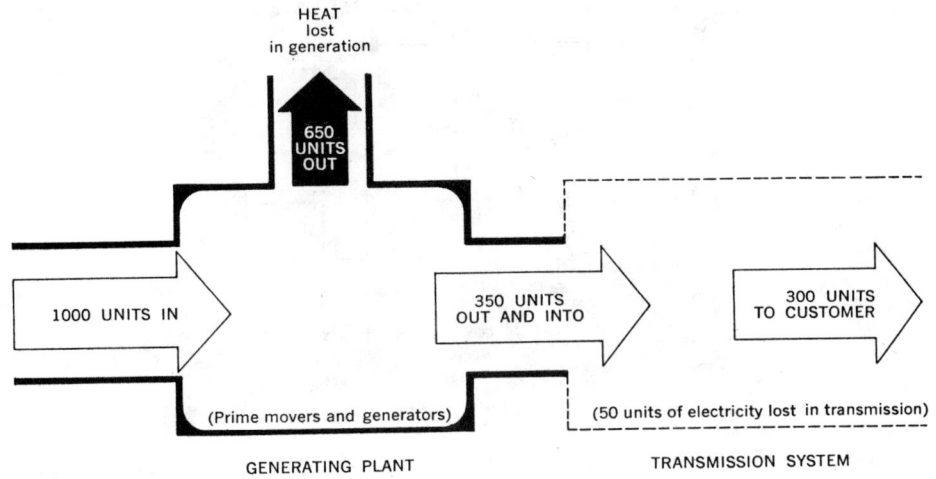

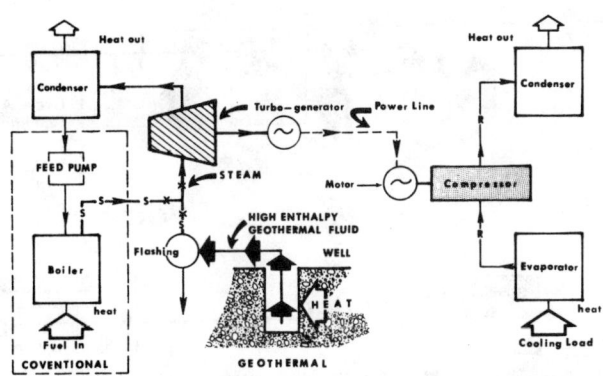

FIGURE 2. THE GENERATION OF ELECTRICAL POWER WITH GEOTHERMAL HEAT IS ONE OF THE LEAST EFFICIENT USES FOR IT. STEAM OR HEATED FLUID FROM THE EARTH CONVEYS NO MAGICAL ECONOMY IN ORDINARY THERMODYNAMIC CYCLES APPLIED TO ELECTRIC POWER SERVICE.

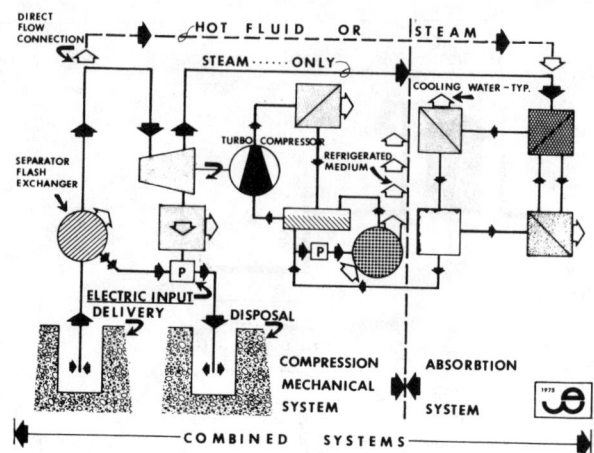

FIGURE 3. STEAM TURBINE MECHANICAL DRIVES MAY BE COMBINED WITH REFRIGERATION COMPRESSORS AND ABSORBTION REFRIGERATION SYSTEMS.

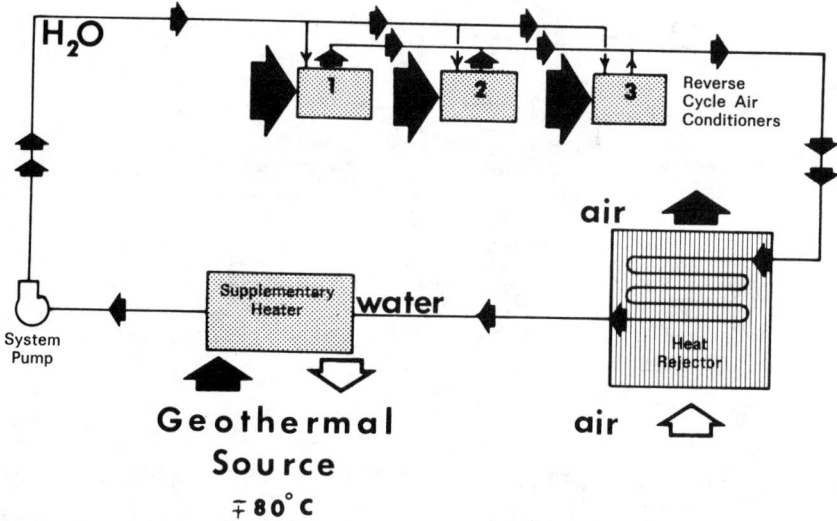

FIGURE 4. MANY HEAT PUMP SYSTEMS MAY BE COMBINED WITH LOW LEVEL GEOTHERMAL SOURCES AND INSTALLATIONS HAVE BEEN MADE ALREADY.

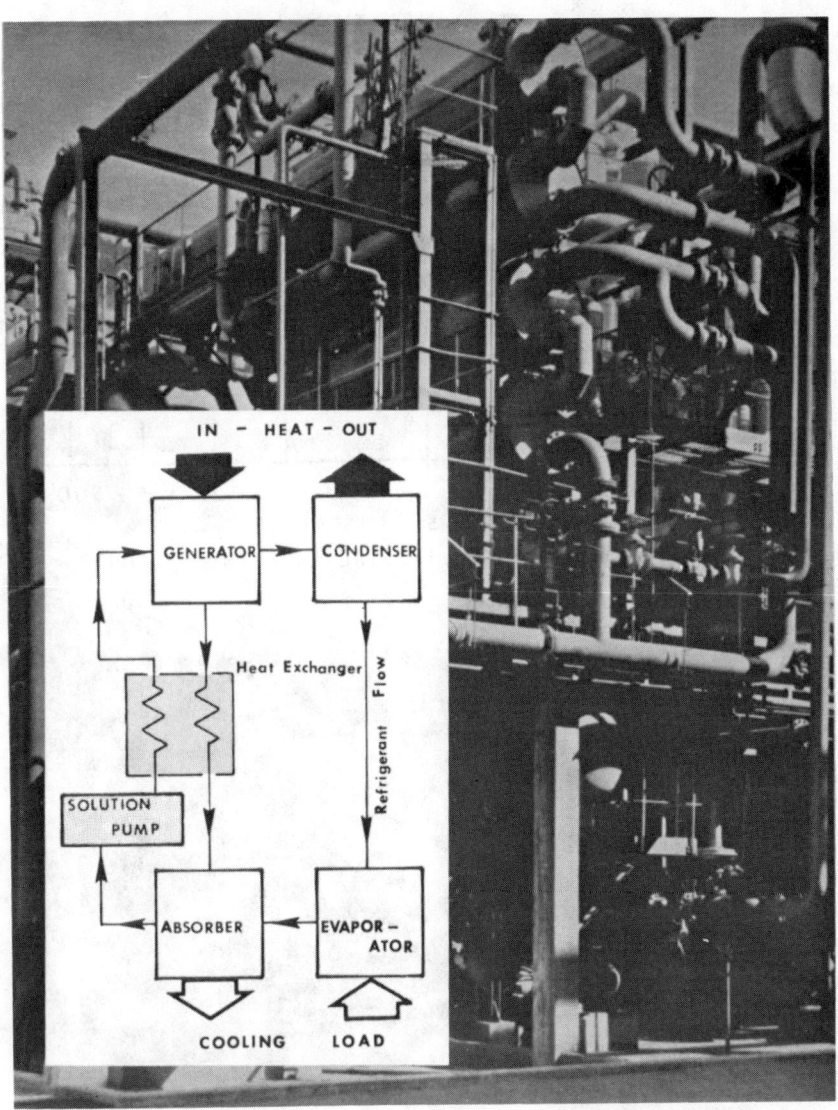

FIGURE 5. THE MODIFIED ABSORBTION SYSTEM FILLS MANY OF THE REQUIREMENTS, BUT THIS EUROPEAN UNIT IS NOT GEOTHERMALLY POWERED.

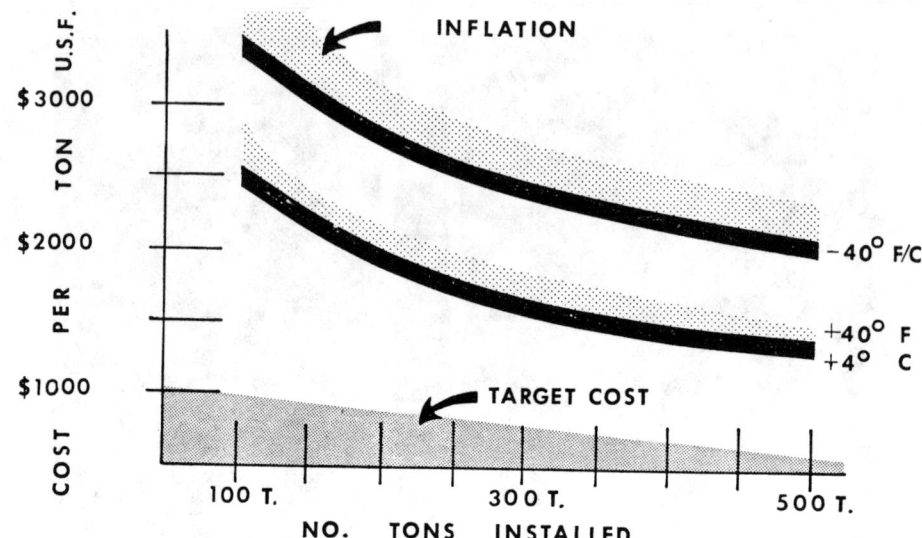

FIGURE 6. ESTIMATED COST FOR INITIAL EXPERIMENTAL UNITS $/TON

FIGURE 7. GEOTHERMAL INDUSTRIAL PARKS MAY BE ABLE TO GIVE SERVICES WHICH SMALLER BUSINESSES COULD NOT PERFORM INDIVIDUALLY.

Heat Transfer Consideration
in Utilizing Solar and Geothermal Energy*

J. W. MICHEL
Oak Ridge National Laboratory
Oak Ridge, Tennessee 37830, USA

ABSTRACT

    Although some forms of solar and geothermal energy will require the application of high temperature technology, most involve only low temperatures. This paper emphasizes the importance of efficient, low cost heat exchangers in the use of low temperature sources and in the waste heat rejection systems of power generation cycles. Results of recent heat transfer development work at ORNL show that for the working fluids which are attractive for use in these cycles the condensation heat transfer effectiveness can be improved by a factor of from 3 to 7 by the use of fluted tubes and condensate drain-off skirts. Data are presented for the condensation of six fluorocarbons, isobutane, and ammonia on a variety of tube surfaces. These results are being applied to the design of a 40-tube heat exchanger to be tested in a geothermal energy test facility.

INTRODUCTION

    While the utilization of any thermal energy source requires the application of heat transfer principles, the use of solar and geothermal energy involves many unique heat transfer considerations. Although some of the applications essentially make use of existing technology, many require adaptations to new working fluids and conversion cycles. Some of the applications involve high temperatures as in the use of solar concentrators; i.e., tower power, or in the use of the magma form of geothermal energy. Most, however, are concerned with the use of lower temperature sources such as the geothermal brines or geopressured fluids or the use of ocean thermal gradients or of unconcentrated solar radiation. It is this lower temperature area that will be emphasized in this paper.

    In the lower temperature applications, heat transfer considerations become of increasing economical and technological significance due to typically low conversion efficiencies and small size installations. One impact of the low cycle efficiency is illustrated in Fig. 1 which shows the amount of rejected heat from a Rankine cycle as a function of the cycle efficiency. Typical geothermal conversion cycles reject up to nine times more heat than a modern coal-based cycle, while an OTEC cycle would correspondingly reject approximately 40 times the amount of heat per net kWe. Thus, in some cases, the heat rejection system may be the most costly system in a power plant operating at low cycle efficiency.

---

    *Research sponsored by the Department of Energy under contract with the Union Carbide Corporation.

SOLAR AND GEOTHERMAL SOURCES

As a basis for discussing the related heat transfer problems, it is important to define the nature of the available solar and geothermal energy resources. Solar energy is usually cyclical (day-night), variable (morning-noon-evening, as well as weather sensitive), and always diffuse. However, in one terrestrial application, ocean thermal gradient, the first two characteristics are not controlling; i.e., solar radiation is already collected and stored, albeit at low temperatures, in the tropical oceans. Geothermal energy sources are available in many forms as summarized in Table 1.

TABLE 1
GEOTHERMAL ENERGY SOURCES

Type	Temperature range	Estimated fraction of available resource
Hydrothermal		
Steam dominated	220–250°C	$\frac{1}{2}$%
Water dominated	<150–360°C	10%
Hot dry rock	150–>400°C	30%
Geopressured	150–200°C	20%[a]
Magmas	700–1200°C	40%

[a]Not including dissolved methane.

In addition to the variation in temperature noted in Table 1, some resources (e.g., hydrothermal) vary in composition (total dissolved solids and gases) from reservoir to reservoir and sometimes from time to time. Most brines are corrosive to many construction materials and have a strong propensity to form scale deposits, especially on heat transfer surfaces. Finally, most of the resources are remote from potential industrial, commercial, and residential users.

LOW TEMPERATURE SOLAR APPLICATIONS

Space and hot water heating are likely to be the largest near-term low temperature solar energy application, are already well-developed and generally make use of conventional heat transfer technology. One potential application in which heat transfer considerations are of great importance and where advancements are required in both performance and costs is the ocean thermal energy conversion (OTEC) system. Design studies have shown that the heat exchangers required for a closed cycle ammonia plant using current technology make up about one-half of the plant cost. Due to the low cycle conversion efficiency (2–3%), heat exchangers (ammonia evaporators or condensers) are large; i.e., for a 25 to 40 MWe module, they may be 50 to 70 ft in diameter with ~10^5 tubes (1 1/2 in.), 50 ft long. Further, the temperature drops are small for optimum operation resulting in large seawater flow rates; e.g., greater than two orders of magnitude more than for conventional high temperature fossil-fired plants.

The heat transfer situation for the design-basis, closed-cycle OTEC plants is illustrated in Fig. 2. Using conventional shell/tube heat exchangers with seawater flowing inside the tubes, overall heat transfer coefficients of ~400 Btu/hr·ft^2·°F can be obtained with titanium tubes including a fouling factor of 0.0003 Btu/hr·ft^2·°F. As indicated in this figure, increasing the ammonia-side coefficient would be of marginal value. Figure 3 shows the effects of increasing the seawater-side coefficient and decreasing the fouling resistance. If the water-side coefficient could be doubled with acceptable pressure losses, then increasing the ammonia coefficient would have a large effect on the overall coefficient, allowing a proportional decrease in the heat transfer surface area and, hopefully, in the heat exchanger cost. At the higher performance levels, the importance of maintaining a low fouling resistance may also be noted.

Experimental Ammonia Heat Transfer Data

Based on the extensive test program with a variety of vertical fluted (Gregorig) tubes at ORNL as part of the OSW Seawater Desalting Project, it was anticipated that high condensation coefficients for ammonia could also be obtained. The principle of operation of the Gregorig-type surface is illustrated in Fig. 4. Surface tension effects of the liquid condensate film on the curved surface act to force the liquid into the troughs, thus maintaining a thin film on the crests while the liquid drains downs the troughs.

Data have recently been obtained at ORNL[1] on ammonia condensation with a variety of 1-in.-OD vertical tubes 4 ft long. The surface geometry of the tubes is shown in Fig. 5. Test with rubber drain-off skirts (Fig. 6) have also been done with Tube F to determine the effect of tube length on performance. Condensing heat transfer coefficients obtained using the Wilson plot technique are summarized in Figs. 7 and 8. In Fig. 7, the coefficients are shown as a function of Q/A where both h and Q/A are based on the total outside surface area. In the region of OTEC interest, Q/A values of 2000 to 3000, it may be noted that a smooth tube gives coefficients of 1000 to 1500, while the fluted tubes give values up to 10,000. It should also be noted that the coefficients are composite; i.e., they include the wall resistance since there is no convenient way to compute the resistance of the fluted tube wall. It may also be noted that the effect of drain-off skirts (shorter tubes) is only evident at very high heat fluxes indicating that due to ammonia's high latent heat and low viscosity flooding of the rills with condensate thus causing a decrease in the thin film area will not be a concern in OTEC condensers.

In Fig. 8, the heat load capability, Q, of the tubes is shown as a function of the composite temperature drop ($t_{NH_3} - t_{H_2O}$). This correlation does not require any assumptions regarding heat transfer area and indicates that substituting Tube E for a smooth tube (A) would allow six times the heat transfer for the same ΔT (~3°F).

Figure 9 shows the improvement in ammonia condensation heat transfer relative to the smooth tube performance. Improvements of about a factor of 7 in the coefficient are indicated in the region of OTEC interest.

Seawater Heat Transfer

The improvement of single-phase heat transfer has been extensively investigated and has recently been summarized relative to OTEC application by Bergles.[2] His analysis was based on enhancement at constant seawater pumping power and may differ somewhat from that based on overall plant re-optimization. The best enhancement technique appears to be with low transverse or spiral ribs. Enhancement of heat transfer coefficients of up to a factor of 2 appear possible.

OTEC Summary

Figure 10 illustrates the potential for heat transfer enhancement as applied to OTEC; e.g., a 100% increase in the overall coefficient appears possible for the ammonia condenser if a water-side enhancement of 1.6 to 1.8 (depending on the achievable fouling factor) is attained. Improvement in evaporator coefficients using Gregorig surfaces as demonstrated at Carnegie-Mellon University[3] appear possible. In all cases, the cost of enhanced surfaces must be factored into the final optimization, but it appears likely that considerable improvement in OTEC heat exchangers can be achieved.

Alternate OTEC Cycles

Second generation OTEC cycles now under study include the open or "Claude" cycle and various water lift cycles. While these cycles can eliminate surface heat exchangers, they do involve the use of flash evaporators and direct-contact condensers. Application of flash evaporators developed during the desalting program and the use of spray or direct-contact condensers seem to offer good possibilities for developing these cycles, but more technology and economic studies are required before meaningful comparisons to the closed ammonia cycle can be made. Deareation of the seawater appears to be a significant problem with these cycles.

GEOTHERMAL APPLICATIONS

While the use of geothermal energy has found and should continue to find wide application for direct heat use in residential/commercial as well as industrial applications, the following discussion will be limited to its use in power conversion cycles. Although there are several types of power conversion cycles in use or under development (i.e., flash, total flow, binary, dual, hybrid, etc.), the binary cycle is receiving perhaps the most interest. Three typical cycles are shown schematically in Fig. 11.

As shown by Milora and Tester[4] for a given temperature of a geothermal heat source, there is probably an optimum working fluid. Figure 12 illustrates this principle where R-115 is an attractive working fluid for a low temperature (~150°C) source; i.e., is capable of capturing a larger fraction of the available energy, while R-22 is correspondingly superior for a 250°C heat source. It should be recognized that this analysis is based on thermodynamic considerations and when economic and safety factors are applied the optimum fluid for a particular temperature may differ from that shown in Fig. 12.

Working Fluid Evaporators

One of the main problems associated with the operation of a binary cycle using moderate to high temperature geothermal brines is that of scale formation on the evaporator heat transfer surface. Several schemes are under development to overcome this problem: (1) use of a hybrid cycle where steam flashed from a brine is used as the heat source for the working fluid evaporator, (2) use of a fluidized sand bed with brine to keep the evaporator surface scoured clean, and (3) use of a direct-contact evaporator where a hydrocarbon such as isobutane is dispersed and bubbled through the brine.

Working Fluid Condensers

As mentioned above, the heat rejection systems for power cycles of relatively low efficiency can be the most costly component of the system. Investigations are in progress at ORNL to improve the condensation heat transfer effectiveness for a variety of candidate working fluids for geothermal power cycles. These fluids, mainly hydrocarbons and fluorocarbons, have typically poor heat transfer characteristics as illustrated in Fig. 13. Also, as with the OTEC ammonia condensers, the use of vertical fluted tubes would be expected to provide improved performance for geothermal cycles. Similar heat transfer experiments to those described for ammonia condensation have been performed for a variety of fluorocarbons and isobutane using many of the tubes shown in Fig. 5. Typical data is shown in Fig. 14 for R-113 with Tubes A, B, C, D, E, F, and F-7 where F-7 is Tube F with 7 rubber drain-off skirts which effectively becomes a tube 1/2 ft long. On this correlation, Tubes F and E show significant improvement over the smooth tube performance and

Tube F-7 shows even more enhancement indicating that R-113 does tend to flood the rills of a fluted tube.

Figure 15 illustrates the difference in condensation performance resulting from the differences in fluid properties with data shown for R-11, R-22, R-113, R-114, R-115, R-600a (isobutane), and R-717 (ammonia). Of the fluorocarbons, R-11 and R-22 give the best performance but are considerably below that for ammonia.

Figures 16 and 17 are plots of the ratio of the heat transfer coefficient for a fluted tube to that for the smooth tube vs heat flux. Figure 16 shows the performance improvement for R-113 by using effectively shorter tubes; e.g., Tube F-7 gives an improvement factor of from 4 to >6.5 over the smooth tube performance. Similarly, the performance of R-600a is improved by the use of drain-off skirts as indicated in Fig. 17.

Work is now in progress to apply these results to the design of heat exchanger to be tested at the DOE East Mesa Geothermal Test Facility. The design of this isobutane condenser shown in Fig. 18 makes use of rubber drain-off tube sheets fitted to aluminum fluted tubes (Tube F). The shell side (isobutane) is also designed to handle noncondensables by vapor sweeping across the tube bank in successively smaller flow areas to the purge point.

SUMMARY AND CONCLUSIONS

In the utilization of solar and geothermal energy for power production, waste heat rejection is often the most important subsystem from an economic standpoint. For many of the applications, working fluids of poor heat transfer characteristics are employed which further complicate the situation. Use of vertical fluted tubes for condensation of these fluids has been shown to provide significant improvement in heat transfer over smooth tubes. The application of condensate drain-off skirts has been shown to provide additional benefits, especially for fluorocarbon and hydrocarbon working fluids.

REFERENCES

1. S. K. Combs, <u>An Experimental Study of Heat Transfer Enhancement for Ammonia Condensing on Vertical Fluted Tubes</u>, ORNL-5356 (in press).

2. A. E. Bergles and M. K. Jensen, <u>Enhanced Single-Phase Heat Transfer for Ocean Thermal Energy Conversion Systems</u>, ORNL/Sub-77/14216/1 (April 1977).

3. R. R. Rothfus and C. P. Neuman, "The OTEC Program at Carnegie-Mellon University Heat Transfer Research and Power Cycle Transient Modeling," pp. VI-55 in the <u>Proceedings of Fourth Annual Conference on Ocean Thermal Energy Conversion, University of New Orleans, New Orleans, Louisiana, March 22-24, 1974</u>, Edited by George E. Ioup (July 1977).

4. S. L. Milora and J. W. Tester, *Geothermal Energy as a Source of Electric Power — Thermodynamic and Economic Design Criteria*, The MIT Press, Cambridge, Massachusetts, 1976.

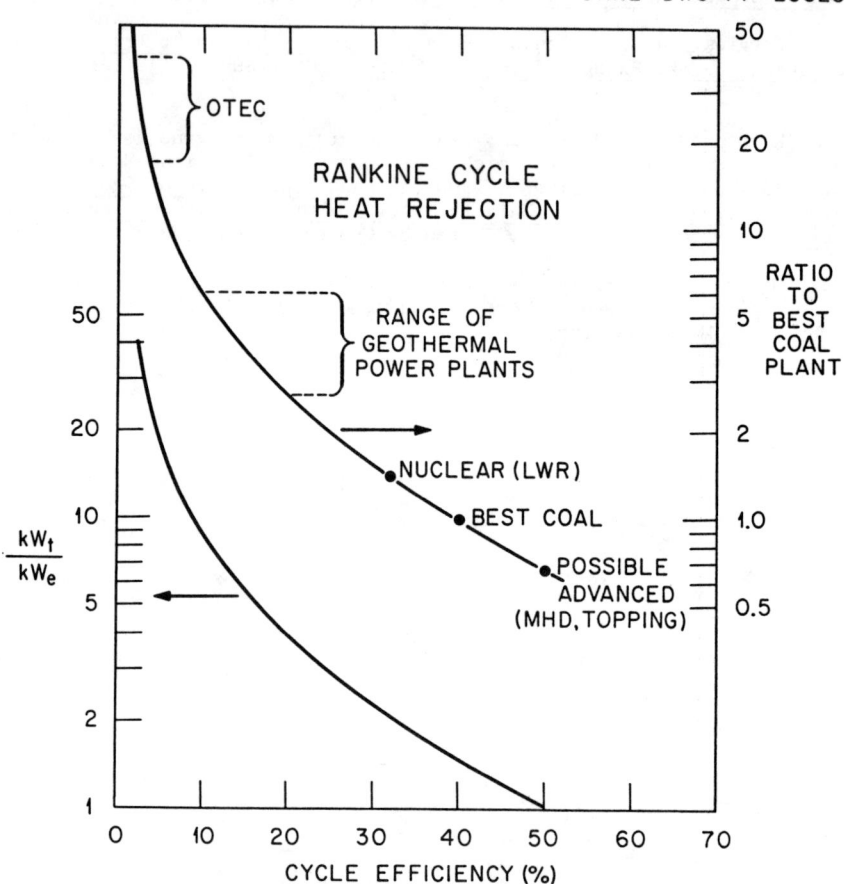

Fig. 1 Rankine cycle heat rejection as a function of cycle efficiency, plot of ratio of kW_t (heat rejected) to kW_e (net electrical power produced) vs. cycle efficiency, and the ratio of this ratio to the value at 40% (best coal plant) efficiency.

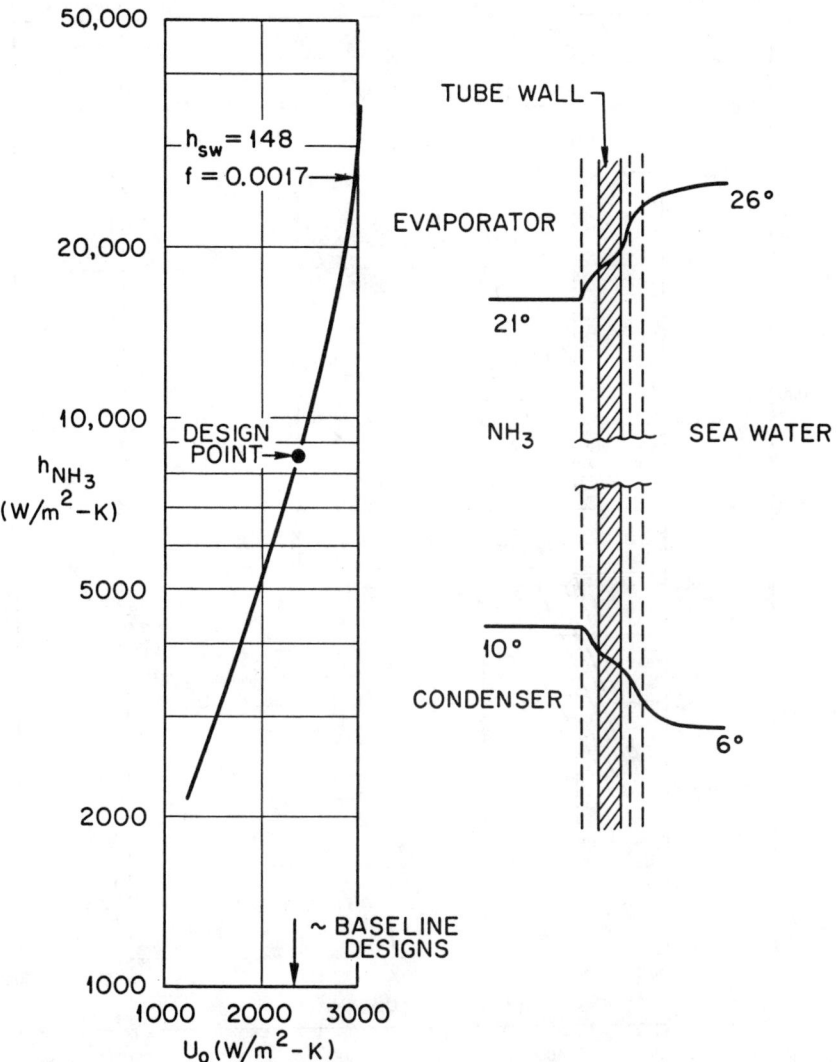

Fig. 2 Effect of changing ammonia heat transfer coefficient on the overall coefficient.

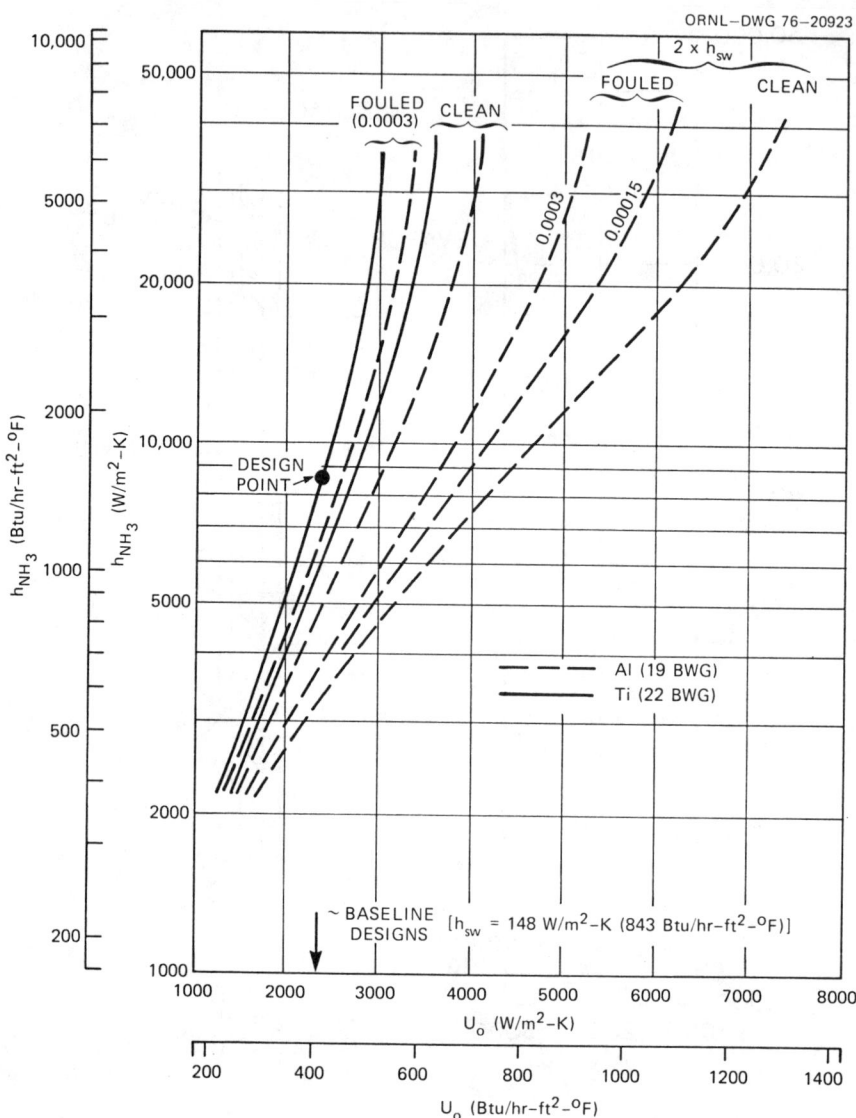

Fig. 3   OTEC Heat Transfer Potentials and Limitations, Plot of Ammonia-Side Heat Transfer Coefficient vs. Overall Heat Transfer Coefficient.

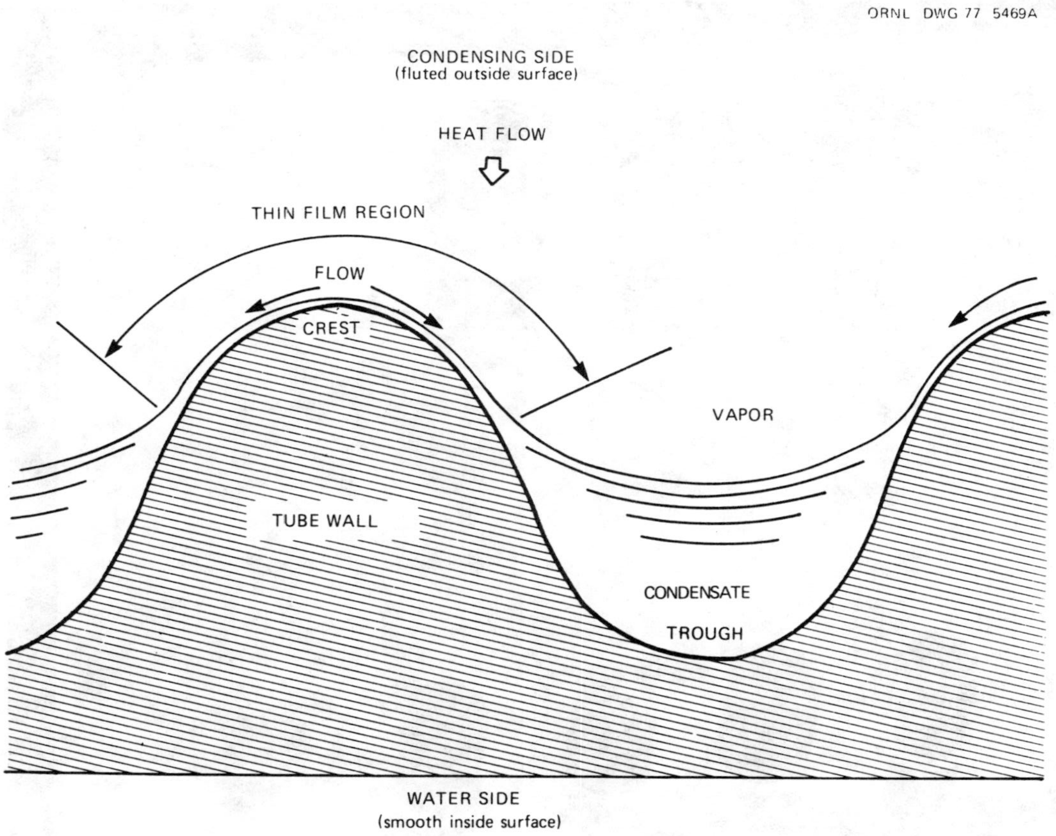

Fig. 4 Fluted tube principle of operation (condensation mode) - surface tension forces acting to push condensate from crests into troughs.

Fig. 5 Examples of experimental fluted tubes.

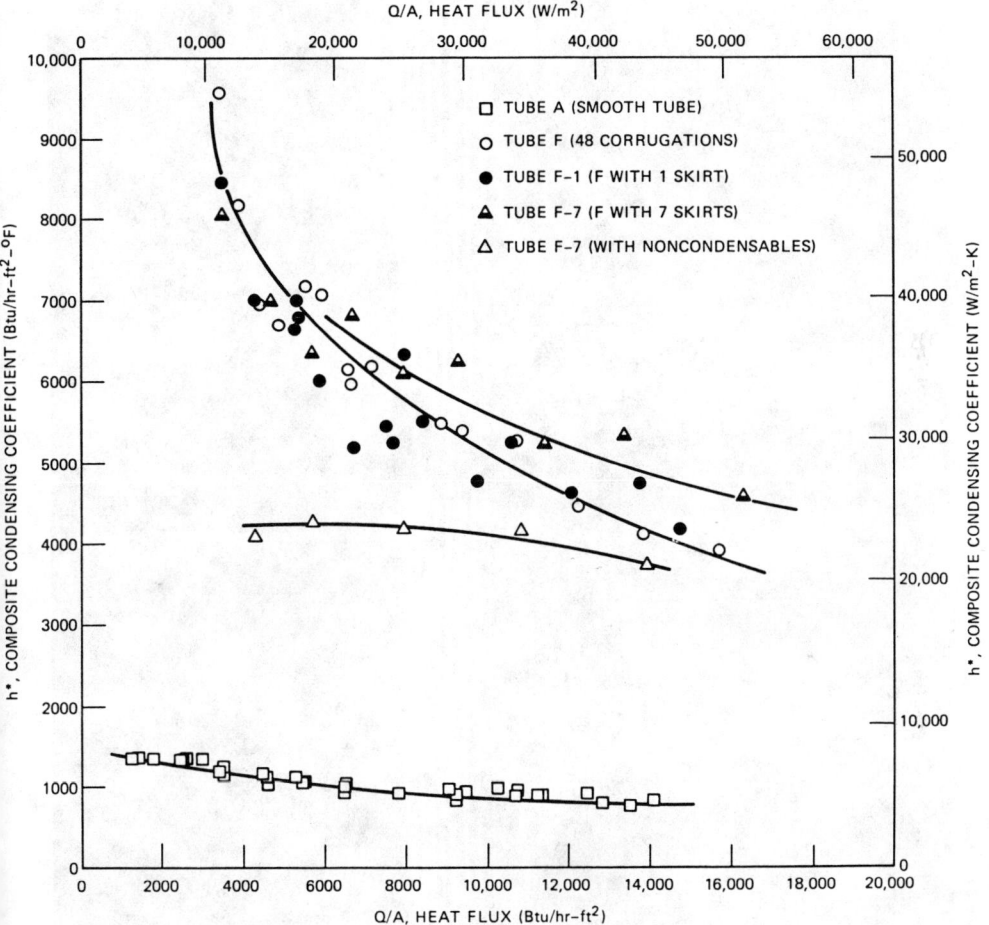

Fig. 7 Ammonia condensation on 1-in.-OD by 4-ft long tubes

Fig. 6 Midsection view of Tube F-1 (Tube F with a neoprene drain-off skirt attached at the midpoint of the condensing length)

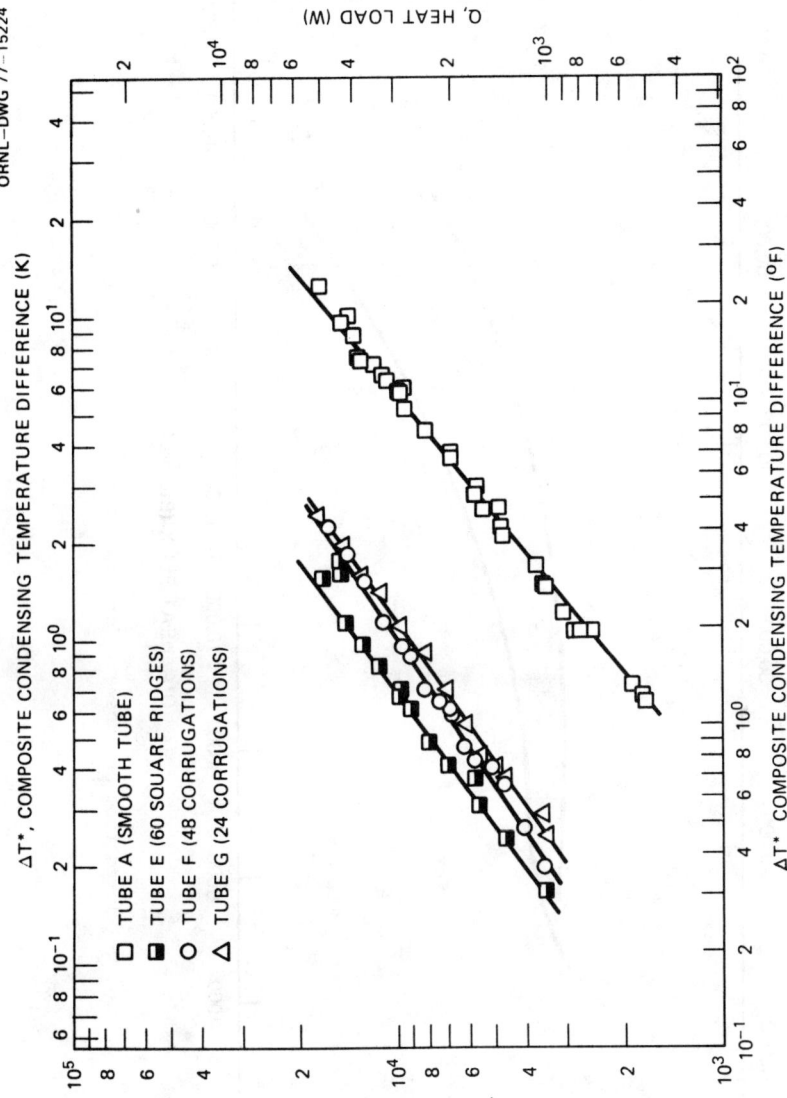

Fig. 8 Heat load vs. composite condensing temperature difference for ammonia.

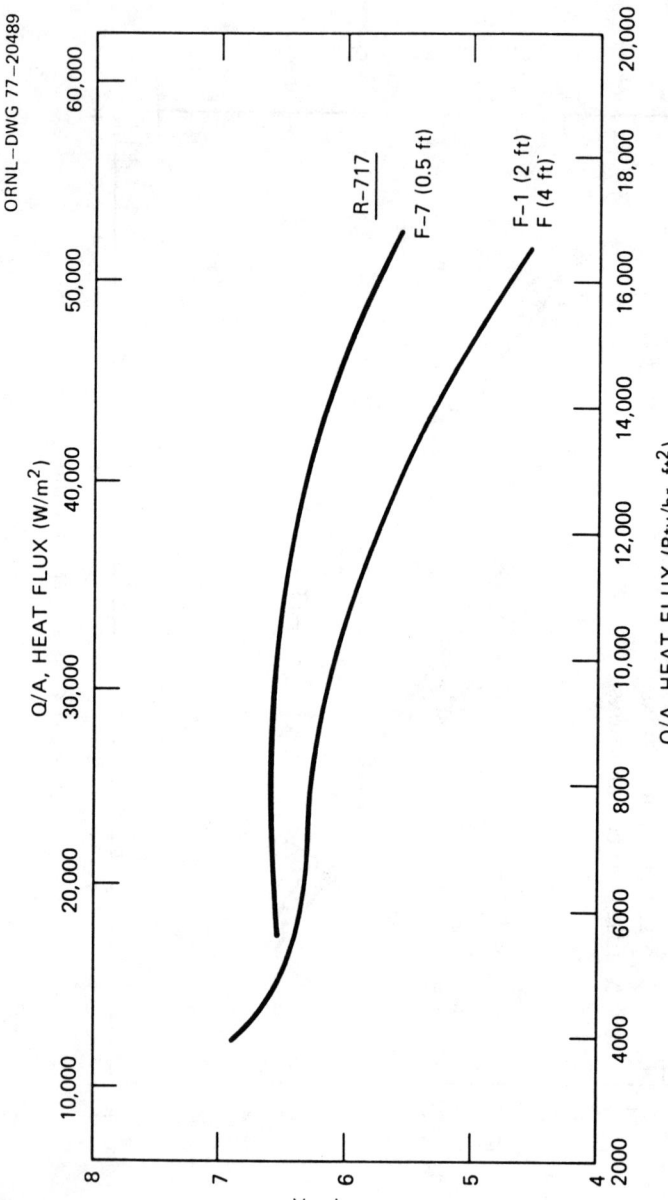

Fig. 9 Fluted tube enhancement for ammonia condensation.

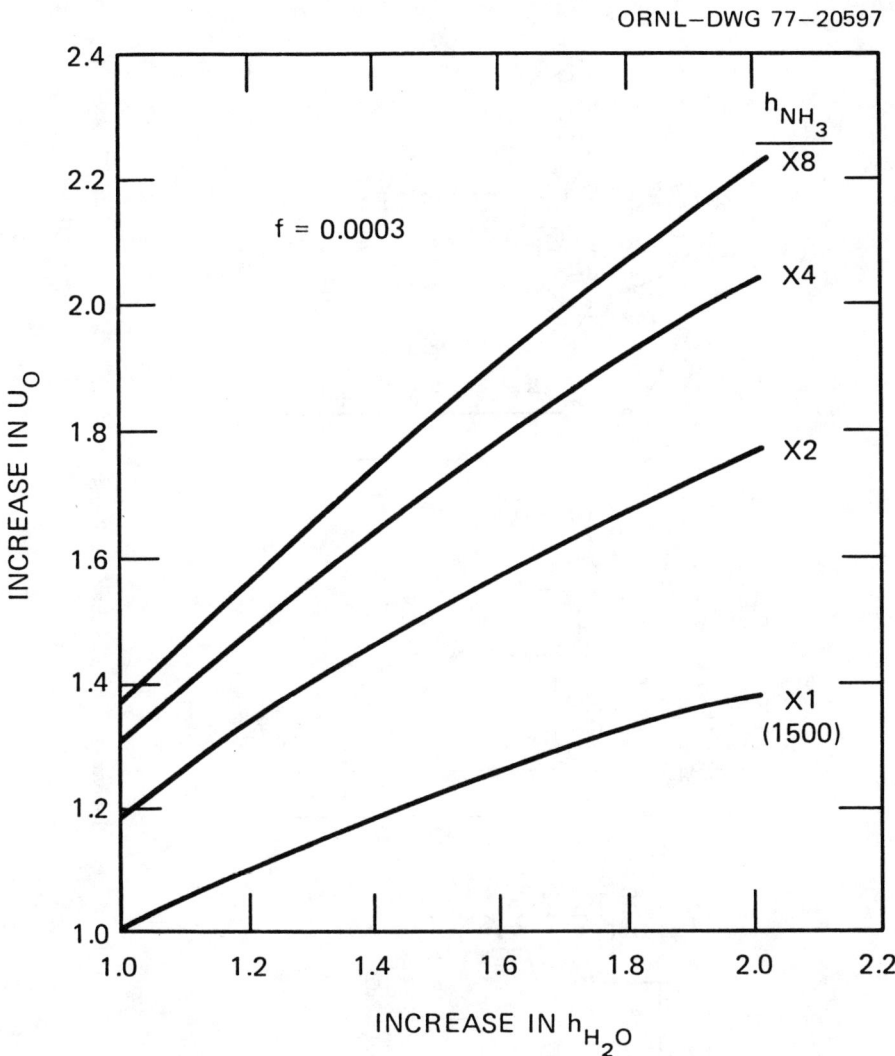

Fig. 10 Relative effects of increasing the ammonia and water coefficients on the overall coefficient.

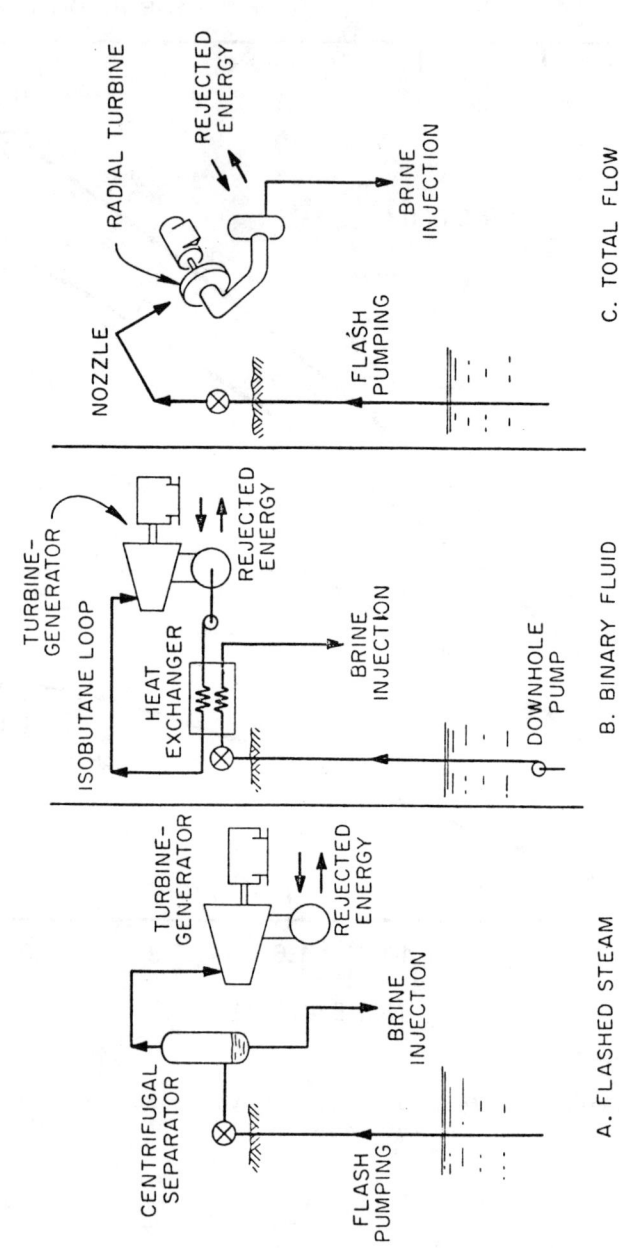

Fig. 11. Comparison of Geothermal Power Cycles.

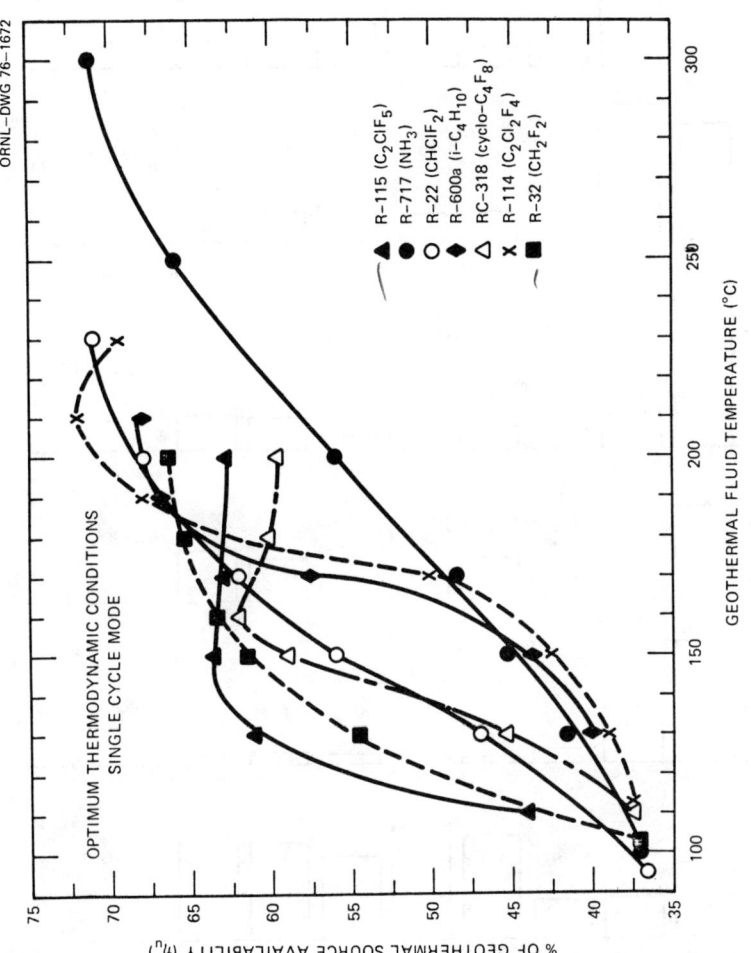

Fig. 12 Effect of working fluid characteristics on the effectiveness of energy use as a function of geothermal source temperature.

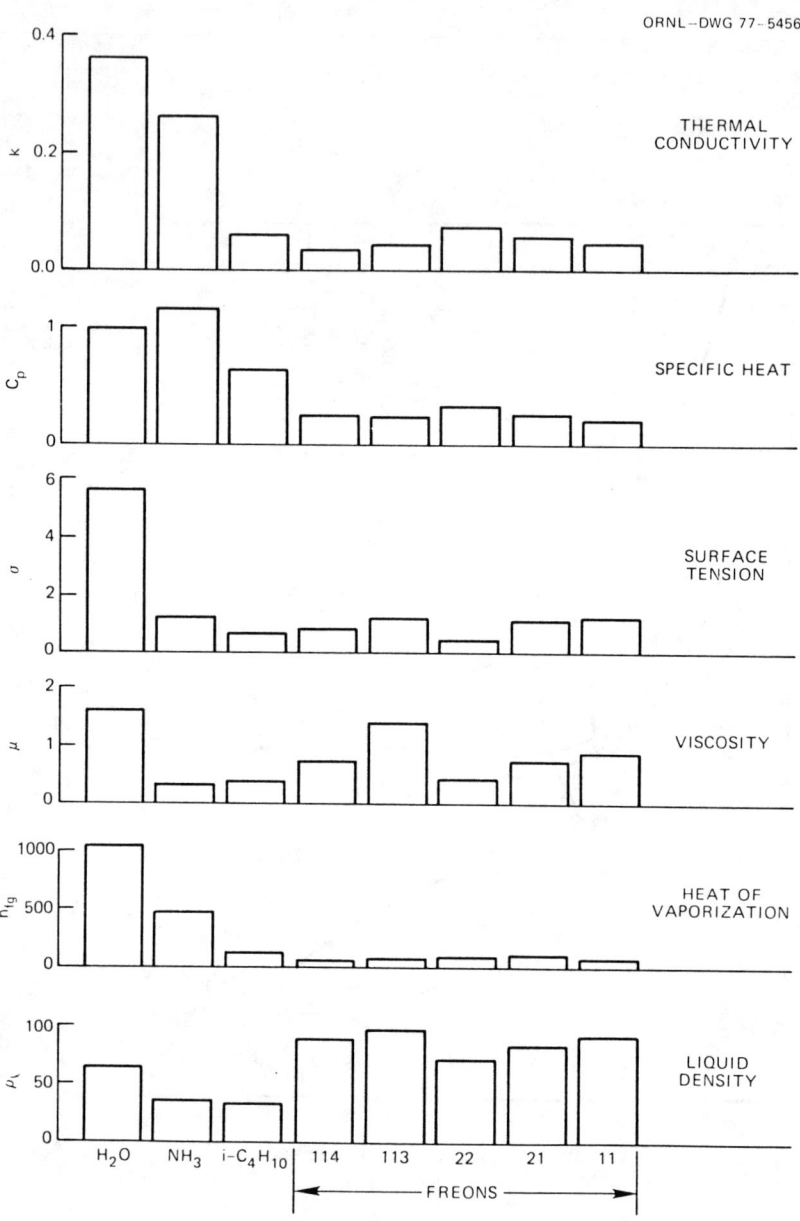

Fig. 13 Comparison of Fluid Properties (at 100° F)

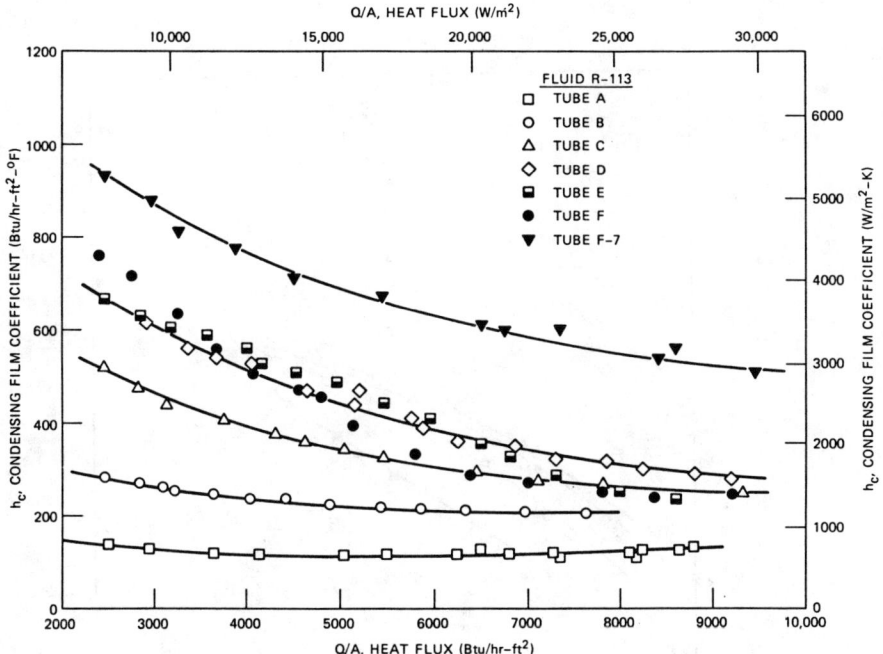

Fig. 14. Condensation characteristics of R-113 on various vertical tube surfaces.

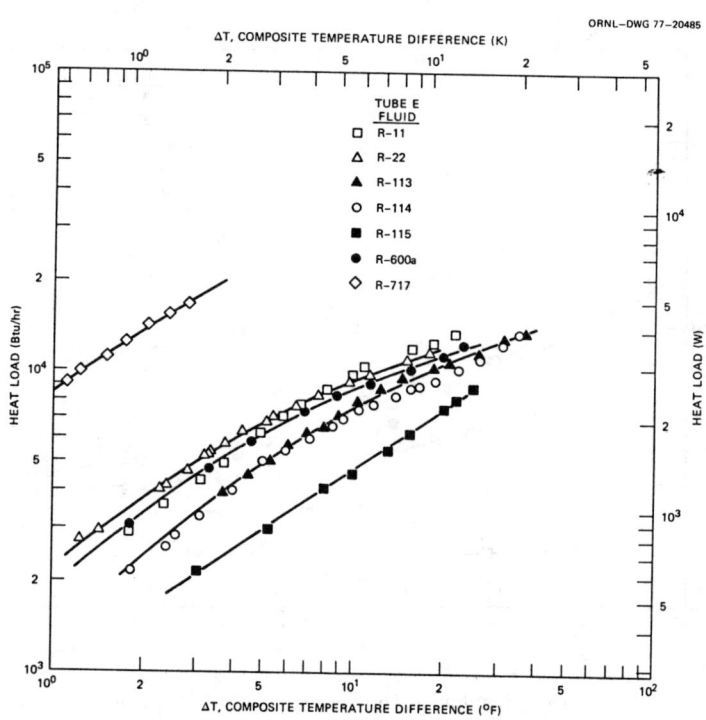

Fig. 15. Comparison of condensation characteristics of various working fluids.

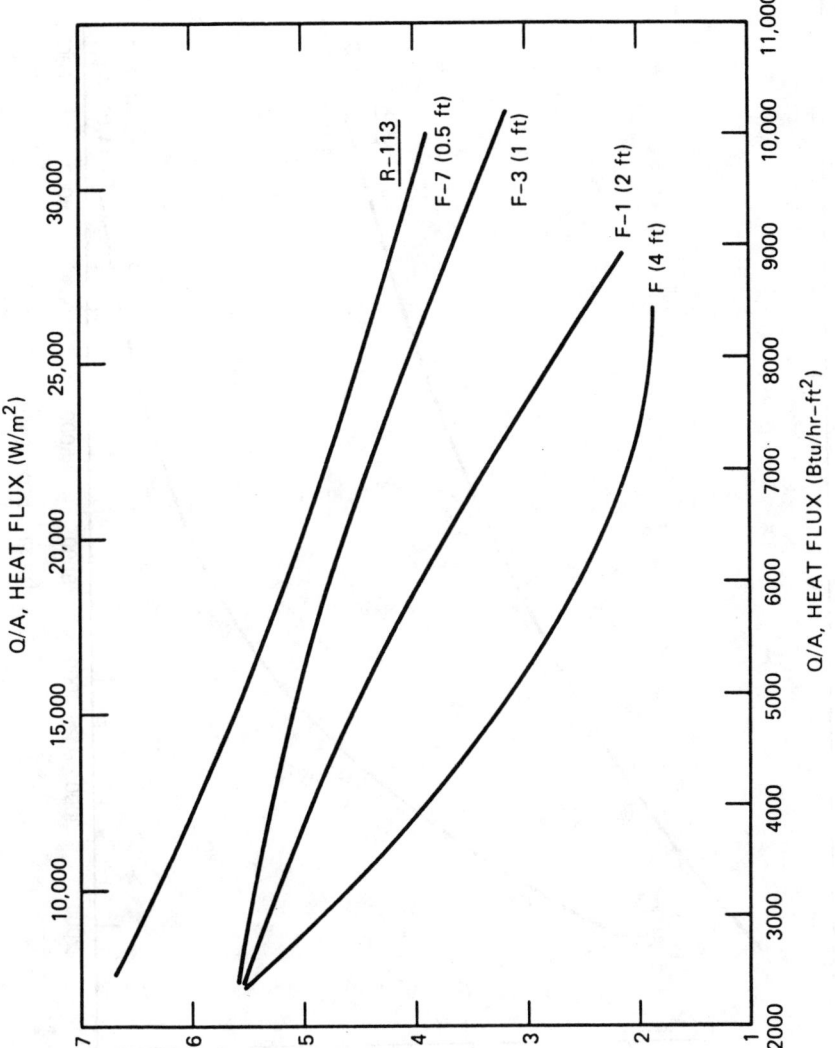

Fig. 16 Effect of tube length on the condensation enhancement of R-113.

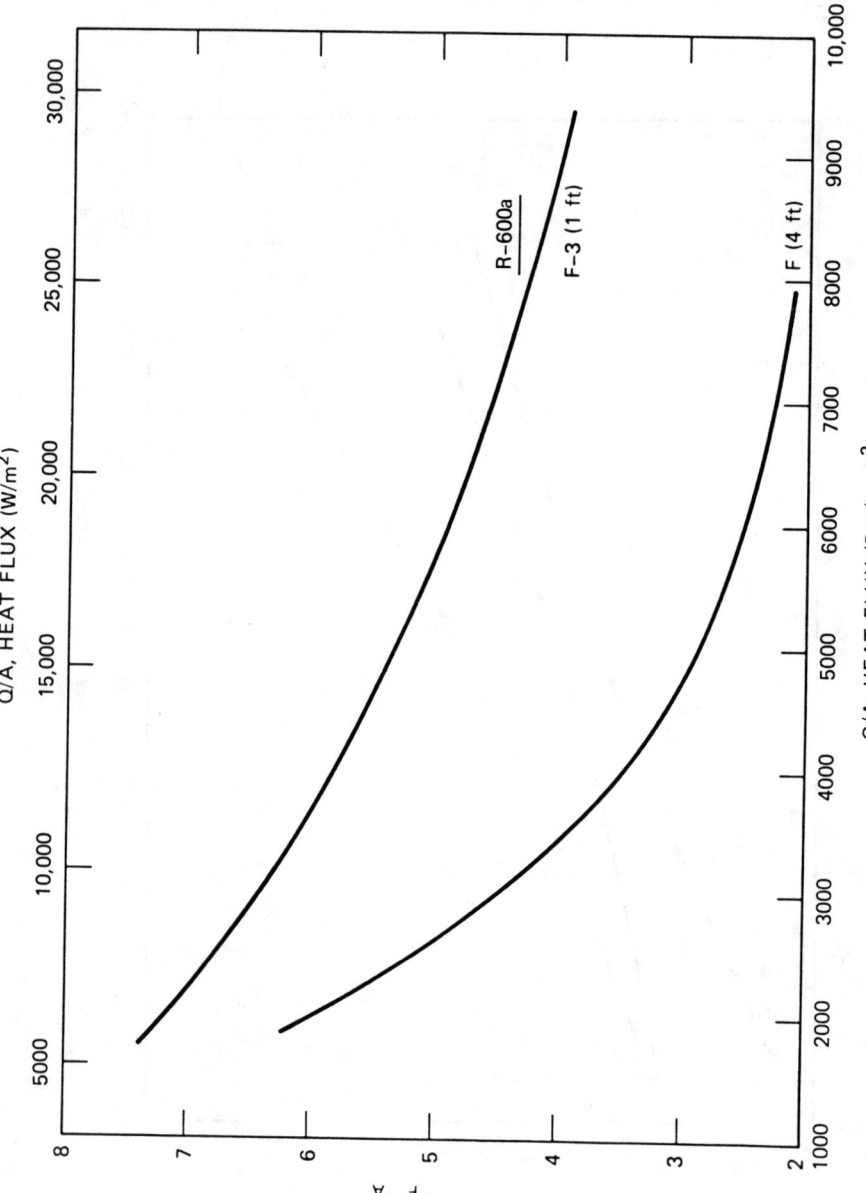

Fig. 17 Effect of tube length on the condensation enhancement of R-600a.

2824

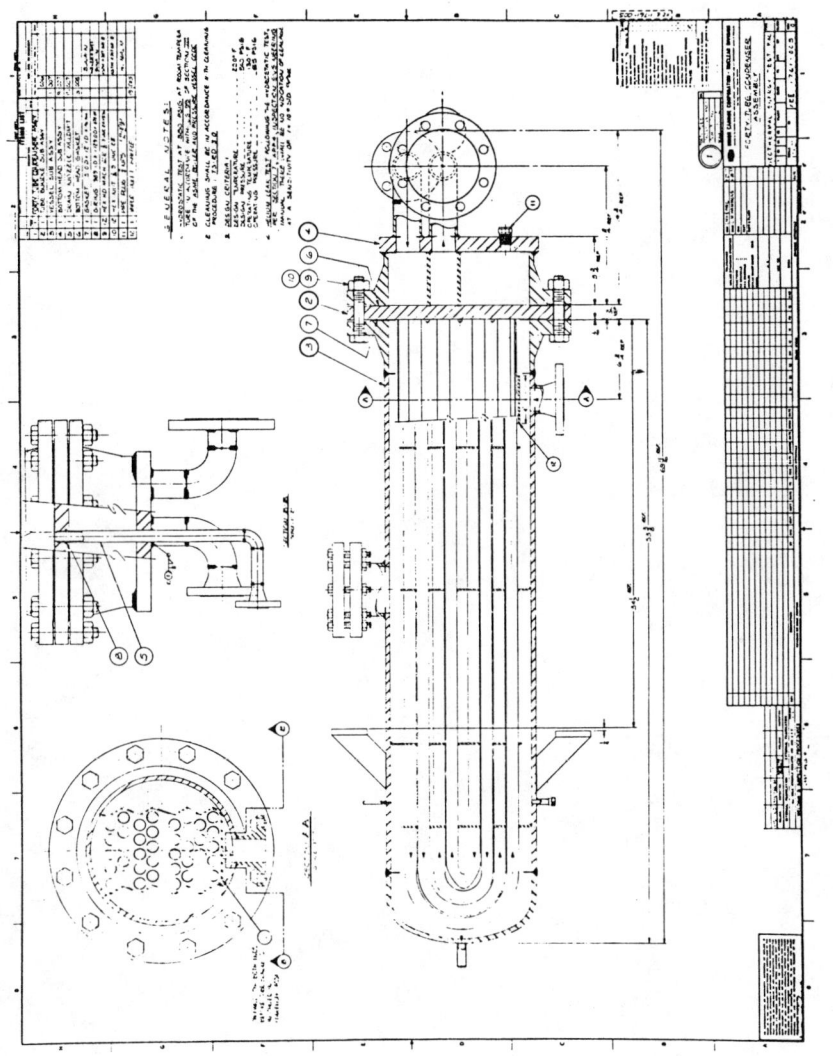

Fig. 18 Experimental vertical fluted tube condenser for isobutane.

The Economics of Upgrading Geothermal Steam by Adiabatic Compression

A. VALFELLS
Department of Chemical Engineering and
Nuclear Engineering and Engineering Research
Institute, Iowa State University
Ames, Iowa 50010, USA
and the
Science Institute of the University of Iceland
Reykjavik, Iceland

ABSTRACT

Because of its relatively low temperature, geothermal heat is much better employed as a source of process heat in the chemical industry, than for electric power generation. Most chemical processes have been designed to use fossil fired steam conventionally produced at 250°C-300°C, whereas geothermal steam is quite often only available at lower temperatures. This raises the question, whether low temperature geothermal steam, which is about the cheapest steam available, can be economically used in a turbine-compressor unit to compress some similar steam adiabatically to higher temperatures. The economics of such a facility, of 12 MW capacity, that is run by steam formed by double flashing at 175°C and 120°C and that compresses 190°C hot steam were investigated, and such compression was found to be economically feasible.

INTRODUCTION

Most geothermal areas in the world provide saturated or superheated hot water; only a few such as the Geysers in California provide dry saturated steam. Because of its relatively low temperature, geothermal steam can only be converted into electrical power with a low efficiency; say 17% with a two stage flashing system.

A much more profitable use for geothermal heat appears to be its utilization as process heat in energy intensive chemical processes. Lindal [1] has made an extensive survey of such processes. However, most chemical processes have been designed to utilize steam at temperatures conventionally generated in fossil fired steam generation plants, often around 200 to 250°C, whereas geothermal steam, produced by flashing, is most economically furnished at temperatures lower than 200°C. This puts a considerable limitation on the number of processes of conventional design that can be operated with geothermal steam. These limits can be extensively widened by modifying the design of many processes so as to use part or all of the required steam at temperatures below 200°C. A case in point is the H_2S-H_2O isotope exchange process for heavy water manufacture which, in its conventional design, requires steam at 300°C. By appropriate modifications of the process design, the steam can be utilized at 170°C or even lower. (See references [2] and [3]). Other processes can only be partially adapted to geothermal steam; petroleum refining is a case in point. Still others cannot be adapted at all.

Since geothermal steam is by far the cheapest form of steam, within the temperature range that it is available, this raises the question whether it cannot economically be upgraded so as to be applicable to processes that require steam at higher temperatures. It is the purpose of this paper to investigate that question.

FLASHED STEAM COSTS.

Geothermal Field Characteristics

The study is based on a field having the characteristics of the Hengill geothermal area in south western Iceland. Fig. 1 illustrates the characteristics of a typical well in the region.

Drilling and Distribution Costs

Based on actual drilling costs at the Krafla geothermal project in northern Iceland, the cost of drilling a single geothermal well was taken as $500,000. Furthermore, using cost data from the same project for the flashing and distribution system ($3.6x10^6 for 15 holes) and a 0.73 power scaling law, the cost for the corresponding system for a single well turns out to be $500,000. The capital costs for the steam supply system then turn out to be $1 million. At 15 per cent annual fixed charges this will be equivalent to $150,000 in annual steam costs, for steam from a single well.

Taking the operating time for a typical well as 7000 hrs per year then the cost per tonne of flashed steam, c_{st}, will be:

$$c_{st} = \frac{1.5 \times 10^5}{7000 \, Q_s} = \frac{21.43}{Q_s} \qquad (1)$$

where Q_S is the hourly steam output of the well in tonnes. Using data from Fig. 1 one can then calculate steam prices as a function of temperature. These are shown in Table I and for the particular well characteristics used here the steam costs may be approximated by the empirical relationship.

$$c_{st}(t) \approx \frac{0.30}{\left[1 - \left(\frac{t-100}{97}\right)^3\right]^{1/3}} \qquad (2)$$

where $c_{st}(t)$ is in \$/tonne of steam and t in °C.

COST OF COMPRESSED STEAM.

Theoretical Considerations

Fig. 2 illustrates the path on a Mollier diagram of the compressed steam for an idealized case. Consider steam flashed at a temperature, t_1 and a pressure p_1. It is desired to transform it to saturated steam af a temperature, t_2, corresponding to a saturation pressure, p_2. The steam at t_1 is adiabatically compressed until it intersects the isobar, p_2, at a temperature, t_1'. The enthalpy increase will be:

$$\Delta h_c = h(p_2, t_1') - h(p_1, t_1) \qquad (3)$$

The optimum value of t_1 will depend upon the desired value of t_2 as well as the characteristics of the geothermal field in question.

It is proposed to compress the steam in a system such as is shown in Fig. 3. The optimum temperature for producing the steam to be compressed depends upon the particular pressure-yield characteristics of the field or borehole in question, as well as the desired terminal pressure. The optimum flashing temperature for the steam driving the compression turbine also depends upon the field characteristics. In the interest of simplicity it may be desirable to use the same flashing temperature for the steam to be compressed and the steam that is to furnish the compression work with a compromise on both optima. In this study, the optima will not be combined. As indicated in Fig. 3, the compression system will be driven by steam obtained in a two stage flashing process at 175°C and 120°C. For the field characteristics in question and a maximum desired saturation temperature of 300°C for the compressed steam, the optimum flashing temperature for the steam to be compressed was found by trial and error to be 190°C. Sample calculations pertaining to these temperatures will be presented here.

Cost of Steam for Compression Work

For the flashing temperatures of 175°C and 120°C for the steam to supply the compression work, the ratio of the steam outputs at the two temperatures will be 1:0.676. This may be determined by calculating the flashing fraction and from the water output of the well, as given by Fig. 1. Equation 2 applies to single flashing and attributes all the cost to the steam flashed at the higher temperature, the steam from the second flashing may be considered "free", giving an average steam cost of:

$$\bar{c}_{st} \approx \frac{0.30}{\left[1 - \left(\frac{175-100}{97}\right)^3\right]^{1/3}} \frac{1}{1.676} = \$.22/\text{tonne}$$

The average enthalphy change of the steam doing the compression work, considering a condensation temperature of about 50°C, will be

$$\Delta h_1 = \frac{h_g(175) - h_g(50) + 0.676\,(h_g(120) - h_g(50))}{1.676}$$

$$= \frac{2771 - 2591 + 0.676\,(2705 - 2591)}{1.676} = 153.4 \text{ kJ/kg}$$

Total Steam Costs

Considering a compression efficiency of 80% the cost of the compressing steam, c_{s1}, per tonne of compressed steam may be calculated as being:

$$c_{s1} = \frac{\Delta h_c}{\Delta h_1} \frac{1}{0.80} \bar{c}_{st} \qquad (4)$$

To get the total steam costs per tonne of compressed steam, i.e. the variable costs, the price of a tonne of saturated steam at 190°C must be added to the above value. From equation 2 this is found to be \$0.51/tonne. The values of Δh_c may then be determined from a Mollier chart as a function of the desired temperature t_2. The results are shown in Table II, where c_s indicates the total steam price..

Fixed Charges

The cost of a turbine-compressor facility and condenser was estimated by using data from the Krafla geothermal project in northern Iceland (a 60 MW facility), with appropriate scaling, and compressor costs estimated from the literature. The size of the facility, that was chosen, was 12 MW. Steam supply (i.e. steam to be compressed) for a larger facility would have to be furnished from more than one well. The cost of such a facility was estimated to be $6,000,000, which at 15% fixed charges and 7000 operating hours per year will give hourly costs of:

$$c_h = 0.15 \times 6,000,000/7000 = \$128.57/hr.$$

Assuming 80% compressor efficiency, the hourly capacity of the compressor will be:

$$Q_c(t_2) = \frac{3600 \times 0.8}{1000} \frac{P_c}{\Delta h_c(t_2)} = 2.88 \frac{P_c}{\Delta h_c(t_2)} \text{ tonnes/hr} \quad (5)$$

where P_c is the compressor power in kilowatts. Given the hourly fixed charges and values of h_c and c_s from Table II, one can calculate the fixed charges per tonne of compressed steam, c_{fc}, and the total cost of compressed steam. The results are shown in Table III, where the total cost is

$$C(t_2) = c_s(t_2) + c_{f_c}(t_2) = c_s(t_2) + \frac{128.57}{2.88} \frac{\Delta h_c(t_2)}{12,000} \quad (6)$$

The results are plotted in Fig. 4, together with the cost of singly flashed steam.

CONCLUSIONS

The cost per ton of steam generated by the burning of fossil fuels, although it depends somewhat upon geographical and political factors (price controls, etc.), is generally in the range from $2.80/tonne to $5.00/tonne. It may therefore be concluded that upgrading geothermal steam by compression is, of itself, economically as well as technically feasible. However, since many geothermal sites are far removed from resources and markets, the freight costs to and from the chemical plant must be balanced against the savings in energy costs, when consideration is being given to whether to utilize geothermal steam for any given process.

REFERENCES

1. Lindal, B., Industrial and other applications of geothermal energy, Geothermal Energy V. 12, p. 135 (1973)

2. Valfells A., Heavy water production with geothermal steam. U.N. Symposium on the Development and Utilization of Geothermal Resources, Pisa, Proceedings (1970).

3. Valfells, A., Optimal recovery of geothermal heat for process steam, U.N. Geothermal Symposium, San Francisco, Proceedings (1975).

ACKNOWLEDGEMENT

This work was carried our partly at Iowa State University, where it was supported by the Engineering Research Institute, and partly at the Science Institute of the University of Iceland, where it was supported by the Icelandic Ministry of Education.

TABLE I

Cost of Single Flashed Geothermal Steam

t_1 °C	Q_s tonnes/hr.	c_{st} $/tonne
100	72	0.30
120	70	0.31
140	68	0.32
160	64	0.33
180	54	0.40
190	42	0.51
197	0	∞

TABLE II

Total Steam Costs as a Function of Final Saturation Temperature

t_2	t_1'	Δh_c	c_{s1}	c_s
°C	°C	kJ/kg	$/tonne	$/tonne
225	267	139.56	0.25	.76
250	324	241.9	0.42	1.05
275	390	365.18	0.67	1.18
300	437	453.57	0.81	1.32

TABLE III

Total Cost of Compressed Steam

t_2	Δh_c	c_s	c_{fc}	C
°C	kJ/kg	$/tonne	$/tonne	6/tonne
225	139.56	.76	0.53	1.29
250	241.90	1.05	0.90	1.95
275	365.18	1.18	1.36	2.54
300	453.57	1.32	1.69	3.01

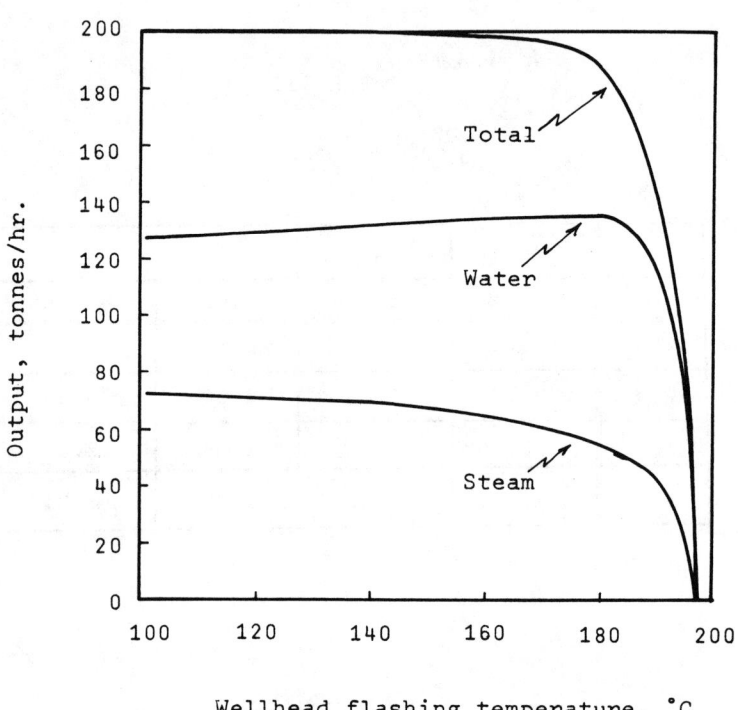

Fig. 1. Geothermal Well Characteristics.

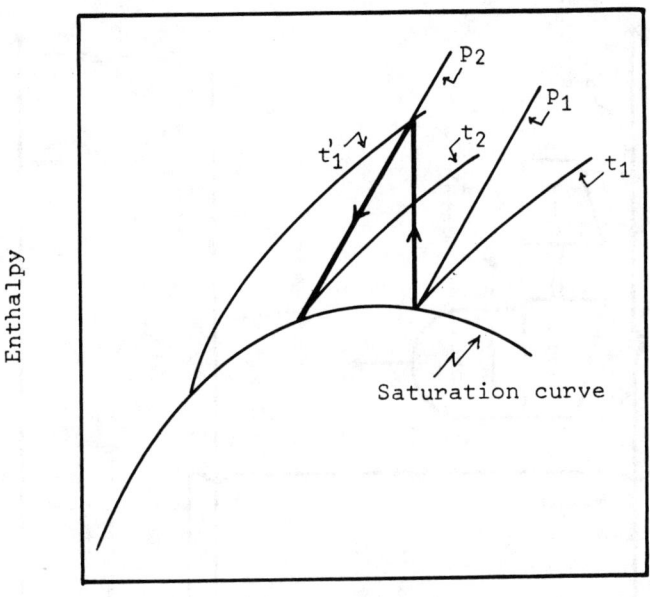

Fig. 2. Idealized Path on h-s Diagram for the Enhancement of Saturated Steam.

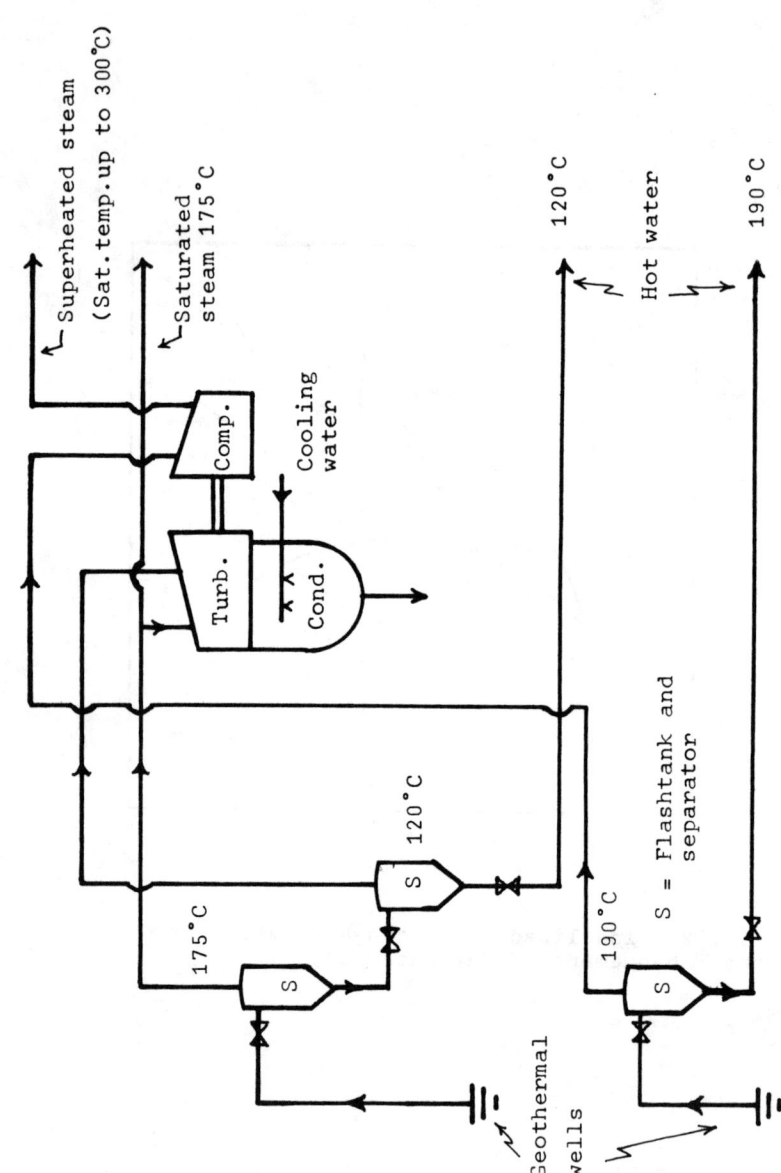

Fig. 3. Schematic Diagram of System.

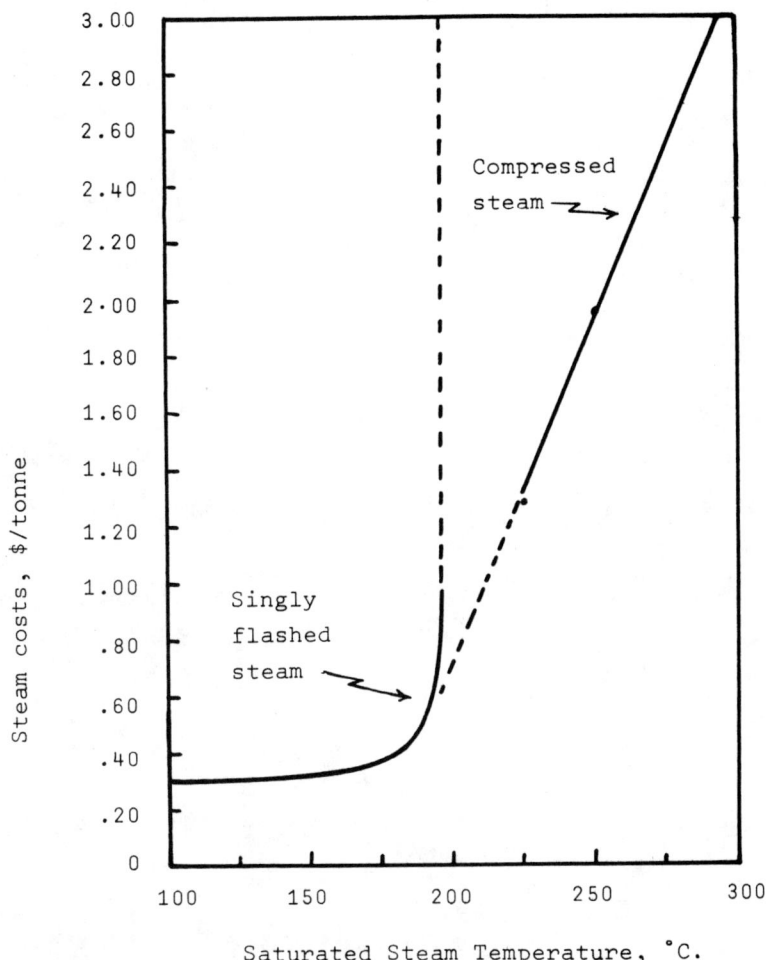

Fig. 4. Cost of Singly Flashed and Compressed Steam.

WATER-BASED ENERGY

Low-head Hydroelectric Power:
A Realizable Alternative

J. I. MILLS and G. L. SMITH
EG&G Idaho, Incorporated
Idaho Falls, Idaho 83401, USA

ABSTRACT

The utilization of hydraulic heads of less than 60 feet to generate hydroelectric power represents a substantial and potentially environmentally innocuous renewable energy source. Statistics indicate that at least an additional 105 million kilowatts of hydroelectric capacity with a corresponding average annual generation of 387 billion kilowatt-hours may have the potential for future development. These data were compiled for sites with a potential greater than 5000 kilowatts, and the potential additional contribution from smaller sites, while not yet fully documented, appears to be substantial. These resources have not been fully developed because of economic and institutional factors together with an historic dependence upon large, centralized steam power plants at the expense of smaller, decentralized, appropriate energy technologies. Changing resource and energy use patterns, together with dramatic socio-economic developments, imply a future rapid expansion of low-head hydropower resource utilization.

INTRODUCTION

The problems of energy availability and the increasing costs of energy have led to a national effort to develop economically and environmentally attractive alternative energy resources. One such alternative involves the utilization of the potential energy in hydraulic heads of less than 20 meters to generate hydroelectric power. This technology, referred to as low-head hydropower, has been relatively well developed in Europe, but in the United States there has been until recently no aggressive program to develop this substantial and potentially environmentally innocuous renewable energy resource.

This historic low rate of development for low-head hydroelectric in the United States is attributed to economic uncertainties, institutional constraints, environmental unknowns and an established energy economy depending upon rapid expansion of large and highly sophisticated centralized power generating stations while largely ignoring the potential of small, disperse, highly efficient and renewable energy sources. Energy planners have been hard-pressed to develop large blocks of centralized power -- either fossil or nuclear -- to meet rapidly growing load demand, and previous economic trends have favored larger plants. However, changing economic patterns together with an increased focus on the potential of smaller, renewable resource energy technologies have dramatically altered the competitive position of low-head hydroelectric power, and this technology can in many instances now compete with power from increasingly expensive fossil-fueled power plants. Many new low-head hydroelectric facilities are currently either under construction or being planned.

Low-head hydroelectric development offers many advantages as an alternative energy technology. Among the most important of these advantages are the facts that low-head plants are non-pollutive, dependent upon renewable resources and amenable to a variety of end uses. Also, low-head facilities can be made quite unobstructive, and unaesthetic perturbations to the national environment can be reduced to a minimum. In addition, the effects upon the natural stream ecology are minor compared to conventional hydroelectric facilities. Another important advantage is that low-head hydroelectric facilities are small and dispersed with each plant designed for maximum effectiveness in its particular environment and for its specific end-use. Low-head systems have a relatively short lead time from inception to on-line production, and this benefit results in greatly reduced exposure to the fluctuating economics and inaccurate demand forecasts that seem certain to plague our future energy planning. Finally, the technology associated with low-head hydroelectric plants is well-established, especially in Europe; and, since the components or technical resources required for the development of the technology are available throughout industry in the United States, the rapid establishment of an industrial base with the potential for mass production and the capacity to support the development of low-head hydroelectric reserves is assured.

The major purposes of this paper are to review the critical economic, environmental and institutional parameters affecting the utilization of low-head hydropower in the United States and to consider the integration of small, decentralized hydropower systems into the national energy economy. Recommendations focusing upon areas requiring additional research are included.

Low-Head Hydropower Resources

The potential low-head hydropower resources of the United States appear to be significant. Statistics [1] indicate that at least an additional 105 million kilowatts of hydroelectric capacity with a corresponding average annual generation of 387 billion kilowatt-hours may have the potential for future development. There data were compiled for sites with a potential specific capacity greater than 5000 kilowatts, and the potential additional contribution from smaller sites, while not yet fully documented, appears to be substantial. In addition to new developments, a recent survey by the U.S. Army Corps of Engineers indicates a total of 54.6 million kilowatts of capacity with a corresponding annual generation of 159.3 billion kilowatt-hours is possible from the rehabilitation, expansion and retrofit of existing dams. Figure 1 provides graphical documentation of these remaining hydroelectric power resources.

The estimates of remaining resources do not reflect potential constraints on development, however; and it should be stressed that economic, environmental, institutional, and geographical and climatological considerations can all impact dramatically on low-head hydropower resources and their development. Nevertheless, the potential for future development of low-head hydropower can still be seen to be significant, and every effort should be made to stimulate the utilization of this important resource.

The total U.S. electric generation capacity together with the percentages of this total contributed by individual technologies are illustrated in Figure 2. These totals include indefinitely postponed units and units scheduled as late as 1995. The percentages contributed by individual technologies, especially coal and nuclear, will no doubt change as a result of political and socio-economic factors as well as changing growth and use patterns. However, this illustration of present and projected future generation capacity does lead to some interesting conclusions.

Presently, hydroelectric power supplies approximately 13 percent of our total generating capacity, although by 1985 projections suggest that this percentage will fall to 10 percent. Currently, only 8 [2] percent of all our energy demand requires electricity for an end-use. The remaining 92 percent represents the energy requirements for mechanical motion and heat. Projections now suggest that by the year 2000, 20-40 percent of our end-use needs would be supplied by electricity if historical growth and technology patterns continue. However, even with this pessimistic scenario, our projected hydroelectric capacity, which excludes the potential low-head hydropower resources below 5 MW, would have the potential of generating, at greater than 90 percent efficiency, one-half to one-fourth of all of our electricity requirements. Hydroelectric power together with industrial inplant electric power generation, or cogeneration, could conceivably supply all of our electricity requirements by the last decade of this century. In 1950, 15 percent of the Nation's total electric power was supplied by cogeneration, although that figure has now dropped to approximately 5 percent [3]. West Germany, an industrialized nation similar in industrial output to the U.S., currently generates through cogeneration about 29 percent of the total electric power. A 1975 study by Paul McCracken suggests that U.S. industry could meet approximately 50 percent of its entire electricity requirements (compared to approximately one-seventh today) by cogeneration [4].

Clearly, it can be neither expected nor suggested that our entire energy economy be re-directed so that hydroelectric power and cogeneration become the exclusive suppliers of end-use electricity, although this might indeed result in the most efficient, economical and environmentally viable system of supply. However, the comparisons do show the significance of insuring that efficient generating facilities of 5 MW and below be given attention equal to that given to the central station steam power plant which is today at most 38 percent efficient.

The problems facing low-head hydropower development are not technological: they are social, economic, institutional and philosophical problems that stem from past associations with conventional, high-head hydroelectric power together with established management philosophies and changing economics-of-scale that have dictated a dedicated affair with large and inefficient centralized plants that too often repay their suitors with waste and pollution.

We are approaching in the United States an increasing awareness that small can be beautiful, and we are poised at the brink of an age that will no doubt see smaller, appropriate energy technologies contribute greatly to our economic survival and the maintenance of our quality of life.

Status of Low-Head Hydropower Technology

The technology associated with low-head hydroelectric developments is well-established, and various hardware options covering a wide range of parametric variations are available. Figure 3 is a comparative portrayal of some of the turbine types commonly used for low-head applications.

A major factor affecting the rapid and economic development of low-head hydropower has been the emphasis upon the very large hydroelectric facility as opposed to small or low-head units. Turbines, generators, and associated equipment designs have been developed for these larger plants; however, versatile, efficient and standardized designs appropriate for low-head installations have been largely ignored. The number of manufacturers of small hydroelectric units has declined from several hundred at the beginning of this century to only a very few. In Europe and China, however, the industry is quite vigorous. Indeed, in the book Water Power Development by Mosonyi (1960) there is an entire chapter devoted to small water power stations in which the specific capacity is less than 100 KW [5].

Increased demand created by vigorous development of resources will, of course, promote increased interest and activity, in the industrial community. In addition, the potential for further reducing the cost of low-head hydroelectric power together with busbar-energy costs for steam stations having risen from an average 15.09 mills/net KWhr in 1974 to an average 25.03 mills/net KWhr in 1977 [6] will further enhance the competitive posture of low-head hydroelectric power.

There are many areas where costs could potentially be reduced. The amount of custom engineering required for each specific site could be dramatically reduced - estimates range from 10 percent to 20 percent [7] through some degree of standardization of equipment and civil works. Turbine and generating equipment, together with the associated foundation walls, equipment embedment, intake and draft tube structures, closure bulkhead, and spillway crests, gates, piers, hoists and miscellaneous controls could be simplified and designed to be functional over a much broader range of site-specific hydraulic and geographical variations. The loss of efficiency resulting from standardization could be recovered by the utilization of automated control and monitoring systems which are presently available. The complexity and cost of these controls can be reduced, if the individual plant is part of an integrated complex of low-head and high-head hydropower facilities operating on a river system.

In addition to a need for standardization of the turbine, generator and associated equipment, development efforts designed to minimize the environmental impact of low-head installations would be beneficial. Specifically, more efficient and economical designs for fish ladders, together with alternative funding strategies, are required.

In summary, the primary objectives of future engineering development efforts should be to increase the economic attractiveness of low-head hydropower developments, to develop expanded applications for low-head hydro resources, and to establish the capability of developing marginal resources that may be optimized by technologies other than those presently available. This development effort should progress systematically from characterization of the resources and the evaluation of new concepts for energy utilizations to the testing of components, subsystems and processes, and finally to scaled testing of systems.

Legal and Institutional Constraints

Next to basic economic considerations, the major factor affecting the development of low-head hydropower resources in the United States is the licensing process. In fact, the length and expense of the licensing process is a major contributor to the less than optimum economic atmosphere presently affecting low-head hydropower development.

In order to simplify and expedite the licensing process, the most important objective must be to insure the separation of the requirements for licensing for large and small hydroelectric facilities. Existing safety requirement criteria include a separation of "large" and "small" dams based upon upper limits of 5 MW specific capacity, 10 meters of hydraulic head and 2000 acre feet of storage. For consistency with these present criteria and the basic definition of "low-head", it is recommended that streamlined licensing procedures be centered around these existing upper limits.

After separation of the licensing requirements for large and small dams has been achieved, it will then become necessary to develop a comprehensive, efficient licensing procedure for smaller, low-head facilities.

One of the more important requirements for casting an effective licensing procedure is to coordinate the overlapping and possibly conflicting authorities that are now involved. Currently, there is growing evidence of increasing interagency cooperation. This cooperation together with a shift of emphasis from more conventional hydropower to small and low-head developments should lead to a simplified and more efficient licensing process.

In addition to, and to compliment, the streamlined licensing procedure, the preparation of aregulatory guidebook to aid individual developers would be beneficial. A step-by-step "cookbook" approach to aid in determination of licensing requirements, procedures and sources of assistance, together with many case studies of low-head hydroelectric facilities covering a wide range of hydraulic, geographical and institutional parameters would appear to be the best approach to this regulatory guidebook. Such a guide would not only simplify, expedite and economize the licensing procedures, but it would reduce local investor uncertainty and provide confidence in the ability of the investor to satisfy regulatory requirements.

Low-Head Hydropower Environmental Considerations

One of the more salient features of low-head hydropower is the relatively minor environmental impact of the technology compared to many competing technologies and especially compared to conventional, high-head technologies. Aesthetically, low-head plants can be made very attractive and compatible with the surrounding environment. The trend to automatic control systems means that few maintenance personnel are required at any given time, and therefore, the human activity around the plant and the traffic to and from are held to a minimum. These facts, coupled with the lack of heat, smoke, or obtrusive noise levels, together with implementation of modern flow-over spillway designs, reduce the aesthetic impact of low-head facilities to little more than that of a bridge. In fact, in Europe some low-head facilities do serve as bridges for automobile, bicycle, and feet traffic.

By definition, the great majority of low-head hydropower developments will be run-of-the-river with very little associated water impoundments. In fact, it has been previously suggested that the definition of low-head include the stipulation that impoundments be no greater than 2000 acre feet. With this limitation in mind, the effects on river profile and the lands associated inundation of agricultural land, wildlife habitation, and lands with scenic or historic value is minimized.

In general, the environmental impact of a retrofit of turbines, generators and associated equipment to existing dams should represent less of an environmental impact than the development of new dams. For this reason, it is recommended that the environmental impact assessment requirements for retrofits be separated from the requirements for new facilities and that this separation be included explicitly in the licensing requirements.

Although the environmental impact of low-head hydropower developments is not significant, there are unique environmental constraints that warrant further investigation and creative solution. Alterations in stream morphology and the potential downstream impairment of a river's natural pollutant absorption mechanisms are questions that are important to address. In addition, the necessity of providing passage facilities for anadromous fish is an area of major concern, and designs incorporating more efficient and economical fish ladders into low-head civil structures would be a welcome contribution to the technology. Present costs for fish ladders range from $3,500 to $93,000 per vertical foot, and this expense is not easily borne by private developers.

There also exist environmental problems associated with expanding on developing transmission facilities for new small and low-head hydropower developments. The requirements for additional access roads as well as costs required for transmission lines and towers pose potential environmental perturbations that should be considered when assessing low-head hydropower sites.

To aid in identification and cataloging of environmental factors associated with low-head hydropower developments, it is recommended that regional assessments of environmental concerns and requirements be conducted in cooperation with local, state and regional government and non-government entities. These assessments should then be cataloged in report form and made available to potential developers.

To compliment these regional environmental assessments, standard reference files for dam records at state levels should be supported to facilitate developers and reviewers of development proposals in assessing and addressing the environmental and other considerations that are required to assess the feasibility of all potential low-head hydropower developments. These files would provide information detailing the hydroelectric generating capacity at all planned, under construction and completed projects. Details of capacity, average annual energy D/C ratio, site-specific institutional as well as environmental factors and other significant features would be referenced. In addition to those projects for which hydroelectric facilities exist or planned, the status of dams which have not been authorized for hydropower additions could be monitored.

Data provided by the general data library, the state-specific dam reference file and individual case studies of low-head developments could be normalized to eliminate or minimize the effects of large parametric variations from one project to another. Based upon this normalization, algorithms could be constructed to allow parametric comparison of a proposed development site with successful sites that have been previously developed, thus providing a general, initial assessment methodology for potential project feasibility.

After all environmental and safety problems associated with low-head hydropower have been identified it would then be helpful to prepare generalized environmental guidelines useful to developers contemplating a specific potential site. These guidelines could provide data on all environmental factors which should be assessed for purposes of initial feasibility determination. Site-specific examples from successful projects broadly distributed geographically could provide valuable base-line information to facilitate the environmental impact assessment procedure.

Low-head hydropower provides a unique opportunity to demonstrate the environmental viability of decentralized, appropriate technologies and every effort should be made by developers to work closely with public and governmental agencies in an attempt to insure minimal environmental impact. In this way, low-head hydropower can become a broadly accepted energy resource, and the potential contributions of this abundant technology to our national energy needs can be realized.

The Economics of Low-Head Hydropower

Presently, the costs of low-head hydropower developments range broadly from about $500 to almost $2000 per KW of installed capacity. This large variation in costs is due to large site-specific parametric variations that primarily affect the costs of reservoir impoundments, the degree of custom engineering required for each site, and the costs of feasibility assessment and licensing procurement. Table I illustrated the approximate costs of energy per KWH for various generation alternatives. The costs of coal, oil, nuclear, and combined cycles are from a report by Seymore Baron [8]. Continuing increases in fuel charges, to which low-head hydroelectric installations are immune, will doubtless continue to boost these energy costs to higher levels.

The initial costs of feasibility assessment and license procurement present a serious deterent to the development of small hydroelectric facilities. Funds expended for these initial expenses, with no guarantee that such expenditures will result in project development represent a "high-risk" investmant with little assurance of obtaining any return. Streamlined and more efficient feasibility assessment methodologies and licensing applications procedures would result in a far more favorable economic climate.

In addition to the standardization of the turbine, generator, and associated equipment, the more economical construction of fish ladders and the utilization of automated control systems, all of which have been discussed previously in this paper, there are other important factors which currently appear to impact negatively on low-head hydropower developemnt. An analysis should be made of the adequacy, availability and expense of liability insurance to determine whether or not current requirements impede the development of existing and new sites. Also, an analysis should procede to determine whether utility market price for the purchase of non-utility electric power production is equitable and sufficient to encourage small development. Finally, determination should be made on whether local utilization of hydroelectric resources has an overall measurable affect on utility production and whether the cost to be charged by a utility for providing stand-by power affects local economies.

Conclusions

Low-head and small hydroelectric power resources in the United States are substantial, and the technology exists to develop efficient and environmentally innocuous generating facilities. Uncertainties about the technology and its economics, expensive and complicated institutional requirements, and a previous dependancy on large, centralized blocks of power have all combined to hinder the development of low-head hydropower resources. However, changing economics together with greater awareness of the benefits of smaller, de-centralized appropriate technologies have resulted in a resurrection of interest in this unique, renewable energy resource.

In response to the growing interest in low-head hydropower in the public and private sector, the Division of Geothermal Energy of the United States Department of Energy has initiated a low-head hydroelectric development program to assist and encourage the acceleration of development of known resources, particularly at those sites where suitable dams currently exist. A second, longer-term objective is to encourage utilities to install hydroelectric generating plants at new dam sites suitable for low-head hydroelectric power production.

The cooperative efforts of the private and public sectors and the Department of Energy, together with changing political and socio-economic factors, seem certain to result in a major contribution, from low-head hydroelectric power to this country's efforts to establish a viable national energy plan.

REFERENCES

1. "Hydroelectric Power Resources of the United States", Federal Power Commission, Washington, D. C., 1976.

2. A. B. Lovins, "Energy Strategy: The Road Not Taken?" Foreign Affairs 55 (October 1976).

3. S. E. Nydick et al., "A Study of Inplant Electric Power Generation in the Chemical, Petroleum, Refining and Paper and Pulp Industries, Thermo Electron Corporation, Te5429-97-76,1976.

4. P. W. McCracken et al., "Industrial Energy Center Study", Dow Chemical Co., NSF, PB-243824, June 1975.

5. "Energy for Rural Development", National Academy of Sciences, Washington, D. C. 1976.

6. "20th Steam Station Cost Curvey", Electrical World, November 15, 1977.

7. "National Low-Head Hydropower Workshop", Sponsored by U. S. Department of Energy, Division of Geothermal Energy, Sept. 1977.

8. S. Baron, "Energy Cycles: Their Cost Interrelationship", Mechanical Engineering, June 1976.

TABLE I

Cost of Delivered Electric Power

Total Delivered Energy (Mills/KWH)	*Low-Head Hydropower	Oil	Coal	Coal-Gas	Coal-Liquid	LWR
	10-20	25.1	24.2	41.7	46.3	27.18

*Based upon limited data from operating or projected United States low-head hydroelectric power stations covering a range of 3 MW to 400 MW.

HYDROELECTRIC POWER RESOURCES

*TOTAL NATIONAL HYDROELECTRIC POWER
CAPACITY FOR SITES WITH CAPACITY
GREATER THAN 5,000 KILOWATTS   170.7 MILLION kW
   PRESENTLY DEVELOPED   57.0 MILLION kW
   UNDER CONSTRUCTION   8.2 MILLION kW
   REMAINING UNDEVELOPED   105.5 MILLION kW

†UNDEVELOPED POTENTIAL AT EXISTING DAMS
CONSIDERING UNIT CAPACITY OF LESS
THAN 5,000 KILOWATTS   26.6 MILLION kW

‡POTENTIAL FOR NEW DEVELOPMENT
CONSIDERING UNIT CAPACITY OF LESS
THAN 5,000 KILOWATTS   173.4 MILLION kW

        TOTAL UNDEVELOPED CAPACITY   305.5 MILLION kW

*FEDERAL POWER COMMISSION
†CORPS OF ENGINEERS
‡APPROXIMATE

Figure 1. Hydroelectric Resources

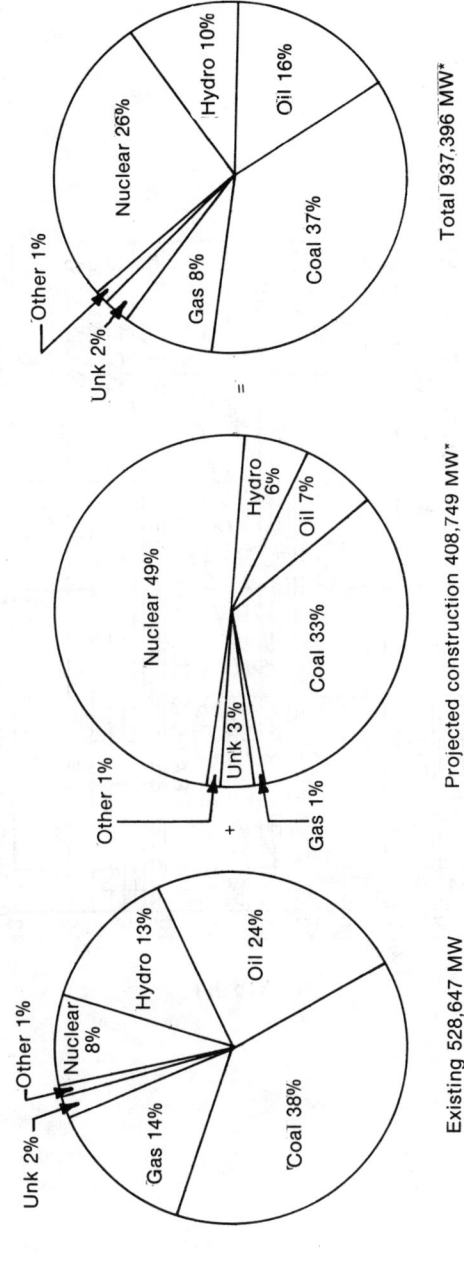

Figure 2. U.S. Generation Distribution

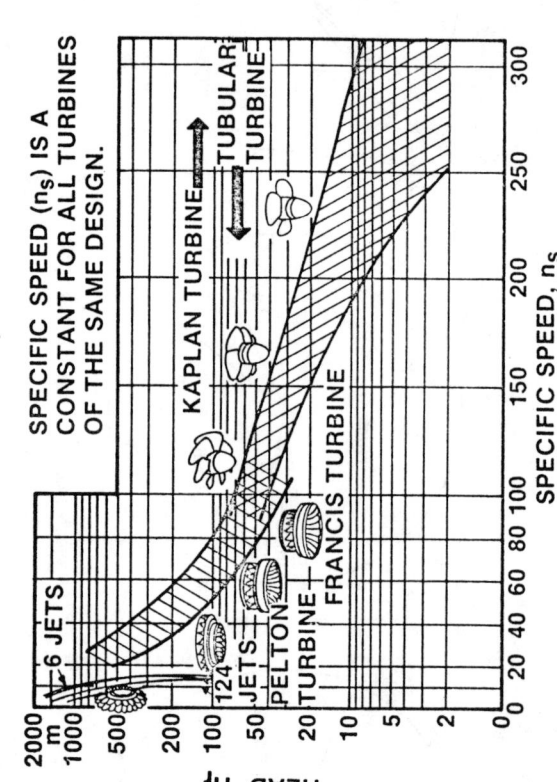

Figure 3. Turbine Utilization

Small Hydropower—Promise and Reality in New York State and the Northeast

R. S. BROWN and A. S. GOODMAN
Polytechnic Institute of New York
Brooklyn, New York, USA

ABSTRACT

President Carter's National Energy Plan and both Houses of Congress in pending legislation recognize hydroelectricity as offering new possibilities of substitution for oil, gas, and coal at local demand centers, especially at smaller existing dam sites. The prospects of expanding production capacity at old and new sites in New York State especially and the Northeast generally will be addressed in this paper. The authors are now in the midst of preparing a state hydropower development and expansion plan under the auspices of New York's Energy Research and Development Authority. Their earlier regional assessment was prepared for the Hydro-Turbine Division of Allis-Chalmers which has since entered the market with several standard lines of turbines for smaller sites. Progress on the New York State assessment and plan will be reported.

INTRODUCTION

Rising energy prices and growing constraints upon all forms of energy production have occasioned a review of old energy producing facilities to see if changed economics, new fuel substitution possibilities, and more modern technologies might make them viable once again. An example might be thermal power plants once fired by coal which shifted to oil or gas ten years ago which are now being considered for use of coal once again. Old abandoned hydroelectric sites are considered here. Also considered are existing dams which have never produced hydroelectric power, existing dams now producing electric power which can be upgraded and new smaller dams. It is assumed that the large sites in the Northeast which were more economical earlier have mostly been developed.

THE BASIC ISSUE

The basic issue is whether the hydroelectric potential of these sites should be ignored or the sites should be revived or adapted for modern levels of hydroelectric production. The authors and their research team have been addressing this issue for several clients, including the largest hydro-turbine manufacturer in the United States, the Allis-Chalmers Hydro-Turbine Division, and the State Government of New York. The geographical areas examined include the nine Northeastern States broadly and the New York State more specifically.

WHY DEAL WITH THE REVIVAL OF SMALL HYDROPOWER NOW?

Recent steep price rises in fossil fuels, increasing opposition and delays in developing nuclear plants, tightened environmental controls affecting thermal power plants, and heightened political desire to reduce import dependency are among the quoted reasons for now exploring the revival of small hydropower. Pinched earlier by these constraints the Europeans are a half decade ahead of the U.S. in applying more efficient, smaller scale technologies in retrofit and upgrading situations. By 1975 and 1976 energy planning authorities in New England and New York State were actively reviewing prospects for expanding all indigenous sources of energy to reduce dependency on outsiders. In 1977 serious assessments of hydropower potential got underway under the aegis respectively of the New England River Basin Commission [1] and the New York State Energy Research and Development Authority [2].

Heightened Enthusiasm For Using Existing Dams For Hydroelectric Purposes

Enthusiasm for reviving old dams in the United States has been greatly stimulated in 1976 and 1977 by publicity emanating from a few members of the Congress and Senate and a few private individuals, notably David E. Lilienthal former head of the Tennessee Valley Authority (TVA) and first Chairman of the Atomic Energy Commission [3]. Both nostalgia and a sense of hope have combined to create a substantial upwelling of interest among potential developers and buffs of all stripes. President Carter's 1977 National Energy Plan in April generated a 90 day study of the potential at 49,000+ dams identified earlier in the 1975 Corps of Engineers national dam safety inventory [4]. A $10 million Low-head Hydroelectric Development Program was approved in the Fiscal Year 1978 budget for the Energy Research and Development Administration which has now been absorbed into the new Department of Energy. Separate pieces of legislation stimulating small hydropower development passed the House of Representatives (August 5, 1977) [5] and the Senate (October 5, 1977) [6]. These measures were still in Conference on December 1st.

DEFINITION OF SMALL LOW-HEAD HYDROPOWER

The first definitions of small hydropower used in the recent revival of interest were keyed to sites with very small potential installed capacity, often under 500 kilowatt or .5 megawatt installed capacity. Then the 5,000 kw figure achieved currency. More recently the pending legislation has embodied a 20,000 horsepower or 15,000 kw (15 mw) standard which would amend the Federal Power Act and would constitute for licensing purposes a new dividing line between major and minor projects. The working definition of "small" utilized at the first national small/ low-head hydropower workshop organized by USERDA in September 1977 used a bottom cut-off of 50 kw as suggested from our earlier work. The working definition of "low-head" was below 58 feet (18 meters) [7].

GEOTHERMAL ENERGY AND HYDROPOWER

ESTIMATES OF POTENTIAL

The Federal Power Commission Estimates

Prior to the report of the 90 day inventory requested by President Carter [8] the most quoted estimates of potential installed hydroelectric capacity were those derived from the Federal Power Commission's quadrennial report for 1976, Inventory of Hydroelectric Resources of the United States (Developed and Undeveloped) [9]. In early 1977 Ronald Corso, Deputy Chief of F.P.C.'s Bureau of Licensing, was saying the nation's developed capacity as 66,000 mw and the undeveloped capacity as 47,000 mw [10]. Corso further estimated if 10 percent of the dams identified in the 49,000 dam safety inventory could be developed for power at only 1000 kw each and operated at a 50 percent plant factor the installed capacity would be 4900 mw producing 22 billion kwh annually. This equals 36 million barrels of oil. However, Corso expected the average size of small developments to be more in the range of 5,000 kw, totalling then as much as 24,500 mw.

An independent estimate using 1976 inventory data was later prepared by A.E. Allen, Harza Engineering of Chicago [11]. Allen used 30 mw as his dividing line for distributing the 1976 list between small and large undeveloped projects. He counted 996 undeveloped projects under 30 mw, including 417 below 10 mw in the nation.

Regionalized to the Northeast the 1976 FPC Inventory showed that the hydroelectric potential was 43.1% developed with 7,572,241 kw (7,572 mw) left to develop at known sites (with or without existing dams). This is elaborated in Table I. Allen counted 170 projects below 30 mw and 81 below 10 mw in the drainage basins of the Northeastern nine states.

A major problem with the F.P.C. data relates to its reliability. In our inquiries we have discovered that most of the back-up data for individual undeveloped sites in the Northeast was discarded years ago and that the published numbers have been carried routinely from one quadrennial report to the next. This leads to a fundamental observation to which we will return repeatedly. Site characteristics change over time and we have found that data in published lists often no longer represent current reality.

The Corps of Engineers 90-Day Report Estimates

The Institute of Water Resources of the Corps of Engineers in preparing the National Inventory of Hydroelectric Resources at Existing Dams made aggregate estimates by river basins and did not examine individual sites. Their estimates were admittedly soft but were considered by them to be a fair approximation of average developable potential. The I.W.R. estimates for the nation are summarized on Table II and the Northeast in Table III.

The Polytechnic Institute of New York Estimate For The Northeastern United States

The Polytechnic approach concentrated on individual sites and it is this approach which we would to discuss here.

At the request of the Allis-Chalmers Hydro-Turbine Division in York, Pennsylvania in February, 1977, the Center for Regional Technology at the Polytechnic Institute of New York prepared a regional marketing assessment of potential for hydro-power development at existing low-head dams in the Northeastern United States [12]. The Center for Regional Technology was joined in this overall task by the Institute's Department of Civil and Environmental Engineering and during the initial phase by the firm of Tippetts-Abbett-McCarthy-Stratton. Nine states are included: Maine, New Hampshire, Vermont, Massachusetts, Rhode Island, Connecticut, New York, New Jersey and Pennsylvania.

After office and map studies reviewed approximately 5,300 existing dams with useful information, the site inventory (Table IV) yielded approximately 1,600 existing low-head dams which are estimated to be capable of producing hydroelectric power within the range of 50 to 5,000 kilowatt (kw) potential. Heads of 10 to 45 feet were examined. A sampling of 25 representative dam sites were examined in more detail, yielding 12 with average heads of 22 feet and power estimates of 1,150 kw which were deemed to have excellent or good prospects for hydroelectric development. The average power estimate would be about 1,250 kw if one site of 78 kw potential were discounted. If this percentage of developable sites holds true then approximately 750 low-head dams could be candidates for hydroelectric development.

THE IMPERFECT ART OF MAKING ESTIMATES

To estimate the potential capacity to generate hydroelectric power it is our firm belief that individual sites with or without existing dams must be dealt with separately as soon as possible. Office searches and field screening that are site-specific must address at least four possibilities for development: uprating existing hydroelectric units; additions to existing hydroelectric plants; all new facilities at existing or restorable dams; and low-head hydroelectric power plants. Without site-specific estimates the aggregation of estimates is at best hazardous, yet it is being done and must be done in some form to meet planning and political needs. Therefore, let us turn to the development of information about sites.

INFORMATION SOURCES

Generalized projections, such as those presented in the U.S. Army Corps of Engineers, Estimate of National Hydroelectric Power Potential at Existing Dams, July 1977, are of little immediate use to potential developers. Of more use is the Corps of Engineers report on the National Program of Inspection of Dams [13]. It is especially useful in those states, such as New York, where the State's own dam safety inspectors generated the basic information on contract for the Corps. Of general use to

us was a report by the New England-New York Interagency Commission
(NENYIAC) on the resources of the New England-New York Region prepared in
1955 [14]. Most States also have their own appraisals prepared over time
which are more or less useful. A notable example is a 1952 Inventory of
Dams in Vermont taken from the Biennial Report of the Vermont Public
Service Commission [15]. Much more work needs to be done to make current
information on sites available to potential developers. Comprehensive
information is being developed within two years by the New England River
Basin Commission for six States and within one year by the New York State
Energy Research and Development Authority. In short, the current rush to
explore development opportunities is almost outpacing the availability of
comprehensive and specific information on sites. The latest Department of
Energy hydroelectric allocations for FY 1978 reduce emphasis on resource
assessment and shift it to feasibility studies for demonstrations. Both
a good and a bad move, this may cause only immediately visible sites to
receive attention. For the commercial market to flourish informational
resources which are accessible to the public must be strengthened, probably at the State level. This lasting resource should not be overlooked
in the rush to develop one or only a few Federally supported demonstrations in a given State.

THE NORTHEASTERN STATES

Initial Screening For The Northeastern States

To prepare our initial inventory of existing dams in the nine Northeastern
States it was personally necessary for our team to go to the office files
of the individual states where very little data is in reproducible form.
For the Allis-Chalmer's project the initial inventory was limited to dams
with a height of 10 to 45 feet and a power potential of 50 to 5,000 kw.
The power potential is a function of the operating head and the discharge
through the turbine.

It was assumed, for this study, that the operating head is equivalent to
the height of the dam. This could vary widely with individual dams and
an on-site survey would be required to determine the actual head. Generally, the operating head would be equal to the distance between headwater
or tailwater or less than the height, but it could be further increased by
extending the draft tube or by using diversion facilities to take advantage
of the lower downstream water levels.

The discharge is generally a function of the drainage area, and for this
study, the discharge for defining installed safety was assumed to be 2 cfs
per square mile. This is also variable and will require more detailed
studies for each site. This discharge relationship was shown graphically
on flow-duration curves for twenty-five selected sites and reproduced on
field report sheets, such as the one attached as Figure 1.

Secondary Screening For The Northeastern States

In the second phase of the study a limited number of sites were selected for further study using the following additional criteria:

1. The site should be near a road to minimize costs of construction access.

2. The site should be near a community or demand center to minimize the cost of transmission.

3. Dams with private or local ownership would be preferred to governmental ownership because of fewer probable development constraints.

4. Flood control and public water supply dams would not be preferred because of possible operational constraints.

The varied characteristics of the 25 sites inspected in the field and the information obtained in the State files and from discussions with personnel in each State yielded a good sampling of the opportunities and problems for small hydroelectric development that are present at existing low-head dams.

Of the 25 sites studied, 12 proved to be governmentally owned, including seven at the municipal or regional level, four by state agencies, and one by a Federal agency (U.S. Army Corps of Engineers). Originally the first office screening of the available ownership records indicated that only six were governmentally owned. Six were associated with local parks, restaurants, real estate developments, summer camps or tourist attractions. Thirteen turned out to be privately owned.

State personnel often suggested that our criteria did not always lead to sites that were as suitable for early development as others they could suggest. We appreciate this and recognize the need for developers to talk with informed State officials at an early time and examine their files. Nothing beats specific discussions and file information although in most States these are spotty at best. The lack of up-to-date centralized information on dams has become painfully apparent in the wake of recent dam disasters. Particularly difficult information to obtain is that concerning ownership and water rights. We often had to turn to the Township's Tax Assessors Office to identify current owners.

More detailed criteria were established for evaluating each site and these are outlined in Table V. In general all or several of the conditions were met at a given site. As mentioned earlier 12 sites with average heads of 22 feet and power estimates of 1,150 kw were deemed to have excellent or good prospects for hydroelectric development.

Presentation of Results For The Northeastern States

The results of the site studies for the Northeast were presented to the client for its own uses in the following form.

1. A general report was prepared that addressed three things: the overview of the project; who are the potential developers by classes; and how Federal actions affect site development.

2. A field reconnaissance review and evaluation.

3. Three appendices containing field reports, site photographs (using color xeroxy of composite photographs), and office file materials.

NEW YORK STATE

The New York State Energy Research and Development Authority aided by the Power Authority of the State of New York has commissioned a $200,000 study of prospects for increasing the state's hydroelectric production. This study concentrates on the potential for small hydropower plants at old and new sites, but does not exclude larger sites where appropriate. This includes sites which formerly produced power or were used for other purposes and those not yet developed.

Emphasis is placed on sites with production potential in the 50 to 15,000 kw range. The lower limit suggests a minimum drainage area of about ten square miles. The 15,000 kw figure reflects the pending changes in licensing requirements that have already been agreed upon in separate, but related measures passed by the House of Representatives and the Senate. After an inventory is prepared by means of initial and secondary screening procedures to be described below, up to 20 representative sites will be examined in pre-feasibility studies. The screening procedures are now underway.

Initial Screening - New York State

The initial data source evaluated for New York State was the State Department of Environmental Conservation's (DEC) set of more than 6,000 files on sites where dams have existed at some point since the early 1900's. This mass of data, including many empty files, had been created for a variety of purposes and is maintained principally for dam safety reasons. A small office with a staff of two inspects as many dams as it can each year but analysis of file information of other purposes can only receive lesser priority. The staff of the Dam Section, Bureau of Facilities and Construction Management, has been very helpful to our research team.

Three years ago in connection with the National Program of Inspection of Dams, the Corps of Engineers subcontracted with DEC for a listing of information for approximately 850 dams that meet the following minimum criteria:

Dam Height	Reservoir Volume	Disposition
less than 6'		None included
6' - 25'	more than 50 ac. ft.	included
more than 25'	more than 15 ac. ft.	included

After initial examination of the condition of DEC files it was judged that:

1. The DEC files for over 6000 dams could not be screened in a reasonable period of time, and the data obtained from such a screening would be rather incomplete for individual projects.

2. It was unlikely that important projects would fail to meet the dam criteria established by the Corps of Engineers and very few smaller projects with existing or potential capacity of 50 kw or greater would fail to meet the criteria.

It was decided, therefore, to concentrate on the approximately 850 dams included in the Corps of Engineers inventory. Data for these dams were obtained from the Corps of Engineers inventory, the files of the DEC, and various other sources.

The approximately 850 dams were further reduced to approximately 250 dams by applying an additional criterion of a minimum drainage area of 10 square miles. As described below an approximate estimate of installed capacity to generate at a flow exceeded 25 percent of the time is as follows:

$$(1) \quad P = \frac{H \times DA}{7}$$

For 50 kw minimum capacity, the corresponding head for a 10 square mile drainage area would be as follows:

$$(2) \quad H = \frac{50 \times 7}{10} = 35 \text{ feet}$$

It was judged that not many sites of interest would fail to meet both the Corps of Engineers criteria and the 10 square mile minimum. A large number of drainage areas had to be determined by outlining drainage areas on U.S. Geological Survey Quadrangle sheets and planimetering.

It has been determined by inspection, that some dams with potential power projects may not be included in the 250 dam list. The assistance of the power utility staff and other individuals involved in the N.Y. State hydropower assessment is being obtained to identify missing projects and furnish data for them.

Next we are preparing tentative and approximate estimates of potential values of capacity, energy, and costs for the 250 sites based upon simplifying assumptions. The following assumptions would be revised as we obtain site specific head and flow data:

H = Head (ft.) - height of dam.

Q_C = Discharge (cfs) - for installed capacity determination, 25 percent availability at 2 cfs/sq mi of drainage area.

Q_A = Discharge (cfs) - for energy determination, at cf/sq mi.

P = Installed capacity (KW), assuming 88% turbine efficiency and 95% generator efficiency.

E = Energy (KWH)

$$(3) \quad E = \frac{P \times 8760}{2}$$

Capacity and energy costs would be determined using simplified assumptions concerning cost of equipment, civil works, etc. such as developed by E. O'Brien of TAMS, New York [16]. Some of his typical capacity and minimum energy cost curves for small hydroplants in the NOrtheastern U.S. are shown in Figure 2.

A selection of approximately 50 sites is being made from the 250 sites analyzed above. This selection involving the following factors:

Power Utility Service Area;
Capacity from 50 to 15,000 KW;
Political Subdivision;
Existing or potential power plants;
Costs of capacity and energy;
Site location with respect to access, transmission lines, and power markets;
Site location with respect to state park, wild river, or other area protected from development;
Current status of studies by power utility, promotor, or other entity.

In addition a variety of sources are being examined for the inventory of potential at sites without dams.

Initial field visits and telephone inquiries have also occurred, especially when considerable interest has been shown in the early development of a particular site. A unique class of sites has emerged from these field contacts, sites with breached or removed dams. They show on many records as existing dams but heavy partial or total replacement may be mandatory before any use can be made for the present or any other purpose.

Secondary Screening - New York State

The characteristics and costs of the 50+ projects with existing dams are being considered in a secondary screening process. Engineering aspects, socio-economic considerations (owner's plans, etc.), legal and environmental considerations are all being employed. There is no good substitute for a combination of (a) actual site plans, (b) telephone inquiries and (c) site visits. These are being sought throughout the project although most visits will occur in later stages.

From the engineering standpoint a variety of data is having to be developed or assumed. USGS quadrangle sheets are sought and used for each site. More accurate information and specific assumptions on head, flow, power, access, transmission, generating facilities, and civil works will be made.

It is important to note that this secondary screening is geared not to the identification of the best sites for further examination in feasibility studies. Instead it is to identify a representative sampling of sites from which extrapolation can be made to the entire population of sites in New York State.

Conceptual Development and Feasibility Assessment

During the second phase of the New York State project an evaluation will be prepared of the potential for increasing hydroelectric production in New York State, including overall benefits of small hydropower projects (economic, fuel substitution, dam safety, environmental enhancement, etc.).

The earlier planning investigations will yield ten to twenty sites which will be reviewed in considerably greater detail. Conceptual designs and associated work will result in better estimates of the unit cost of producing power. Greater emphasis be placed on the socio-economic, political, legal, institutional and environmental problems involved in development of specific sites. This phase will also include an evaluation of the potential of multipurpose developments associated with representative sites, such as industrial parks, commercial centers, rural development projects, recreation, and water supply.

Follow-on Activities in New York

One important component of this project will be to help developers of New York State sites prepare projects for possible State and Federal funding through full-scale feasibility studies and demonstration projects.

THE NATIONAL IMPACT OF SMALL HYDROPOWER DEVELOPMENT IN THE NORTHEAST AND NEW YORK STATE

Much of the nation's undeveloped or abandoned hydroelectric potential at smaller sites exists in the Northeast. Pending federal assistance programs should place substantial emphasis on this region and the Mid-Atlantic region. Development of sites in the Northeast should merit early attention because the rising costs of alternative sources of energy should make small hydropower competitive sooner. This is especially true if the OPEC nations raise the price of oil 10-25% in 1978 and achieve a doubling of prices in the 80's, as has been recently reported.

In conclusion, compromise legislation for small hydropower development was approved late last week by the Energy Conferees in Washington, D.C. It will take the form of a combination of feasibility studies and loans funded in equal annual shares over three years with a total program of about $30 million for studies and $300 million for loans. We hope it will be put to good use.

GEOTHERMAL ENERGY AND HYDROPOWER

REFERENCES

1. New England River Basin Proposal "A Special Study to Investigate the Feasibility and Implications of Hydropower Expansion in New England," May 1976.

2. Proposal to NYS ERDA by the Polytechnic Institute of New York, "Hydropower Restoration and Expansion in New York State," May 1977.

3. "MacNeil/Lehrner Reports on Public Broadcasting Service," Lilienthal (May 27, 1977), Representative Richard Ottinger (August 8, 1977), Piece by Lilienthal in Smithsonian (July 1976, Vol. 7, No. 4, Pg. 108 and September 1977) and the New York Times (OP-ED Page, December 28, 1976.

4. National Energy Plan by Executive Office of the President Energy Policy and Planning (August 29, 1977, Pg. 73).

5. U.S. House of Representatives 95th Congress 1st Session Bill H.R. 8444 In the Senate of the United States, September 7, 1977, Chap. 6, Small Hydroelectric Natural Energy Power Projects introduced by the Honorable Richard L. Ottinger.

6. U.S. House of Representatives 95th Congress 1st Session Bill H.R. 4018 In the Senate of the United States October 6, 1977 Public Utilities Regulatory Act Chap. 6 Small Hydroelectric Power Projects introduced by the Honorable John A. Durkin.

7. "Small Hydro Workshop, New England Center in Durham, New Hampshire (September 6-9, 1977) Sponsored by US ERDA". Panel Reports published by Aeta Corp.

8. U.S. Army Corps of Engineers Estimate of National Hydroelectric Power Potential at Existing Dams, July 1977, Institute for Water Resources.

9. Hydroelectric Power Resources of the United States Developed and Undeveloped, January 1, 1976, Federal Power Commission.

10. Ronald A. Corso, "A New Look at Hydropower," Federal Power Commission.

11. Allen, A.E., Harza Engineering Company, Chicago, Illinois, "Development of Small Hydroelectric Plants," Summary of Remarks for the Conference at University of New Hampshire, September 6-9, 1977.

12. Brown, Ruben S., Director Center for Regional Technology, Polytechnic Institute of New York, "Potential Hydro-Power Development at Existing Low-Head Dams in the Northeastern United States," August 31, 1977, prepared for Allis-Chalmers, Hydro-Turbine Division.

13. National Program of Inspection of Dams, Vol. IV, May 1975 by Dept. of Army Office of the Chief of Engineers, Washington, D.C. 20314.

14. <u>The Resources of the New England-New York Region</u>, New England-New York Interagency Commission (NENYIAC), 1955.

15. <u>1952 Inventory of Dams in Vermont</u> from the Bienneal Report of Vermont (Public Service Commission).

16. O'Brien, Eugene, "Small Hydroplants for the Northeast," <u>Electrical World</u>, August 15, 1977, p. 61-2.

TABLE I

SUMMARY* OF DEVELOPED AND UNDEVELOPED HYDROELECTRIC POTENTIAL

JANUARY 1, 1976

Geographic Division and State	Developed			Undeveloped			Total Potential		Percent Un-developed
	Installed Capacity Kw	Average Annual Generation 1000 Kw	Installed Capacity Kw H	Average Annual Generation 1000 KwH	Installed Capacity Kw	Average Annual Generating Capacity 1000 KwH	Installed Capacity Kw	Average Annual Generating Capacity 1000 KwH	
United States	57,036,438	271,124,113	112,259,510	401,067,925	169,295,948	672,192,038	66.3%
Northeast	5,738,201	31,501,128	7,362,991	19,663,519	13,101,192	51,164,647	56.2%
New England	1,506,462	6,046,791	3,274,231	7,542,441	4,780,693	13,589,232	68.5%
Maine	536,497	2,739,862	1,704,403	4,421,541	2,240,900	7,161,403	76.1%
New Hampshire	417,992	1,250,980	683,100	998,500	1,101,092	2,249,480	62.0%
Vermont	198,308	827,400	433,350	816,500	631,658	1,643,900	68.6%
Massachusetts	233,245	836,129	270,178	753,500	493,423	1,589,629	54.8%
Rhode Island	1,500	4,000	0	0	1,500	4,000	0
Connecticut	128,920	388,420	183,200	552,400	312,120	940,820	58.7%
Middle Atlantic	4,231,739	25,454,337	4,088,760	12,121,078	8,320,499	37,575,415	49.1%
New York	3,779,775	23,651,645	1,286,160	3,253,800	5,065,935	26,905,445	25.4%
New Jersey	3,564	15,400	265,000	1,001,500	268,564	1,016,900	98.7%
Pennsylvania	448,400	1,787,292	2,537,600	7,925,773	2,986,000	9,713,065	84.9%

*Constructed from Draft Table 4 of the 1976 Federal Power Commission Report on the Hydroelectric Power Resources of the United States

Center for
Regional Technology
Polytechnic Institute of
New York
February 1, 1977

TABLE II

CONVENTIONAL HYDROELECTRIC CAPACITY POTENTIAL AT EXISTING DAMS

	Existing Capacity (MW)	Rehabilitation Potential (MW)(9%)	Hydro Expansion Potential (MW)	Hydro Installation Potential (MW)	Small Dam Potential (MW)	Total Regional Potential (MW)
New England	1,427	127	188	223	2,432	2,970
Mid-Atlantic	1,290	116	565	521	5,580	6,782
South Atlantic-Gulf	5,753	518	3,342	874	4,244	8,978
Great Lakes	4,008	360	253	143	644	1,400
Ohio River	1,465	132	19	1,414	1,873	3,438
Tennessee River	3,658	329	30	0	75	434
Upper Mississippi River	581	52	80	199	4,378	4,709
Lower Mississippi River	724	18	88	25	2,582	2,713
Hudson Bay	13	1	0	0	51	52
Arkansas-White-Red River	1,839	165	236	245	2,318	2,964
Texas-Gulf	393	35	11	43	460	549
Missouri River	3,370	303	1,037	486	250	2,076
Rio Grande River	65	6	20	130	184	340
Upper Colorado River	359	32	15	24	465	536
Lower Colorado River	2,847	256	0	59	87	402
Great Basin	530	48	1	0	85	134
Columbia-North Pacific	22,342	2,010	10,681	628	757	14,076
California-South Pacific	7,050	634	970	514	*	2,118
Alaska	123	11	36	46	25	118
Hawaii	18	2	1	0	30	33
Puerto Rico	0	0	0	0	10	10
Virgin Islands	0	0	0	0	0	0
TOTALS	57,855	5,155	17,573	5,574	26,530	54,832

*No estimate was available for the California-South Pacific Region

Prepared by U.S. Army Corps. of Engineers - Institute for Water Resources, Estimate of National Hydroelectric Power Potential at Existing Dams
July, 1977 - p. 8

TABLE III

CONVENTIONAL HYDROELECTRIC ENERGY YIELD POTENTIAL AT EXISTING DAMS

	Existing Energy Yield (10⁶ KWH)	Rehabilitation Potential (9%)	Hydro Expansion Potential	Hydro Installation Potential	Small Dam Potential	Total Regional Potential (10⁶ KWH)
New England	5,719	515	502	517	11,685	13,219
Mid-Atlantic	5,201	477	1,945	792	15,279	18,493
South Atlantic-Gulf	14,521	1,307	9,228	1,255	21,846	33,636
Great Lakes	24,754	2,228	578	580	2,423	5,809
Ohio River	5,505	495	100	4,680	2,849	8,124
Tennessee River	16,112	1,450	139	0	371	1,960
Upper Mississippi River	3,006	270	260	1,077	8,991	10,598
Lower Mississippi River	424	38	136	104	8,110	8,388
Hudson Bay	68	6	0	0	72	78
Arkansas-White-Red River	5,019	452	269	322	5,525	7,568
Texas-Gulf	1,074	97	6	61	1,095	1,259
Missouri River	15,294	1,376	4,632	1,106	580	7,694
Rio Grande	234	21	49	311	788	1,169
Upper Colorado River	1,634	147	80	49	593	869
Lower Colorado River	10,541	570	1	289	526	1,394
Great Basin	1,976	178	0	0	179	358
Columbia-North Pacific	127,182	11,446	14,865	2,949	2,495	31,755
California-South Pacific	33,400	3,006	3,264	2,304	*	8,574
Alaska	493	44	141	239	112	536
Hawaii	104	9	8	0	40	57
Puerto Rico	0	0	0	0	129	129
Virgin Islands	0	0	0	0	0	0
TOTALS	272,261	24,141	36,203	16,635	84,688	161,667

*No estimate was available for the California-South Pacific Region

Prepared by U.S. Army Corps. of Engineers - Institute for Water Resources, Estimate of National Hydroelectric Power Potential at Existing Dams, July 1977, p. 9

TABLE IV

POTENTIAL HYDRO-POWER DEVELOPMENT EXISTING LOW-HEAD DAMS IN NORTHEAST

INVENTORY OF EXISTING DAMS

State	Total No. of Dams Listed by State	No. of Abandoned Power Sites Identified	No. with Useful Information Reviewed	No. Suitable for Power Development	No. Selected for Further Study
Pennsylvania	2,324	N.A.	2,324	95	5
New Jersey	1,129	16	632	28	4
New York	6,352	200	108	200	2
Connecticut	3,522	N.A.	367	59	2
Rhode Island	521	82	83	50	3
Massachusetts	2,704	222	270	190	2
Vermont	355	100	314	92	3
New Hampshire	3,000	293	182	97	2
Maine	1,010	N.A.	1,010	800	2
	20,917	913+	5,290	1,611	25

*Estimate developed by PINY/TAMS from reliable but incomplete data sources.

Prepared by Polytechnic Institute of New York and
Tippetts-Abbett-McCarthy-Stratton
30 April 1977

TABLE V

ESTABLISHING CRITERIA FOR SITE EVALUATION

1. **Excellent** Opportunities for Development:

A. Modern dam and structures, or structures appear in excellent condition
B. Appurtenant structures appear sound and in good condition
C. Public access to site is limited and can be limited
D. No conflicting uses
E. Sizeable drainage area and head
F. No floodway problem
G. Adequate space for hydropower installation

2. **Good** Opportunities for Development;

A. Site structures in good condition
B. Public access to the site is limited
C. Appurtenant structures are compact with respect to the site and in good condition
D. Previous use for hydropower
E. No conflicting uses
F. No floodway problem
G. Adequate space for hydropower development
H. Adequate drainage area and head

3. **Fair** Opportunities for Development;

A. Structures are in good condition
B. Public access is not limited at the dam area
C. Minor appurtenant structures are present
D. There may have been/or has not been previous use of site for hydropower
E. Space available for hydropower development

4. **Poor** Opportunities for Development;

A. A very old structure
B. Important appurtenant structures of unknown condition are present
C. The site has a present public use with unlimited access close to the dam such as for a park, recreation, great historic or scenic features or real estate development

Prepared by Polytechnic Institute of New York

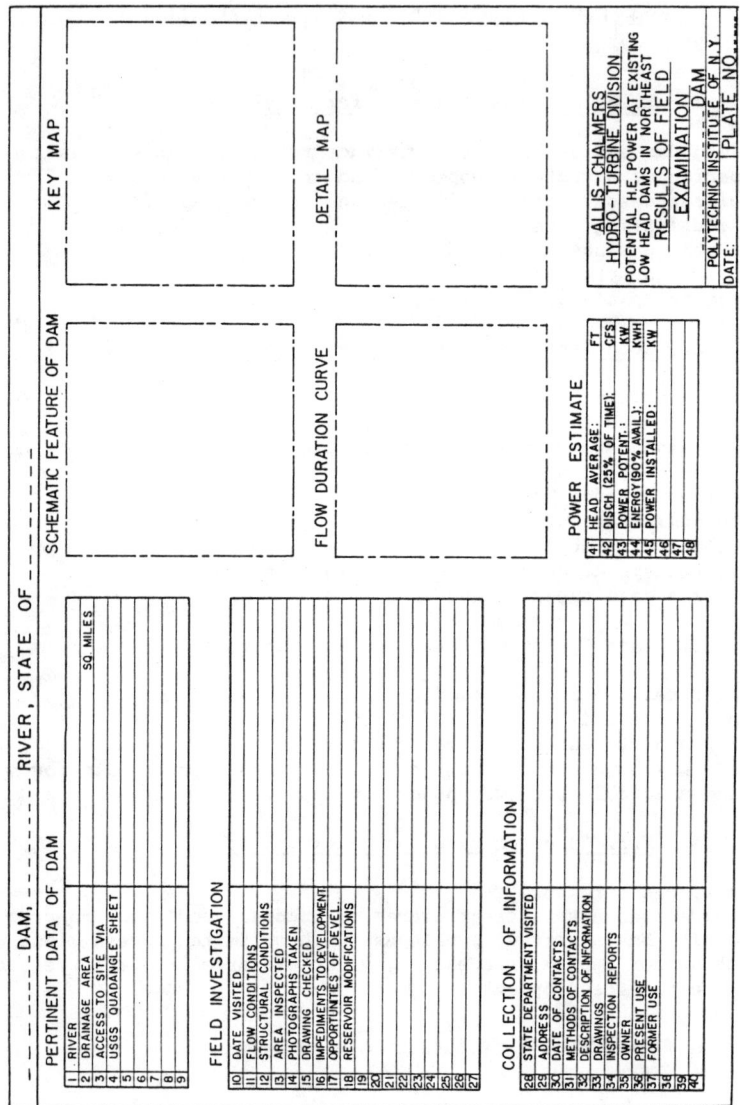

Figure 1. Field Report Sheet

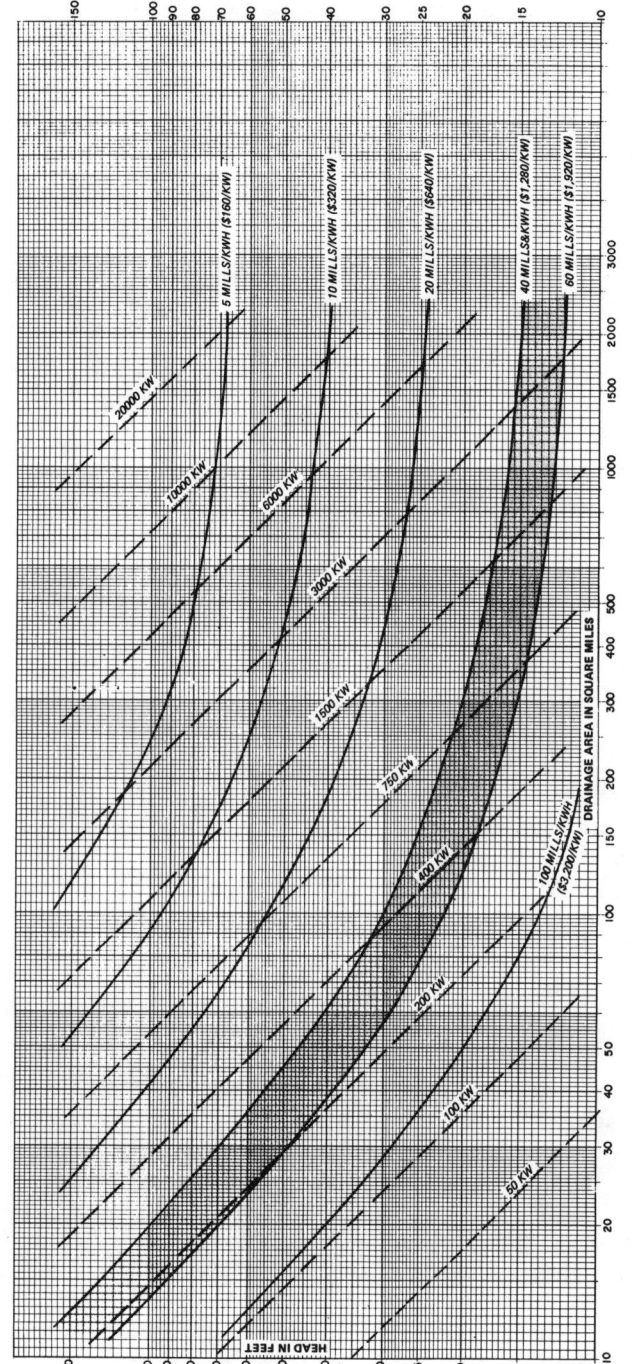

Figure 2. Typical Capacity and Minimum Energy Cost for Small Hydroplants in Northeastern United States

Energy from Sea and Air by Large-span Tensioned Foils

D. Z. BAILEY
Bailey Engineering
East Greenwich, Rhode Island, USA

ABSTRACT

Kinetic energy in ocean and tidal streams, and in atmospheric winds may be extracted in large quantities by stretching hydro (aero) foils horizontally over large distances across moving streams. These foils can be made to span very large distances by designing them to sustain large tensile loads and arranging them in "catenary like" arcs such that the major forces are devoid of bending moments. These tensile loads are applied to the system by placing massive concrete anchorages at opposing points across the stream similarly to those of a suspension bridge. By this means the earth absorbs the complex compression forces assumed in presenting this array to the energy flux. Supplementary structures stabilize the catenaries and extract the mechanical energy. For smaller systems of 1 to 10 Mw the axial thrust might be absorbed by one or more barges. George's Banks is a likely site for wind systems in the Northeast. Wind velocities and "real estate" values favor such a marine location where the array may stretch for miles.

INTRODUCTION

Conventional systems for extracting kinetic energy from winds and water currents such as wind mills, turbines, and water wheels have been in existence for many years. However, these systems are limited in size by constraints of mechanical design. Concurrently, during the latter part of the above development cycle, suspension bridges have reached a high level of technical sophistication. By combining the two technologies and providing suitable controls, restraints, and power conversion equipment large cross-sectional areas may be presented to the stream in order to capture the large but diffuse kinetic energy. This is achieved by placing the major structural elements in tension. For the larger systems the earth assumes a sizable compression load in supporting this man-made "spider's web" of tensile elements thereby playing a crucial role in the mechanical strength loop. It does so at no cost since the site must be provided for any system. For the smaller systems inexpensive concrete or other cost effective materials may be used for barges since size is an advantage, not a drawback, in handling the marine environment.

With the judicious addition of stabilizing elements and power take-off elemants this mechanical strength loop can be stretched over great distances.

ARRANGEMENT OF THE SYSTEM

In order to obtain energy in a most efficient manner with a lift device (aerofoil or hydrofoil) translation of the foil should be in a plane perpendicular to the wind or water-current velocity vector. In addition the added restriction that the earth or other cheap horizontal structure provide the required axial tension demands that the foil should move in a vertical direction. Mechanically this is not accomplished easily. In the interest of mechanical simplicity rotary motion was chosen for primary study. This system is a horizontal cross axis machine. Figures 1 and 2 are views of such systems.

Figure 1 is a perspective view of such a rotary system consisting of three bays, four barges and rings for stabilization and power take-off, and three cables with aerofoil shapes continuously enveloping the cables within the bays. Figure 3 is a cross-sectional view of the system taken at the center of the central bay. As mentioned above power is developed when the travel is primarily vertical (perpendicular to the flux). As a corellary to this fact lift, either negative or positive, is developed when the travel is essentially horizontal. Sectors are shown in figure 3 where lift may be generated and where power is generated. Of course the sector boundaries are not sharply defined since there are both horizontal and vertical components of the rotational velocity. Figure 2 shows a 3 Mw system at a rated velocity of 14 meters per second (32 mph). This would have, typically, a span of approximately 190 to 200 meters (650 ft.) and a ring diameter of 25 meters (82 ft.).

ARRANGEMENT OF THE ANCHORAGES

Massive concrete anchorages need to be provided for the larger systems tensioned by earth compression in order to translate this tension into the earth. In addition these anchorages must accommodate the downstream directed forces associated with the energy extraction. Since they are fixed in the earth a two anchorage system must operate in bi-directional winds. If they are multi-directional, three or more anchorages need to be provided to form a delta or triangular configuration in plan view. Figure 4 shows three span assemblies of three bays each. Obviously, with the wind vector as shown span A would produce the majority of the power; span B a small portion; and span C would produce essentially no power. For the large spans contemplated the spans would be sufficiently separated to avoid interference problems between spans. This configuration allows a significant cross-section to be presented to the flow for any flow direction.

For the smaller systems the barges would orient themselves to the wind as shown in figures 2, 5, 6 and 7. As shown in figures 5 and 6 it may be advantageous to have steel or concrete stabilizers or fixed rudders to aid in wind alignment. For a land system the compression structure might be placed on a crane or mining shovel bearing and race. A light cantilever structure or a car on a track with stabilizer as shown in figure 7 would orient the system to the wind. An imaginary line drawn between pairs of anchorages forms an approximate axis about which the dynamic system rotates. A rotating assembly, the cable cone, is placed in bearings at each anchorage such that the cable cone rotates about the described axis.

ARRANGEMENT OF THE CABLES

Three or more cables are strung between each pair of cable cones in a manner similar to that used in the construction of suspension bridges. These cables accept the high axial tension determined by the fluid dynamic forces, and their geometry and length in relation to the distance between cones. In order to provide the required translational velocity the cables are forced out from the axis by cable rings.

ARRANGEMENT OF THE CABLE RINGS - LARGE SIZE SYSTEM

Cable rings are located in vertical planes adjacent to the anchorages, called the master rings, and then periodically along the span to form bays. The cables fan out from the cable cones to the master rings and then proceed from ring to ring essentially parallel to one another. Figure 8 is an axial view showing representative geometry of such a ring. The forces exerted on the three or more cables are resolved and combined at the rings to produce torque and vertical forces while the cable tension forces are carried through the ring to the next bay and thence to the anchorage. The vertical forces are absorbed by means of trolleys or levipads affixed to barges and riding on the rims of the rings, while the horizontal forces are assumed by increased cable tension. As can be seen in Figure 4, 5, 6, and 7 the downstream forces are assumed by increased tension in the bays toward the anchorage such that the midspan barges or rings sag downstream in a "catenary like" configuration. Torque at the rings is transmitted to generators on the barges, and the power is conveyed ashore by electric cables.

ALTERNATE ARRANGEMENT OF RINGS AND POWER-TAKE-OFF

In order to eliminate or reduce the number of ring trolleys, power take-offs and associated equipment, an alternative system is envisioned as shown in Figure 9. Toroidal buoyancy structure is provided on the circumference of the rings to maintain static lift of the system during calm and light wind periods. Preliminary calculations indicate that the spinning resistance of the toroids in the water detracts significantly from the power generation. However, as detailed subsequently, aerodynamic lift can be provided to raise the toroids free of the water. Also, there is no reason why the toroids should not be helium filled should such become cost effective. Torque is carried from bay to bay by skewing the cables. In

Figure 9 each successive ring to the right would lag in its rotation
behind its left-hand neighbor, thereby transfering the torque from bay
to bay to the master ring near shore where the power is converted into
electricity. Alternatively the power may be taken off at the cable cone
shaft provided the tension in the system and the cable cone diameter are
sufficiently large to prevent the cables from twisting upon each other as
the torque is applied. This method would be used for 1 to 10 Mw.

ARRANGEMENT OF THE FOILS

Aerofoils or hydrofoils are placed surrounding the cables and extending
essentially continuously from master ring to master ring. Figure 5 shows
an axial view in section through one foil and a control boom and foil.
These foils are free to rotate on the cables and have a degree of flexibility in bending as the cable forms a catenary between rings. However,
they must be sufficiently stiff in torsion to maintain the desired angle
of attack as induced by control foils spaced as required along the span.
These control foils operate in a similar manner to the tail surfaces of
an airplane. It is likely that the control foil will be negatively
loaded with respect to the main foil in order to provide inherent stability. In such a case the cable and cable bushing would be placed forward of the aerodynamic center of the main foil.

DESIGN BASIS

The fundamentals of design equations were presented in a paper submitted
to Marine Technology Society Journal. They are as follows:

A certain amount of downstream sag of the bays must be allowed in order to
achieve reasonable cable tensions and anchorage forces. Although this sag
makes the flow three-dimensional, the hydrodynamic model assumes the flow
past the foils to be two-dimensional. It is further assumed that the deflection, $\bar{y}$, defined below is small in relation to the radius, R, so that
foil positions at each section fall on a circle of constant radius
(Figures 3 and 11). The flow is analyzed according to the momentum
balance equation (1) adapted from Wilson & Lissaman [3]. Their analysis
was developed for propeller systems with the plane of rotation perpendicular to the stream. In the foil system of Figure 12, the planes a, b
are assumed to play the role of the propeller plane in their analysis.
This is plausible because the propeller plane enters the argument only
for the definition of a pressure drop equivalent to the net downstream
force divided by the cross-sectional area. This appears to be the simplest basis for preliminary design computations, and is plausible in view
of the flow fields for the hydrodynamically similar giromill presented by
Brulle & Larsen [4].

A. General Design Equations

Momentum Balance [3] (see Figure 12)

$$T = \rho A V_a (V_\infty - u_1), \qquad u_1 = 2V_a - V_\infty \tag{1}$$

Foil Force Balance (Figure 11)

$$\underline{f} = \tfrac{1}{2}\rho v_R^2 \, c(C_D \, \underline{i}_R + C_L \underline{i}_L)\rho \tag{2}$$

Cable Force Balance (Figure 13)

$$\frac{F}{\bar{\tau}} = \sinh(Fx/b\bar{\tau}), \qquad \bar{\tau} = af, \qquad F = bf \tag{3}$$

The mean horizontal hydrodynamic force or thrust, T, is computed from the foil force balance equation (2) by averaging over a period. The hydrodynamic load along any ribbon foil is taken to be uniform and normal to the line joining the cable bearings of adjacent rings. From elementary statics the foil sections must adopt the catenary shape, equation (3), along the force line, F, in Figure 11. The combined system of bays will then also adopt a catenary shape in the horizontal plane.

The system of equations (1), (2), (3) must be solved simultaneously, and it soon becomes obvious that solutions depend crucially on the strategy used in the control of the angle of attack of the foils as the system turns through a revolution. In most of the calculations carried out to date the angle of attack has been modulated to maximize power production while minimizing cable and anchorage forces.

B. Detail Design

1. **Method of Analysis** The foils (or blades) are spaced at equal intervals around the rings. Initially, the index foil, m = 0, is assumed to be at the location θ_o = -180. While in this condition the aerodynamic forces are resolved into their various components, tangential, radial, vertical, and horizontal for an assumed velocity at the rotor area, not free stream. This, of course, could have been non-dimensionalized, but was not in the current program. A tension factor, a, is assumed for the central bay whereby the parameters of the bay may be determined. Of course, in a rotating system all the main cables within one bay must be of equal length. It is assumed in this analysis that the rings are parallel and that, therefore, the true straight line length, ℓ, (see Figure 13) for each of the foils is the same within each bay (although this length may, and does, vary from bay to bay). At the first ring from the centerline the horizontal (downstream) components of the forces acting on the adjacent lengths, b, of the foil surfaces on each side of the ring are summed, as are the tension forces due to the first bay. It should be noted that the tension factor, a, is a unit of length and is the same for all foils

in the bay, but the forces per unit length, f, are not the same. These forces must be balanced by an increased tension factor in the second bay. The reaction of the horizontal forces is assumed by a turning angle, δ. Since the tension in the bays increases from bay to bay, the present procedure increases the cable length, b, of successive bays in order to keep the deflection at the center of the bay equal from bay to bay. The coordinates of the rings are found in this manner and the process is repeated for the remaining bays. An iterative procedure is then used to adjust the sag (the position of the central barges in relation to the outboard barges) to the proper value by assuming a new value for the central bay tension factor. With system components determined, the foil system is rotated incrementally to find average and maximum values of the various forces. In the removal of energy, a downstream thrust is applied to the roils (the above mentioned horizontal force). This average upstream thrust reaction to the flow is used to find the undisturbed upstream velocity and, thereby, the force and power coefficients in the manner prescribed above.

2. **Determination of Aerodynamic Coefficients** Empirical formulae based on experimental results equate angle of attack to the standard airfoil coefficients as follows:

$$\alpha \leq 8° \qquad C_L = 0.1063\alpha \qquad (4)$$

$$\alpha > 8° \qquad C_L = 0.1063\alpha - 0.026(\alpha - 8)^2 \qquad (5)$$
$$+ 1.6(10)^{-3} (\alpha - 8)^3$$

$$C_D = 0.014\left[(0.1063\alpha)^{1.8} + 0.4738\right]^{0.56} \qquad (6)$$

This closely approximates a NACA 65_1 - 012 airfoil with standard roughness and a Reynolds number of $6(10)^6$; [5].

Induced drag is taken to be zero, since the systems are of extremely large aspect ratio.

3. **Characteristics of the Catenary**

Referring to Figure 13, it can be seen that the paramount characteristic of a catenary is the y axis intercept, a. At the ring plane $x = \ell/2$ and the following pertain:

a. Relationship between b and ℓ :

$$\frac{b}{a} = \sinh\frac{\ell}{2a} \qquad (7)$$

b. Tension in the cable at the centerline, y axis:

$$\bar{\tau} = af \qquad (8)$$

c. Tension in the cable at the ring (maximum cable tension):

$$\tau = af \cosh \frac{\ell}{2a} \qquad (9)$$

d. Deflection of the cable with reference to the ring support points, $\bar{y}$:

$$\bar{y} = a(\cosh \frac{\ell}{2a} - 1) \qquad (10)$$

e. Cumulative aerodynamic force at the ring (in the ring plane):

$$F = bf \qquad (11)$$

f. Carry-through tensile load perpendicular to the plane of the ring (same as the centerline tension):

$$\bar{\tau} = af \qquad (12)$$

4. Interbay Force Balance

As mentioned above, the next outboard bay must assume the downstream thrust of all the bays inboard. Let t be the horizontal component of f which is found by solution of equation (2) and the velocity vector diagram (see Figure 11).

Then:

$$(a_n \Sigma f_m)^2 = \left[(b_n + b_{n-1}) \Sigma t_m \right]^2 + (a_{n-1} \Sigma f_m)^2 \qquad (13)$$

$$\delta_n = \frac{(b_n + b_{n-1}) \Sigma t_m}{a_{n-1} \Sigma f_m} \qquad (14)$$

where the summation m, is for all foils in the bay.

5. Modulation of Angle of Attack

In order to distribute the load more uniformly over the rotation cycle the angle of attack is modulated as a function of $|\theta|$. The flip points where $\alpha = 0$ are at $\theta = \pm 180$ and 0. The aerodynamic forces are of the same magnitude, but, in general, directed downstream; hence, the absolute value of θ may be used for modulation (see Figure 51). At $|\theta| = \theta_j$ the aerodynamic force, f, becomes a maximum. This value, f_{max}, is found and stored for use in modulating the angle of attack.

$$\alpha = 0 \quad \text{for} \quad \theta_h < |\theta| < 180 \quad (15)$$

$$\alpha = \alpha_{max} \frac{180 - |\theta|}{180 - \theta_i} \quad \text{for} \quad \theta_i < |\theta| < \theta_h \quad (16)$$

$$\alpha = \alpha_{max} \quad \text{for} \quad \theta_j < |\theta| < \theta_i \quad (17)$$

$$\alpha = \alpha_f \quad \text{for} \quad \theta_k < |\theta| < \theta_j \quad (18)$$

$$\alpha = \frac{(|\theta|)}{(\theta_k)} \alpha_f \quad \text{for} \quad 0 < |\theta| < \theta_k \quad (19)$$

where α_f is the angle of attack which produces f_{max}.

6. Sizing of Cables

An allowable stress of 87,000 pounds per square inch was used for the main cables of the Verrazano-Narrows Bridge based on an average minimum tensile strength of 225,000 pounds per square inch [6]. In the present analysis a factor of safety of 5 is used giving an allowable stress of 45,000 pounds per square inch. For a compaction ratio (the ratio of steel in the cross-section to the total cross-sectional area of the cable) of 0.742 the effective allowable stress is 33,400 pounds per square inch or $23.5(10)^6$ kgm^{-2}.

C. Design Results

A computer program employing the solution scheme presented above was used to obtain the coefficients shown in the graphs, Figures 14, 15, and 16. Figure 14 shows power coefficient, C_p, as a function of V_T/V_∞ with c/R as a parameter and with constant maximum angle of attack. Figure 10 shows similar curves with maximum angle of attack as a parameter and with c/R fixed. The practical operating regime is to the left of the maxima since, on the right, power falls off rapidly and forces increase correspondingly.

The curves of Figures 14 and 15 correspond to the kind of design curves normally presented for conventional systems. Designs based on such curves tend to ignore structural criteria in the optimization of the system. For the tensioned ribbon foil system a very useful set of design curves is presented in Figure 16. Here C_p is plotted against C_A, the anchorage coefficient, and C_F, the cable tension coefficient. The control strategy for the angle of attack has been chosen to maximize power and minimize the anchorage and cable forces. In these curves, sag is held at 25%. When sag is reduced to 10% the forces double.

Referring to Figure 16, a design point was chosen such that:

$$C_P = 0.451; \quad C_A = 0.454; \quad C_F = 0.210.$$

For a rotor mechanical power equal to 1.3 times the electrical power, the electrical power in Mw, P, is:

$$P = \frac{9.8067(10)^{-6}}{1.3} \left(\frac{\rho}{2} A V_\infty^3 C_P\right) \quad (20)$$

where $9.8067(10)^{-6}$ is the conversion factor from kgms^{-1} to Mw; whence:

$$A = 4.7(10)^6 \frac{P}{V_\infty^3} \quad (21)$$

$$\text{anchorage load} = \frac{\rho}{2} A V_\infty^2 C_A \quad (22a)$$

$$= 132{,}600 \frac{P}{V_\infty} \left(\frac{C_A}{C_P}\right) \quad (22b)$$

$$= 133{,}500 \frac{P}{V_\infty} \quad (22)$$

$$\text{cable load} = 132{,}600 \frac{P}{V_\infty} \left(\frac{C_F}{C_P}\right) \quad (23a)$$

$$= 61{,}700 \frac{P}{V_\infty} \quad (23)$$

For an aspect ratio of 15:

$$\frac{B}{2R} = 15 \quad (24)$$

$$B = \sqrt{15A} \quad (25)$$

Table 1 is constructed from the above formulae and coefficients. It is obvious from the above that the anchorage load is about 2.2 times the cable load.

The most striking conclusion to be drawn from Table 1 is the very large power levels which are attainable with cables which fall easily within the scope of existing technology. In fact, systems have been designed for a wind energy system in the Aleutian Islands and an underwater system in the Gulf Stream with power levels in excess of 10,000 Mw. Even for these huge systems the cable sizes are about the same as those employed in large suspension bridges.

SUBSEQUENT DESIGN ANALYSIS

As indicated above, in the sub-section, Alternative Arrangement of Rings and Power-take-off, recent design work has been focussed on simplifying structure while maintaining the advantages of a large-span tensiled system. Rough calculations indicated that the buoyancy tores would contribute substantially to power loss through spinning drag in the water. This would be true especially if one modulated the angle of attack through an orbit for maximum aerodynamic power without regard to the vertical forces so produced. As a result the emphasis has been to investigate the concept of modulating the angle of attack to provide significant vertical lift with a minimum of fluctuations.

Design for Vertical Lift Management - Normally one would change angle of attack from radially inward to radially outward at the top of orbit (Figures 3 and 11), $|\theta| = 180°$, and back to radially inward at the bottom of the orbit, $\theta = 0$. However, the aerodynamic forces in the vicinity of top and bottom, where the resultant velocity is approximately in the horizontal plane, are predominantly radial for an efficient aerodynamic device with high lift-to-drag ratio. Hence, these forces contribute little and more importantly detract little from power output. In the original analysis the foils advanced into the wind at the bottom of their travel and down wind at the top, Figure 11. Initially it appeared that the top foil should advance into the wind for the purposes of lift management. This may or may not be the best scheme; however, it is the scheme employed in the systems of Figure 17 and 18. For programming and viewing Figure 11 one may rotate the system through 180°. Figure 3 shows the foils so arranged. Lift is provided by phasing angle of attack management behind physical foil position. The angle of attack would lead if the bottom foil were advancing into the stream. Figure 17 shows a phase lag, $\bar{\omega}$, of 16° and Figure 18 shows an $\bar{\omega}$ of 4°. It should be noted that these coefficients are based on the assumptions of the original analysis including that method of accounting for conservation of momentum. That is why some of the curves on the graphs (Figures 17 and 18) move to the right as the phase lag increases.

Influence of Vertical Lift on Power - There is more vertical lift and less energy extracted as $\bar{\omega}$ increases, for a given V_T/V_a, and it is energy extraction that slows down the stream at the machine. Therefore, as $\bar{\omega}$ increases, the system may rotate faster before blocking the stream to the same extent. For this reason substantial vertical lift may be obtained without significant reduction in power. Indeed, it would appear intuitively that the lift reaction on the air stream would draw down upper air to provide increased power. This might be particularly beneficial for the configuration of Figure 6.

GEOTHERMAL ENERGY AND HYDROPOWER

Angle of Attack Management - For the initial investigation, the results of which are shown in Figures 17 and 18, the angle of attack is modulated straight line as follows:

$$\alpha = \alpha_{max} \frac{\omega}{\omega_1} \qquad 0 < \omega < \omega_1 \qquad (26)$$

$$\alpha = \alpha_{max} \qquad \omega_1 < \omega < \omega_2$$

$$\alpha = \alpha_{max} \frac{180 - \omega}{180 - \omega_2} \qquad \omega_2 < \omega < 180 \qquad (27)$$

$$\alpha = \alpha_{max} \frac{\omega - 180}{\omega_3 - 180} \qquad 180 < \omega < \omega_3 \qquad (28)$$

$$\alpha = \alpha_{max} \qquad \omega_3 < \omega < \omega_4 \qquad (29)$$

$$\alpha = \alpha_{max} \frac{360 - \omega}{360 - \omega_4} \qquad \omega_4 < \omega < 0 \qquad (30)$$

The values of ω_1; ω_2; ω_3 and ω_4 are adjusted such that there is the least variation in vertical lift coefficient as the system rotates through a cycle. Though the search is preliminary and the angle of attack management elementary the vertical lift can be maintained at all times. The variations in the curves are due to the selections of ω_1; ω_2; ω_3; and ω_4.
A one degree difference can change the value of a cusp a measureable amount. For the larger $\bar{\omega}$ the minimum lift in the cycle does not fall off substantially from average lift. Anyone familiar with float plane operation on a calm water surface understands that it is not easy to "break" the water surface and employs "tricks" to get air borne. In these systems a "trick" which might be employed to lift the tores off the surface at "start-up" is to apply power momentarily.

Weight Management - The dynamic system can be exceedingly light. There are two major strength elements - the tensile cables which distribute the aerodynamic forces and the master rings which assume the major compression loads of the dynamic system. These compression loads are associated with spreading the tensile loads to the desired operating diameter of the foil system. A triangular arrangement such as is shown in Figure 8 can accommodate these forces efficiently. Indeed, if the axis of the cable cone is sufficiently high the ring and tore may be eliminated leaving the skeletal triangular truss. It is important to note that the cable system is at the neutral axis and that it is highly loaded by downstream displacement. There is, therefore, no stress reversal at the least loaded portion of the orbital cycle. Although the stresses vary throughout the orbital cycle, these factors tend to mitigate fatigue in the most highly fatigue-prone element of the structure. The remaining dynamic structure need cope with only the local loadings of the individual bays. The foil envelope may be fabric supported by rib structure as required stiffened locally to accept the torque of the control foils by a tensioned wire truss. The tores also may be of fabric.

RESULTS (SUBSEQUENT ANALYSIS)

For the system shown in Figure 2 two conditions are presented to illustrate the use of Figures 17 and 18. It should be emphasized that this is a preliminary attempt to acquire vertical lift. This system of three foils has a span between master rings of 200 m (660 ft); the cable length between master rings is 205.5 m; the ring diameter is 25 meters (82 ft); and the aerofoil chord is 2 meters (79 inches). The cross sectional area, A, is 5000 m². The density of air is 0.125 kg m^{-4}s².

Lift-off Condition; V_∞ = 4.5 ms^{-1} (10 mph) - The additional assumptions that V_T/V_∞ equals 2.8 and ω equals 16° are made for low speed lift-off. Then, from Figure 17 the following pertain:

$$C_P = 0.51$$
$$C_F = 0.69$$
$$C_V = 0.87/4 = 0.22$$
$$C_A = 4/3(0.89) = 1.19$$

This gives:

Mechanical power = 280 kw
Cable force = 4400 kg
Vertical lift force = 1390 kg
Anchorage force = 7500 kg

Design Condition; V_∞ = 14 ms^{-1} (32 mph) - Let V_T/V_∞ = 2.65 and $\bar{\omega}$ = 4°.

Then, from Figure 18 the following are:

$$C_P = 0.52$$
$$C_F = 0.65$$
$$C_V = 0.045$$
$$C_A = 1.15$$

GEOTHERMAL ENERGY AND HYDROPOWER

Whence:

$$\text{Mechanical Power} = 4{,}400 \text{ kw}$$
$$\text{Cable Force} = 40{,}000 \text{ kg}$$
$$\text{Vertical Lift force} = 2{,}800 \text{ kg}$$
$$\text{Anchorage force} = 70{,}000 \text{ kg}$$

One may compare these results with Table 1, which is based on Figure 16. Note that the forces are very much higher. This fact illustrates the necessity of using the proper angle of attack management scheme as the system approaches design stresses. It is anticipated that when high C_V is not required (as is the case for the higher V_∞ values) the scheme employed will closely approximate that used for Figure 16. Using the same coefficients that were employed in the construction of Table 1, we obtain:

$$\text{Mechanical Power} = 3{,}800 \text{ kw}$$
$$\text{Cable Force} = 12{,}900 \text{ kg}$$
$$\text{Anchorage Force} = 28{,}000 \text{ kg.}$$

CONTROLS

Control dynamics is a formidable task which has not been addressed as yet. However, microprocessor and sensor technologies are expanding at a rapid rate. There are at least two loops in the control system: maintenance of desired angle of attack of the foil systems and determination of desired angle of attack. Desired angle of attack may be influenced by many factors. One necessary input is position of foil in orbit. Many other inputs may be necessary or desirable such as power setting, system stresses, rotational velocity, vertical lift, and anemometer inputs. Controls for the wind energy systems contemplated must begin at a level of sophistication greater than those of conventional systems. However, it would appear that costs of control systems would fall within reasonable limits quickly as size and quantity of units increase. Also, there are advantages to the present system. For example, resonance frequencies and stress levels can be changed by axial movement of the cable cones in the cable cone bearings.

SUMMARY

Within the limitations of the theoretical assumptions set forth a basis for the design of HCATF-WECS, horizontal cross-axis tension-field wind energy conversion systems, is presented. Curves are presented in terms of conventional aerodynamic coefficients applicable to all wind speeds and sizes but to specific ratios of physical parameters, C/2R, $2\bar{D}$ or percent sag, as noted and specific angle of attack management. Since siting variations and applications give an infinite variety of sizes emphasis is placed on these curves. However, Table 1 and the characteristics of a 3 Mw system are presented to illustrate curve usage and an estimate of the physical parameters. The larger systems supported by earth-compression may extend indefinitely provided power take-off and an upstream reaction force are supplied periodically.

GEOTHERMAL ENERGY AND HYDROPOWER

NOMENCLATURE

T	net average, horizontal, hydrodynamic force (kg)
ρ	fluid density (standard air 0.125 kg m^{-4}s^{2})
A	projected cross-sectional area = 2RB (m^2)
B	overall projected length of foils, master ring to master ring (m)
2b	cable length spanning one bay (m)
R	radius of rings (m)
V_∞	undisturbed stream velocity (ms^{-1})
u_1	downstream velocity (ms^{-1})
V_a	mean velocity at the central plane (ms^{-1})
V_R	relative velocity (see Figure 11) (ms^{-1})
V_T	tangential velocity (see Figure 11) (ms^{-1})
c	chord of the foils (m)
C_D	drag coefficient
C_L	lift coefficient
$\underline{i}_R, \underline{i}_L$	orthogonal unit vectors in the direction of the drag and lift forces (see Figure 11)
$\underline{f}$	hydrodynamic force per unit length of foil (kg m^{-1})
bf	magnitude of $\underline{f}$ (kg)
$\bar{\tau}$	cable tension at center of catenary = af (kg)
ℓ	length of one bay (m)
θ	foil orbital position angle
C_P	power coefficient = Power/$(1/2)\rho V_\infty^3 A$
C_A	anchorage force coefficient = anchorage force/$(1/2)\rho V_\infty^2 A$

C_F cable force coefficient = cable force$/(1/2)\rho V_\infty^2 A$

C_H horizontal downstream force coefficient = horizontal force$/(1/2)\rho V_\infty^2 A$

C_V vertical force coefficient = vertical force$/(1/2)\rho V_\infty^2 A$

t   horizontal component of f, force per unit length (kg m^{-1})

a   y axis intercept of catenary (see Figure 13)(m)

P   electrical power output (Mw)

θ phase angle of foil

α angle of attack (degrees)

ω angle of attack phase angle = $\theta - \bar{\omega}$

$\bar{\omega}$ angle of attack lag behind foil position

BIBLIOGRAPHY

(1) Bailey, D. Z., U.S. Patent 3,978,345

(2) Bailey, D. Z., U.S. Patnet 3,407,770

(3) Wilson, R. E., P. B. S. Lissaman, "Applied Aerodynamics of Wind Power Machines," July 1974, NTIS, U.S. Dept. of Commerce, PB-238 595.

(4) Brulle, R. V., H. C. Larsen, "Giromill (Cyclagiro Windmill) Investigation for Generation of Electrical Power," Proceedings of the Second Workshop on Wind Energy Conversion Systems, 1975, Mitre Corp. NSF-RA-N-75-050 MTR 6970 pp. 452-460.

(5) Abbott, I. H., A. E. vonDoenhoff, L. S. Stivers, Jr., "Summary of Airfoil Data," NACA Rept. 824 (1945), p. 210.

(6) Brumer, M., H. Rothman, M. Fiegen, B. Forsyth, "Verrazano-Narrows Bridge: Design of Superstructure," J. of the Construction Division, ASCE, Vol. 92, No. CO2 Proc. Paper 4742, March 1966, p. 30.

ACKNOWLEDGEMENTS

Professor Bruce Caswell of Brown University has been most helpful, particularly in helping me analyze the fluid flow. I am most grateful for his help and support of my project.

Table 1

Nominal Electric Power Mw	Design Rotor Power Mw	Area m²	B m	2R m	C m	Cable Load kg	Cable Diameter cm (in.)
Design Velocity 6.71 ms⁻¹ (15mph)							
0.1	0.13	1,560	153	10.2	0.41	920	0.70 (0.27)
1	1.3	15,600	484	32.2	1.29	9,200	2.2 (0.87)
10	13	156,000	1,530	102	4.08	92,000	7.0 (2.7)
100	130	1,560,000	4,840	322	12.9	920,000	22 (8.7)
500	650	7,800,000	10,800	721	28.8	4,600,000	49 (19)
Design Velocity 8.94 ms⁻¹ (20mph)							
.1	0.13	660	99	6.6	0.27	690	0.60 (0.24)
1	1.3	6,600	315	21.0	0.84	6,900	1.9 (0.75)
10	13	66,000	995	66.3	2.65	69,000	6.0 (2.4)
100	130	660,000	3,150	210	8.4	690,000	19 (7.5)
500	650	3,300,000	7,040	469	18.8	3,500,000	43 (17)

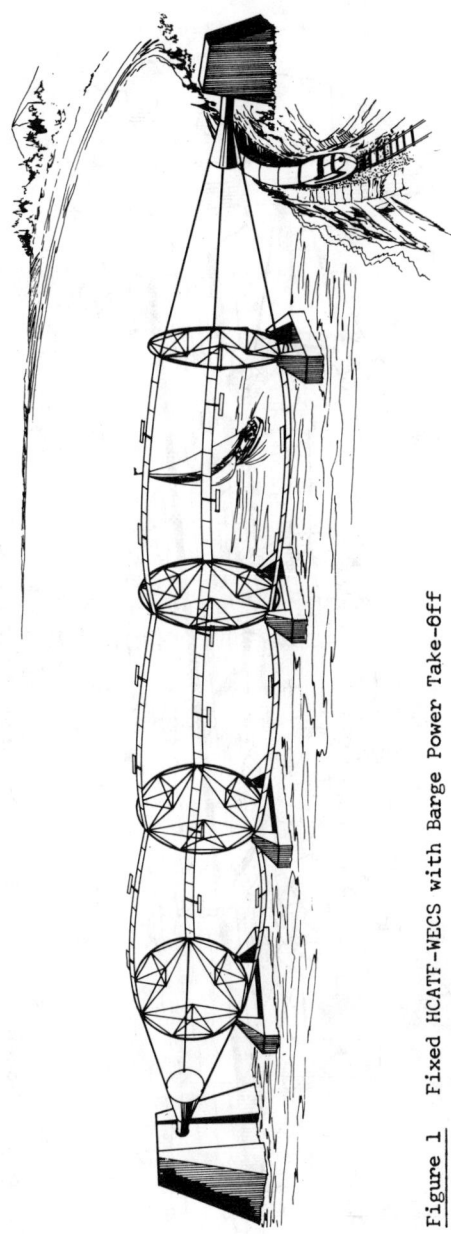

Figure 1 Fixed HCATF-WECS with Barge Power Take-Off

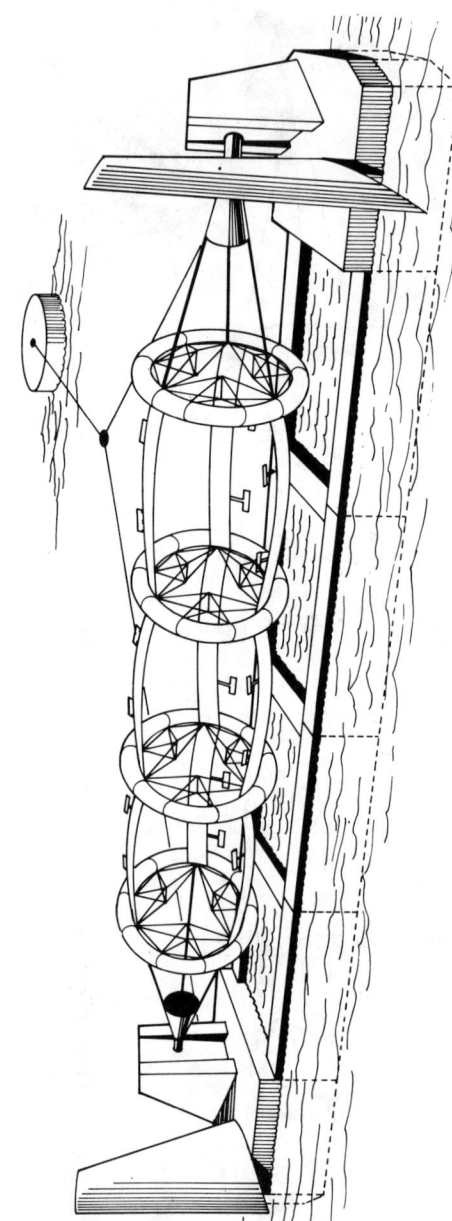

Figure 2　Floating HCATF-WECS with Anchorage Power Take-Off

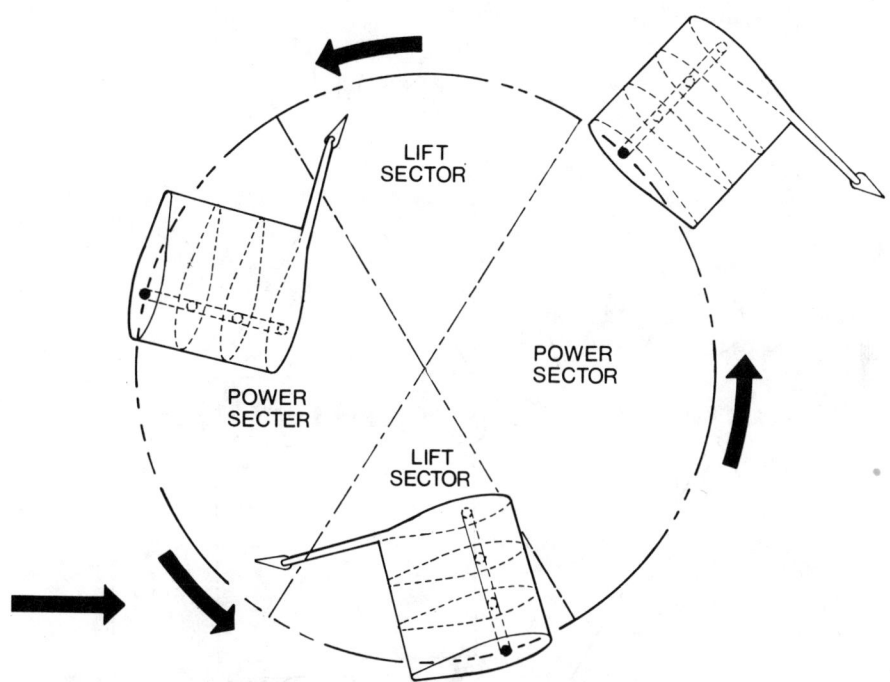

Figure 3 Cross-sectional View of HCATF=WECS Showing Foil Displacement and Lift and Power Sectors

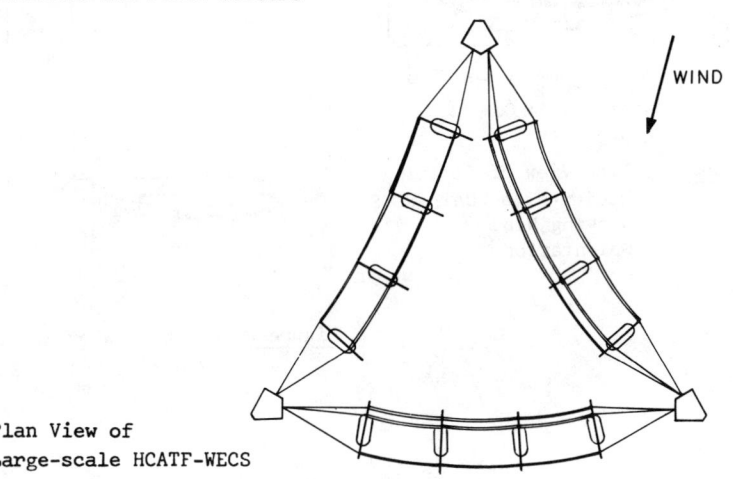

Figure 4 Plan View of Large-scale HCATF-WECS

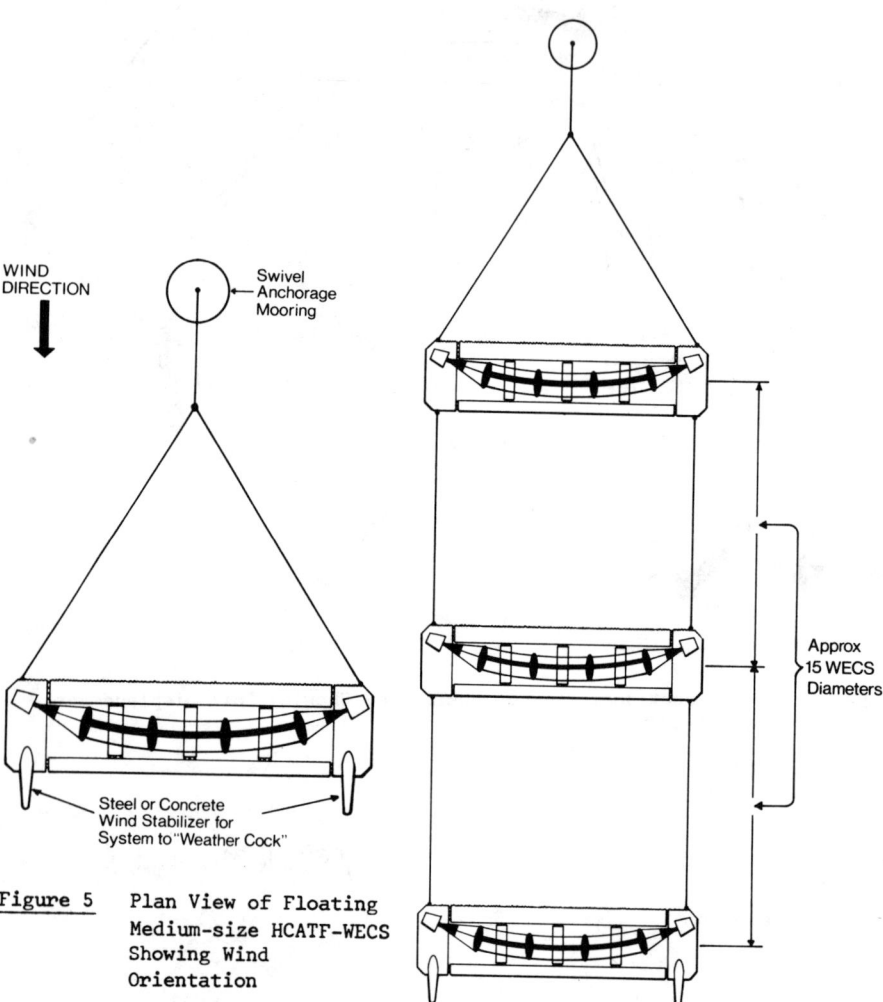

Figure 5 Plan View of Floating Medium-size HCATF-WECS Showing Wind Orientation

Figure 6 Plan View of Multi-unit Floating HCATF-WECS

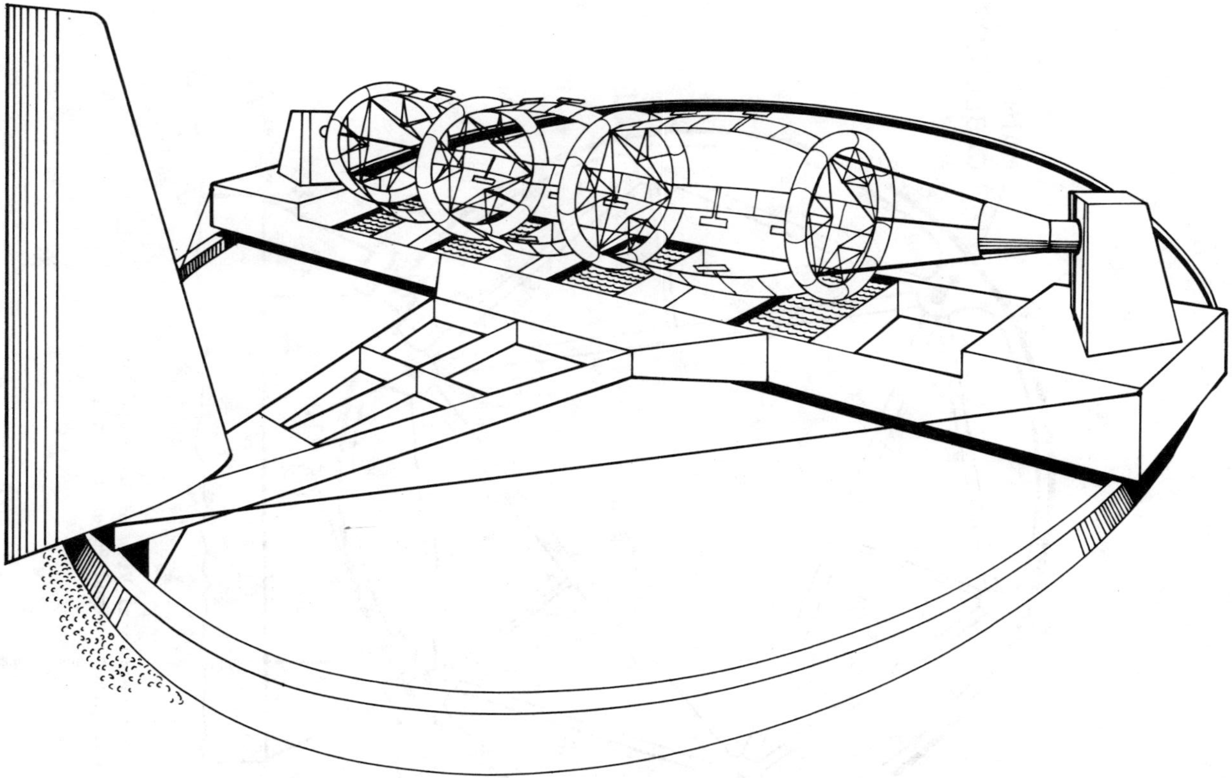

Figure 7   Three-quarter View of Rotating Land HCATF-WECS

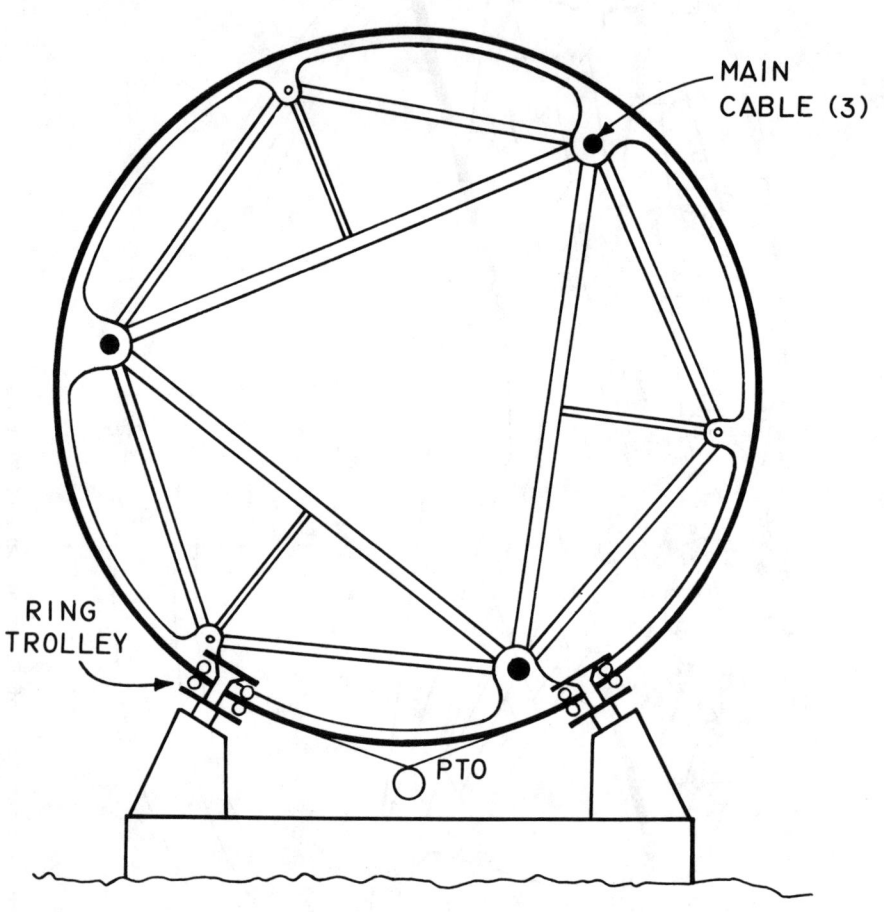

Figure 8 Cross-sectional View of Large HCATF-WECS Showing Barge Stabilization and Power Take-off

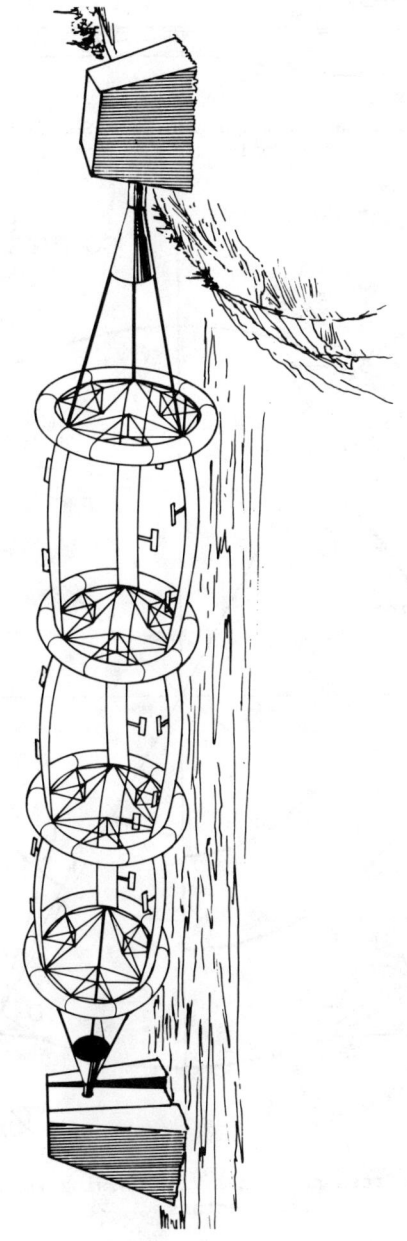

Figure 9 Three-quarter View of Large HCATF-WECS Showing Tore Stabilization and Torque Carry-through

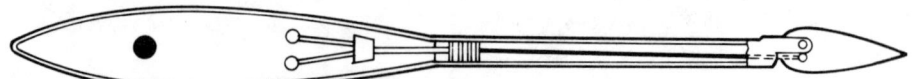

Figure 10 Cross-sectional View Through Boom Showing Angle of Attack Control Foils

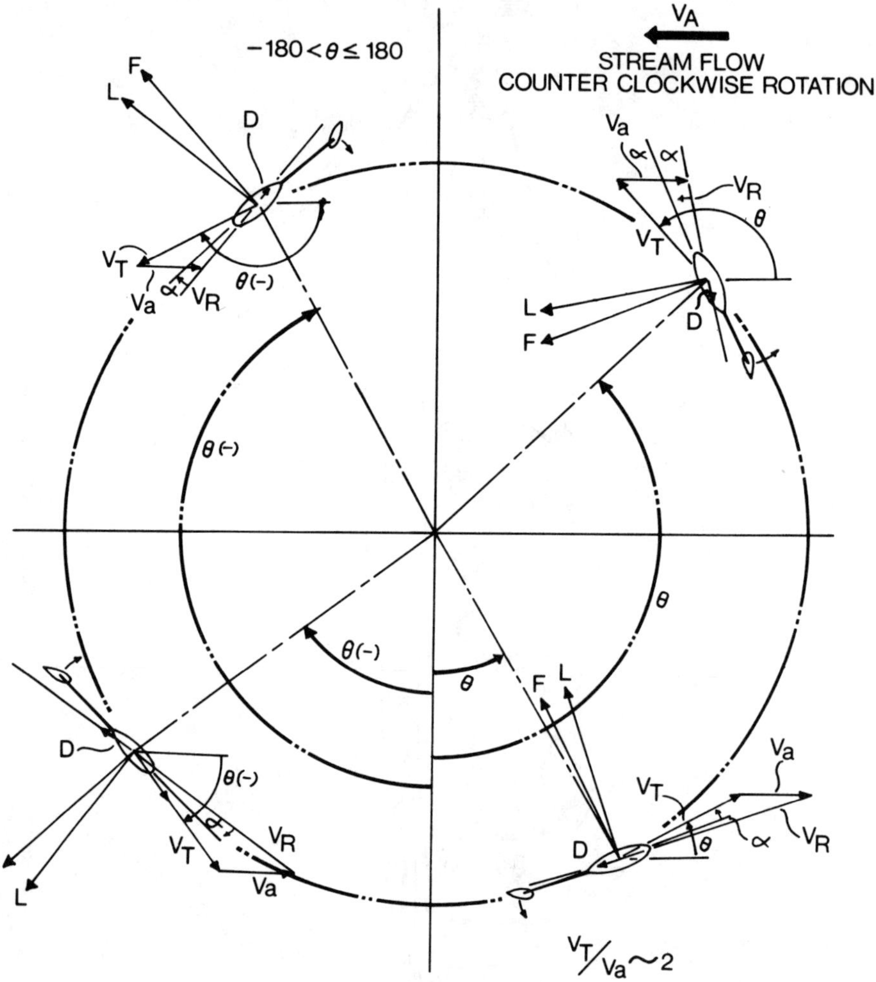

Figure 11 Schematic Cross-sectional View Showing Velocity Vectors and Aerodynamic Force Vectors

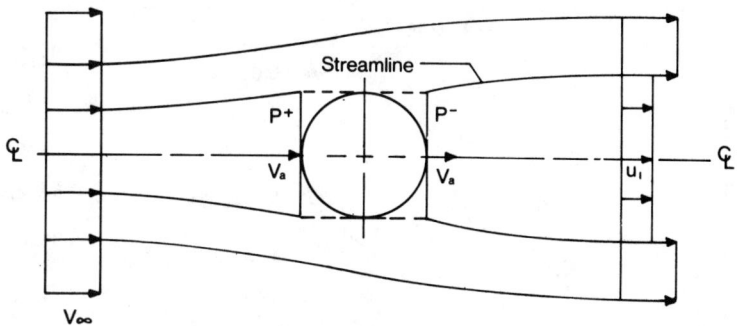

Figure 12 Idealized View of Flow Field

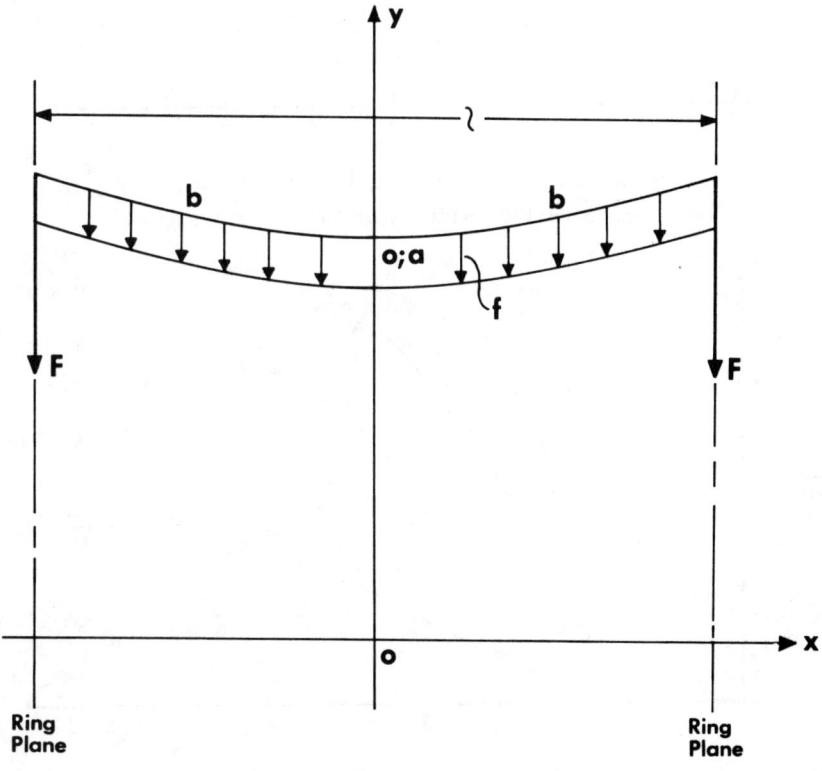

Figure 13 Foil Catenary Within the Bay

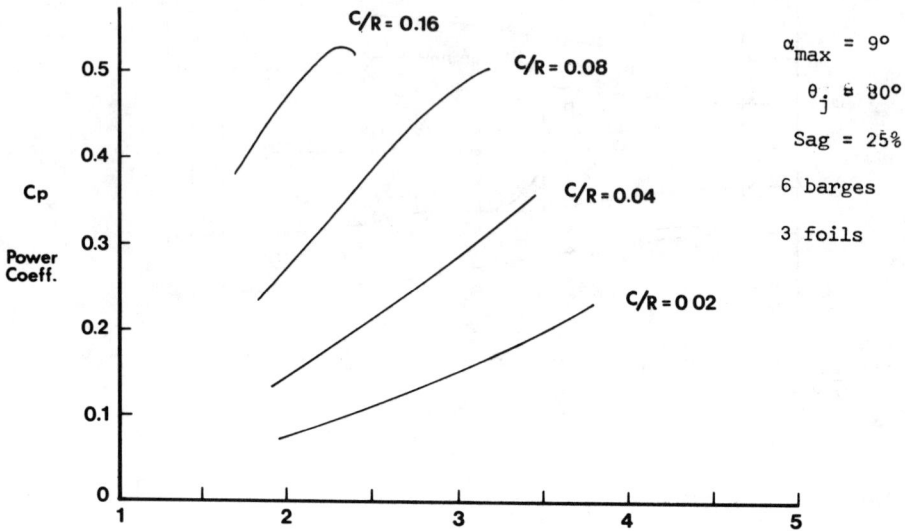

Figure 14   Power Coefficient, C_P, Versus Velocity Ratio, V_T/V_∞, for Various Foil Chord to Ring Radius Ratios, C/R.

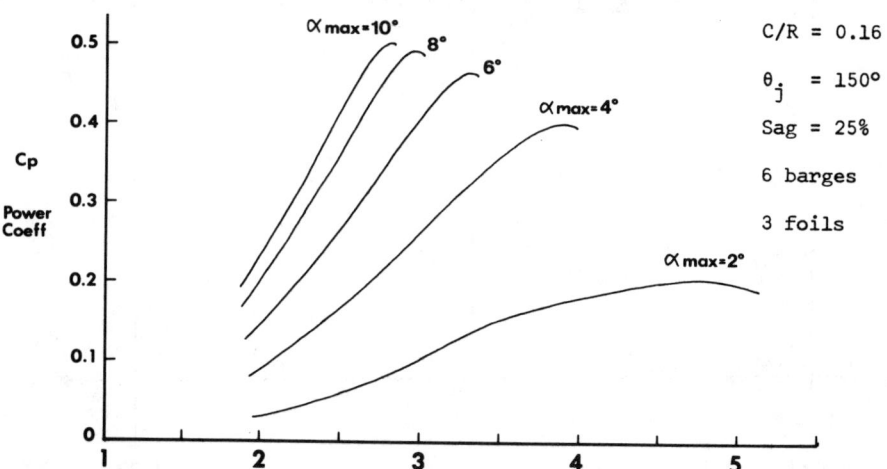

Figure 15   Power Coefficient, C_P, Versus Velocity Ratio, V_T/V_∞, for Various Maximum Angles of Attack

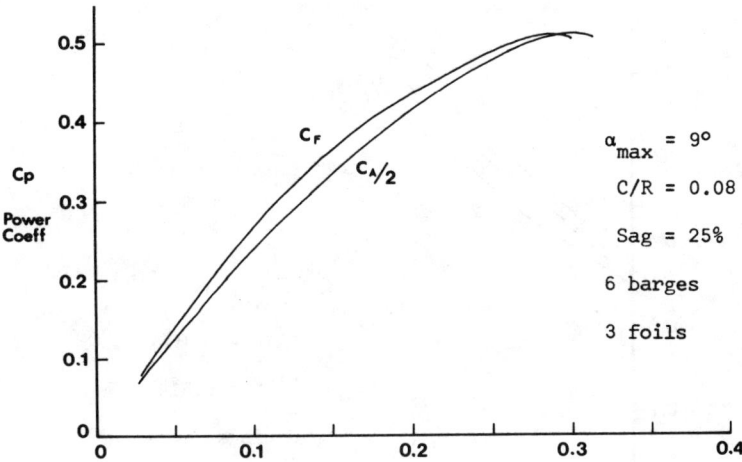

Figure 16 Power Coefficient, C_p, Versus Anchorage, C_A, and Cable Force, C_F, Coefficients

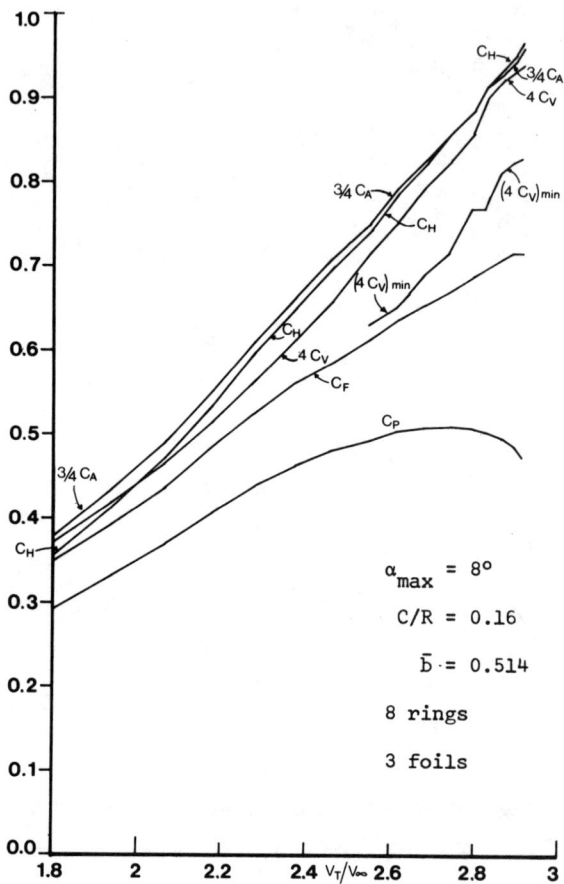

Figure 17 Power, C_P, Anchorage, C_A, Horizontal, C_H, Cable, C_F, Vertical Lift, C_V, and Minimum Vertical Lift, $(C_V)_{min}$, Coefficients Versus Velocity Ratio V_T/V_∞ for Phase Lag, $\bar{\omega}$, of 16°

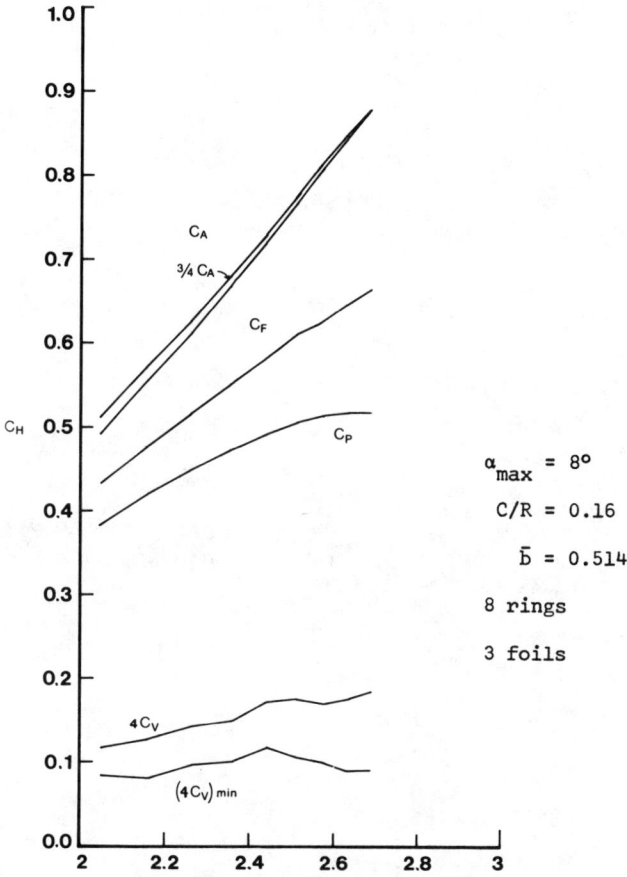

Figure 18 Same as 17. Phase Lag, $\bar{\omega}$, 4°

Salinity Power Station at the Swedish West-Coast: Possibilities and Energy Price for a 200 MW Plant

ALLAN T. EMREN and STURE B. BERGSTRÖM
University of Gothenburg
Department of Physical Chemistry
Fack, S-40220 Gothenburg, Sweden

ABSTRACT

The most favourable position for a salinity power plant in Sweden is at the west-coast. The total outflow of the rivers along this coast-line is 1000 - 1500 m /s. From five methods of energy production, the electrochemical one was considered to be most promising in this case. The stacks should be put together in units producing 20 kW each. The units should be placed in a frame of steel tubes about 250*1000*4 m^3 which also includes water channels. The entire construction is thought to be floating just above the water-surface. to avoid shellfishes and similar animals the input water has to be filtered to 0.1 mm. The adhesion of microbes is probably unavoidable. Thus the stacks have to be cleaned at regular intervals. The energy cost is estimated to be 3.4-4c/kWh. Of this the capital costs of the stacks give a contribution of 1.2-1.8 c/kWh.

INTRODUCTION

In 1976 the Board for Energy Source Development decided that the large scale possibilities of salinity power had to be studied. Thus a group of 12 scientists was established. The scientists should cover the disciplines most important in connection to salinity power. The members of the group have been:

Lars Bergdahl   Department of Subaqueous Constructions CTH
Sture Bergström   Department of Physical Chemistry GU
Lennart Claesson   B. Ludvigson Engineering Office Gothenburg
David Dyrssen   Department of Analytic Chemistry GU
Allan Emren   Department of Physical Chemistry GU
Jan Åke Frost   Department of Electric Plant Technique CTH
Lars Gertmar   Department of Electro-Engineering CTH
Lars Gunnarsson   Department of Marine Microbiology GU
Rolf Halde   Department of Chemical Apparatus Technique CTH
Karsten Pedersen   Department of Marine Microbiology GU
Elbadri Sherif   Department of Electric Plant Technique
Kjeld Thorborg   Department of Electro-Engineering CTH
Chairman of the group has been Allan Emren.

PRINCIPLES

Mixing fresh and salt water seems to be the only endothermic process which has been proposed for producing energy. If fresh and salt water are mixed in a tank, the mixture will have a temperature somewhat lower than that of the original solutions. The process then does not release heat, but, on the contrary, consumes a certain amount of heat energy. The fact that energy can be released depends on nature striving towards the greatest possible degree of disorder. It is obvious that there is greater disorder in a system where the salt is dispersed in the total amount of water than in one where the molecules have been separated so that salt is to be found only in one of two tanks. This striving by nature to obtain maximum disorder is usually called the second law of thermodynamics. Accordingly energy can be obtained from a system with a large heat content (e.g. oil) or from one with a large degree of order (e.g. fresh and salt water) or from one with a combination of both (e.g. a drawn bow).

By constructing obstacles of various kinds, one can force the molecules to do work on their way towards forming a homogenous mixture. If these obstacles are made too big, the impelling force is not strong enough but the process goes in the opposite direction. Consequently a plant for obtaining energy from differences in salinity can be worked in reverse I.e. one adds energy and brackish water at one end and fresh and salt water come out the other. Accordingly the difference between a salinity power plant and a plant for removing salinity is mainly one of purpose. This influences the technical design, but the main components are the same. For gaining energy one can imagine one of the following methods:

Steam turbine

Fresh water has a somewhat higher vapour pressure than salt water so one could place a steam turbine in the opening between a tank with fresh water and one with salt water. At room temperature, however, the vapour pressure difference is very small, which leads to a big machine. Further more a considerable amount of energy is consumed by evaporation where a corresponding amount is set free when the vapour condenses in the salt water. This will lower the temperature of the fresh water and raise that of the salt water so that the difference in vapour pressure disappears. One could avoid the problems by letting a heat exchanger raise the temperature to the critical point of water, 647 K (Fig.1). Then no heat is consumed during evaporation and the difference in vapour pressure between the solutions is about 5 atms. Owing to the high density of the vapour (0.2g/cm) the turbine can be made compact. There might, however, be difficulties in finding a heat exchanger material that is cheap enough and still stands the highly corrosive environment.

Freezing machine

Fresh water with a pressure of 200 atms can be cooled by salt water to -2 degrees without turning into ice. If this cold water is passed through a turbine some of it will freeze and the volume will increase. Consequently the volume of water going out through the turbine is larger then the one that was pumped into it, so that work is saved. The obtained ice is taken down into the salt water in order to maintain the low temperature. The difficulty in obtaining energy according to this method is to find turbines and pumps effective enough at the pressures in question. Fig. 2 shows the principle for such a machine.

Concentration cell

If a membrane letting through only positive or only negative ions is available, the salinity energy can be directly converted into electrical energy. (Fig. 3).

From the figure one can see that there is a net current of positive charges to the right and a net current of negative ones to the left. This is due to the chance movements of the ions. A positive ion for instance, is as likely to move to the left as to the right. But the movement to the left is blocked by the membrane A so that only move to the right is possible for that kind of ions. The contrary applies to ions. At the electrodes the ions are discharged and their charges converted to current in an electric cable. This type of apparatus has been studied experimentally in England /1/, Israel /2/, USA /3/ and Sweden /4/.

Mechanochemical energy production

A muscle is built up of polymers which have different lengths in different chemical environments.This can be imitated by making polymers which have different lengths in salt and fresh water. Then a piston engine /5/ /6/ or a turbine-like machine /7/ /8/ can be constructed. Flodin /5/ has shown that such a machine can be economically possible. It might however be afflicted with certain leakage problems, the same surface having to be in contact with fresh and salt water in turn (fig. 4).

Osmosis

If one has a membrane letting through water but not salt and places salt water on one side and fresh water on the other, the fresh water will flow through the membrane into the salt water. Then the pressure of the salt water will increase (if it is enclosed).

The process will stop when the pressure difference over the membrane reaches about 25 atms. With a conventional water turbine the pressure can be transformed into work (fig. 5).

A great deal of theoretical and practical work has been done in USA /9/, Israel /10/ and Sweden /11/ concerning such machines. As late as the summer 1976, most people working in the salinity energy area considered this method the most promising. Since then discoveries have been made /12/ /13/ which indicate great practical difficulties with this method. The best prospects of a successful result of this method are in areas where there is a supply of fresh water (<0,1 % salt) and salt water with a great content of salt (>5%).

CONDITIONS FOR THE WORK OF THE PROJECT GROUP

The purpose of the work was to investigate the possibilities of using salinity power on a large scale. Therefore a theoretical study was made concerning a plant of concentration cell type with a power of 200 MW. A fictive plant has been constructed, after which the costs of delivered energy have been calculated, its influence on its surroundings inquired into and till now unsolved problems investigated. It has not been possible to optimise the plant. Thus on every possibility of choice the corresponding parameter has jammed.

A suitable position for the plant seems to be the sound between Björkö and Hisingen, south of the mouth of Nordre Älv. The distance to the mouth is relatively small and a deep natural channel (30 m deep) directs salt water there with a salinity of 3.2 percent. The deep channel ends near the power plant. Consequently none of the important shipping passages will be disturbed. Small tonnage should be able to pass without trouble also in the future. The following parameters were fixed from the beginning:
Output power: 200 MW
Fresh water flow 390 m/s; 0.06% salt in and max. 2.0% out.
Salt water flow 1500 m/s; 3.0% salt in and at least 2.5% out.
Membrane characteristics:
Resistivity: 0.2 cm
Selectivity: 95 %
The goal has been to produce a costs estimate, as realistic as possible considering the expected commercial technological level attained by 1990. The calculation should if possible be correct within 20%. Further an investigation into problems and questions, which can not yet be answered, is planned.

OUTPUT POWER AND PUMP POWER FOR CONCENTRATION CELLS

A pair of membranes with fresh and salt water layers will have an electromotive voltage which is described by

$$(1) \quad E_o = \frac{(\alpha_+ + \alpha_-) RT}{F} \ln \frac{a_s}{a_b}$$

where $\alpha_\pm$ is the selectivity of cation and anion membranes respectively. R is the gas constant, F Faraday's constant and T the absolute temperature. a_s is the salt activity in the salt water and a_b in the brackish water. The resistance of a cell can approximately be obtained from

$$(2) \quad RC = R+ + R- + \frac{1}{\Lambda}(\frac{hs}{cs} + \frac{hr}{cb})$$

where R+- are the resistivities of the membranes, Λ is the equivalent conductivity of the dissolved salts. hs and hr are the layer thicknesses of the two solutions, cs and cb their concentrations and A is the area of each membrane. RL is the load resistance. Then the output power is

$$(3) \quad W = \frac{E_o^2 \, RL}{(RC+RL)^2}$$

(3) has maximum for RL=RC.

When the cell is working the salinities are changed and the solutions must be renewed sufficiently rapidly. In an energy-producing unit the pump power necessary for the exchange must not exceed the output power of the cell. The flow of ions through a membrane is

$$(4) \quad \emptyset_i = \frac{I}{F\alpha}$$

To maintain a stationary state the waterflow by the membrane must be

$$(5) \quad \emptyset_w = \frac{\emptyset_i}{\Delta c}$$

where ΔC is the concentration difference between ingoing and outgoing water. In the work of the project group $C = 0.01$ M was chosen. The velocity of the water flow will be

$$(6) \quad V = \frac{\emptyset_W}{hL}$$

where L is the effective edge length of the membrane. In the work of the project group, it was taken for granted that all membranes are square. The pump power at laminar flow between two level parallell disks is approximately

$$(7) \quad W_p = \rho\emptyset_W(5\ V^2 + \frac{12\nu VL}{h^2})$$

where ρ is the density of the water and ν its viscosity. Table 1 shows an example of a simulation where the salt water was kept at 0.48 M NaCl and the membrane qualities have been chosen according to todays laboratory values (the commercial technological level of 1990). Water load means the number of m water which must pass every m of the membrane surface during one second to maintain the calculated concentration, voltage and power conditions. The temperature (T) is given in K and salinities in moles/litre (1% salt 0.17 M). WP is the pump power and WOUT is the output power from the pile. Consequently the WP must not exceed WOUT.

In the shape described above the water must stream equally fast through all piles. $0.5*10-4$ m/s is a suitable choice of water load. The result is shown in table 1 where every pile is divided into imagined strips with cross section $3*3$ cm^2. 16 of those strips form a real pile which consequently will have an effective cross section of $12*12$ cm^2. The electrodes will force upon the stripe a voltage which is an average of the voltages of the individual strips. The numbers of the piles is counted from the fresh water inlet. For the calculations changes in the salinity of the salt water have been neglected. The voltage refers to the voltage of the whole pile.

ENERGY PRODUCING UNIT (20 KW)

A membrane with a profile according to fig. 6 is produced and cut into square pieces with an edge length of about 15 cm. Production can be accomplished by some kind of polymer material being pressed through tuyeres, milled or cast so that the required profile is obtained. The film formed in this way is chemically treated to give the desired ion exchanging proper-

GEOTHERMAL ENERGY AND HYDROPOWER

ties. Another possibility is making the film out of raw materials such that it becomes ion exchanging without subsequent treatment. The laths of the film are covered with glue after which membranes of both kinds are put alternatively on top of each other with the laths perpendicular to the direction of those on underlying membranes. In this way a pile segment according to fig. 7 is built. Having reached a height of about 5 cm (400 membranes), 5 - 10 mm are cut away at each corner and the section is suplied with a frame of plastic. This frame is given a shape facilitating connection with water inlets and outlets. It is also furnished with insulators or attachments for them. Thus a manageable pile segment is obtained the effective membrane surface of which is 8 - 12 m^2. 20 - 40 such segments are screwed together to a complete pile and are provided with one electrode at each end. 24 of these piles form a unit with a power output of 20 kW. Fig. 8 shows the building of the pile and Fig. 9 the complete unit.

THE SITUATION AND STRUCTURE OF THE POWER PLANT

The plant is beeing built as a large platform and placed between Björkö and Hisingen according to fig. 10. From "the fiord" of Nordre Älv a shield (type floating barrier) is built to shut in the fresh water. Owing to the fact that fresh water floats on salt water the bay can be used as a regulating store with a constant top water surface.

The floating barrier is estimated to be about 9 km long. It will shut off the to meter. The power plant consists of 10000 units which are imagined to produce 20 kW each. Since each unit is about 4.5 m^2 the power plant requires an area of about 45 000 m^2.

The power plant is constructed as a floating structure of units kept together. The units consist of electricity producing units - "membrane piles" - and of channels of plastic or plastic coated plate. The units are kept together by a frame of steel tubes and are suspended by floats. The power is imagined to be supplied to the which involves some problems as the power is very large. A small harbour construction and administration buildings are also required. A switchboard plant for the power transmission is needed.

The power plant is to contain 10000 units. These are supposed to be arranged in 200 rows of 50 units according to the figures on the opposite page. The rows are placed parallell with a separation of one or two meters. in this way channels for the salt water are formed. The fresh water is transported through pipes on top of the row of units.

In order to feed the power into the public net the voltage must be increased. This might be done in two steps, one at the inverter and one at the output of the high voltage wire. Reactive power is consumed by inverters because the A.C. fed to the A.C. net is not in phase with the net voltage. Overtones are generated since the A.C. is not a sine wave. Reactive power generation and overtone reduction is done with a net filter. Its dimensions are dependant on the caracteristics of the feeding point and demand for quality in the A.C. net.

Further, the A.C. side should be sectioned so that redundancy is available for the handling of fault conditions and service.

COSTS OF THE 20 KW UNIT

In estimating the costs we have assumed a membrane structure like that in fig. 6. The cost of the film is estimated at 4 c per square meter, which includes machine and manufacturing costs. Chemicals will be used in activating the film and also in surface treatment of the distance rods. The chemical costs are estimated at 45 dollars per kg of activated surface, which gives 55 c/sq meter membrane. The costs of putting glue on the rods has been set equal to the printing costs of newspapers which would give 0.06 c/sq meter. Cutting the membranes and putting them together has been compared to works, which gives a cost of 5.5 c/sq meter. In the membrane profile described above, 25% of the surface is inactive. These facts give a membrane cost of 90 c/sq meter. The 20 kW unit is assumed to be constructed from plastic which costs 2.1 $/sq meter. Each unit will need 270 kg plastic at a cost of 600 $. The cost of the electrodes is estimated to be 215 $. Labour costs are estimated to be 11 $ per hour, and 106 hours are assumed necessary to assemble a unit. This makes 1170 $ for the work. A unit will use 7600 sq meters of membrane and it will cost 6840 dollars. The electrical connections to the electrical system of the power plant are estimated to cost 220 dollars. Finally the unexpected costs are assumed to be 255 dollars per unit. Thus the total cost of a unit is 9300 dollars.

ANNUAL COSTS

A summary of the capital costs is shown in table 2. Annual capital costs calculated from 10 years life and 10 % interest will be 22.5 million dollars per year.

Running costs:
Buildings: 5700 dollars/year.

Salaries: 80000 dollars/year.
Administration: 13000 dollars/year.
Cleaning of filters: 5.3 million dollars/year.
Maintainence: 11 million dollars/year.
Unexpected costs: 4.1 million dollars/year.

Thus the total annual costs are 43 million dollars. With an availability of 70 percent this gives an estimated energy price of 3.5 c/kWh.

TECHNICAL BIOLOGICAL PROBLEMS

Growth of bacteria in natural waters, mostly occurs on the surfaces of solid materials and on interfaces of different kinds. Microbes which grow in this way (microbial fouling) stick very strongly to the surface and they are difficult to remove by washing. The adhesion is of various kinds depending on the kind of microbe and surface. It can be for example adhesion by charges, by chemical adhesion and so on. Often the fouling by microorganisms in sea water is a prestate to macro fouling. The adhesion of bacteria to the membranes will evenually prevent water from entering the stacks. The magnitude of this problem, and the possibilities of solving it, are very difficult to foretell. Only practical tests in the sea can give accurate information.

Most stationary macro organisms are free swimming during some period of their lives. The positions on which they settle are often carefully chosen with respect to light, water current, salinity and temperature. They will give macro fouling at the position where they settle. Probably a 0.1 mm filter and a salt water inlet at more than 20 m depth will prevent macro fouling in the membrane system. However fouling problems will occur in the salt water system before the filters, mainly due to molluscs, balanoids and achides. In short periods (0.5-1.5 years) they may give 0.2 to 0.3 m thick layers on free surfaces in the sea.

ENVIRONMENTAL EFFECTS

The natural variations of the water level are about +/- 0.5 m during a day and about +/- 1 m during a year. In the bay of Nordre Älv, the upper first meter can be considered fresh water (<0.5% salt). If the interface of fresh and salt water is forced to move between one and two meters during 24 hours, a very small change has been made. Therefore animals and plants will hardly be effected.

A problem will come from the considerable quantities of water which are brought up from the bottom. During the summer, the temperature of the surface will be decreased from 288 K to 283 K. During the winter it will be increased from 273 K to 276 K. For ships this will mean a decrease of difficulties from ice, but more fog during winter and spring when the wind comes from the east. During autumn and winter, there will be an increased tendency of fog over the land, when the wind comes from west.

Besides being colder, the bottom water has higher concentrations of nitrogen and phosphorus than the surface water. Rough calculations indicate that the quantities of fish caught on the west-coast will be increased by 2500 - 15000 tons per year if the power plant is built.

RESULTS AND DISCUSSION

The estimated energy price 0.035 $/kWh may be somewhat too optimistic. The temperature dependence of activity coefficients, conductance and ion selctivity is neglected. The simple rule that 10 degrees decrease in temperature reduces reaction rates by half, will lead to a doubling of the cost for the unit. the result would be that the energy price rises to 0.041 $/kWh. Pattles experiment in the temperatur interval 293 K to 306 K gives evidence for that the factor 1.5 is better. Furthermore no optimisations have been made in this study, which will make the calculations too pessimistic. To day we can state that the energy price will be between 0.035 and 0.040 $/kWh when the stacks can be made for a price of 1.22 $ per squaremeter of working membrane area. The estimated energy cost is of the same order of magnitude as that of wind or wave energy.

Unsolved problems

The temperature dependence of the output power should be studied both theoretically and experimentally in the laboratory. The most important problem seems to be fouling on the membranes. This will stop the flow and perhaps destroy the membranes. With fouling we mean microufoling The filters will take care of everything else. Pairs of membranes with electrodes that are forcing current through them should be placed in waters with different grades of impurities, to study biological problems and aging. One or two stacks with an output power between one and ten watts should also be placed in natural surroundings for the same purpose. An investigation should be made of how the temperature of the surface water will be effected by a salinity power plant. One will have to consider the existing and generated sea water streams and also the temperatures in different layers of the sea. The change in climate too should be investigated. The frequency of fog and rain is probably effected. Several problems concerning electrical installations occur from the combination of low voltage, high power and a great number of power producing units. DC to AC converters for this type of power station are not yet constructed. They must be made in such a way that one unit can be disconnected for service.

ACKNOWLEDGEMENTS

This work has been supported by the the Board for Energy Source Development. We are indebted to Mr J Bergman, Prof. P Flodin and Mr P J Svenningsson for valuable discussions and suggestions.

REFERENCES

1. R E Pattle, Electricity from Fresh and Salt Water - Without Fuel, Chemical & Process Engineering; Vol. 36, No 10 p. 351 (1955).

2. C Forgacs, Private communication.

3. J N Weinstein and F B Leitz, Science, vol. 191, p. 557 (1976).

4. S Bergström and A Emren, Förstudie beträffande saltkraft, (in swedish), report to the Board for Energy Source Development , NEPÖ/O 770520

5. P Flodin and P Lagerkvist, Mekanokemisk energiutvinning ur saltgradienter, (in swedish), report to the Board for Energy Source Development , NEPÖ/O 770501.

6. O. Elmqvist, Private communication.

7. W Kuhn et al., Fortschr. Hochpolymforsch., Bd 1, p. 540 (1960).

8. A Katchalsky et al., Elementary Mechanochemical Processes, in A Wassermann, Size and Shape Changes of Contractile Polymers. Pergamon Press, London, 1960.

9. H H G Jellinek, Osmotic work from ERDA's Wave and Salinity Gradient Energy Conversion Workshop. Univ. of Delaware 24-26 may 1976.

10. S Loeb, M R Bloch, Salinity Power, Potential and Processes, Especially Membrane Processes. Presented at the Joint Oceanographic Assembly Edinburgh, Scotland, September 1976.

11. S Lindhe, Private communication.

12. S Loeb, Private communication.

13. A Emren, Concentration polarisation in pressure retarded osmos, report to the Board for Energy Source Development, NEPÖ/O 770912.

FIGURE CAPTIONS

Fig. 1. Steam turbine.

Fig. 2. Freezing machine.

Fig. 3. Concentration cells.

Fig. 4 Mechanochemical engine.
A. The polymer fibres are shrunk and the cylinder is washed with fresh water.
B. The outlet valve is shut, the fibres absorb water and expand. The balance wheel keeps them stretched.
C. The salt water and outlet valves are opened and the fresh water in the cylinder is exchanged for salt water.
D. Power stroke. The salt water ventilator is shut and the fibres shrink at which they take the balance wheel along with them.

Fig. 5. Osmotic turbine.

Fig. 6. Membrane structure.
The height of the laths are about 0.1 mm and the distance between them about 0.5 mm. The thickness of the membrane between the laths is 0.01 - 0.02 mm. Each membrane is given the shape of a square with sides of about 15 cm. So the figure shows only about 4 mm of a membrane.

Fig.7 Part of a pile segment.
By placing the laths of the cation and anion membranes perpendicular to each other, a stable structure with separate channels for fresh and salt water is obtained. The pile segments will have the approximate dimensions 15*15*5 cm. Every such segment will then contain 450 membranes (see fig. 6). When the membranes have been joined, the segment is provided with water connections and details to facilitate fastening to other segments.

Fig. 8. Membrane pile
20 - 40 segments according to fig. 7 are coupled together to a complete pile which is supplied with one electrode at each end. The total length of the pile is 1 - 2 m. It contains about 300 m of membrane. A typical value of the delivered power is 800 W. The fresh water respectively salt water is pumped through the pile in inlets situated perpendicular to each other.

Fig. 9. 20 kW Module.
The unit consists of 24 piles of the type shown in fig. 8 built together. Every dark square is a pile seen from one end. The fresh water is distributed from a vertical tube to

the left and is taken out through the tube to the right. The
salt water rooms are directly connected to horizontal salt
water channels.

Fig. 10
The power plant is situated between Björkö and Hisingen. Salt
water comes in from the south through a deep channel which
ends near the power plant. A floating barrier shuts off the
fiord of Nordre Älv from Stora Överön and directs the fresh
water to the south towards the power plant.

TABLE I

```
RUN
SIDE LENTH (M)? 0.03
PERMSELCTIVITY? 0.95
SALT WATER (M)? 0.48
FRESH WATER (M)? 0.005
PILE LENGTH (M)? 2
T=? 279
WATER LOAD? 6E-5
MEMBRANE RESISTIVITY (OHM*CM**2). AN, KAT? 0.2 , 0.2
THICKNESS (MM). FRESH, SALT? 0.1 , 0.13
THE MODULE HAS 40.46296 PARALLELL PILES
```

C(FRESH) M	EFFECT W/M**2	VOLTAGE V	WP/WPUT	W TOT
0.00500000	2.124229	1661.429	0.01961246	1319
0.01071695	3.048469	1383.918	0.01366632	3223
0.0205665	3.78768	1146.649	0.01099917	5596
0.03533675	4.112754	949.6296	0.01012979	8174
END OF PILE 1				
0.05470201	4.000203	790.5694	0.01041481	10681
0.07732693	3.602651	664.5733	0.01156408	12936
0.1015665	3.0975	565.3206	0.01344999	14872
0.1260662	2.598688	486.6621	0.01603168	16491
END OF PILE 2				
0.1499428	2.156157	423.5274	0.01932195	17830
0.1727066	1.782642	372.0792	0.02337056	18933
0.1941293	1.474448	329.5165	0.02825555	19841
0.214137	1.222491	293.811	0.03407908	20588

```
PUMPING EFFECT SUBTRACTED IN W TOT
TOTAL MEMBRANE AREA=7600 M**2

READY
```

TABLE II
Costs of the plant.

Part	Unit	Quantity	Cost (1000$)
wires	m	9 000	50
tarpaulin	m^2	18 000	200
Concrete anchor	m^3	2 000	111
Gates for small boats		4	2
weights	kg	45 000	20
Connection to external power network		1	4 500
Building (office)		1	1 100
Small boats etc.		1	450
Tubes (inlet)	m^2	64 000	350
Tubes (outlet)	m^2	64 000	350
Pumps		200	2 200
Salt water filters		2 000	450
Anchor cables (power station)	m	1 000	22
Concrete anchors	m^3	240	13
Internal electrical system			4 500
20 kW units		10 500	100 000
Canal walls	m^2	600 000	5 300
Steel frame	kg	2 000 000	2 200
Hulls	m	10 000	1 100
Fresh water filters		10 000	1 100
Shipping and assembling			11 000
Unexpected costs			9 982
Total cost			145 000

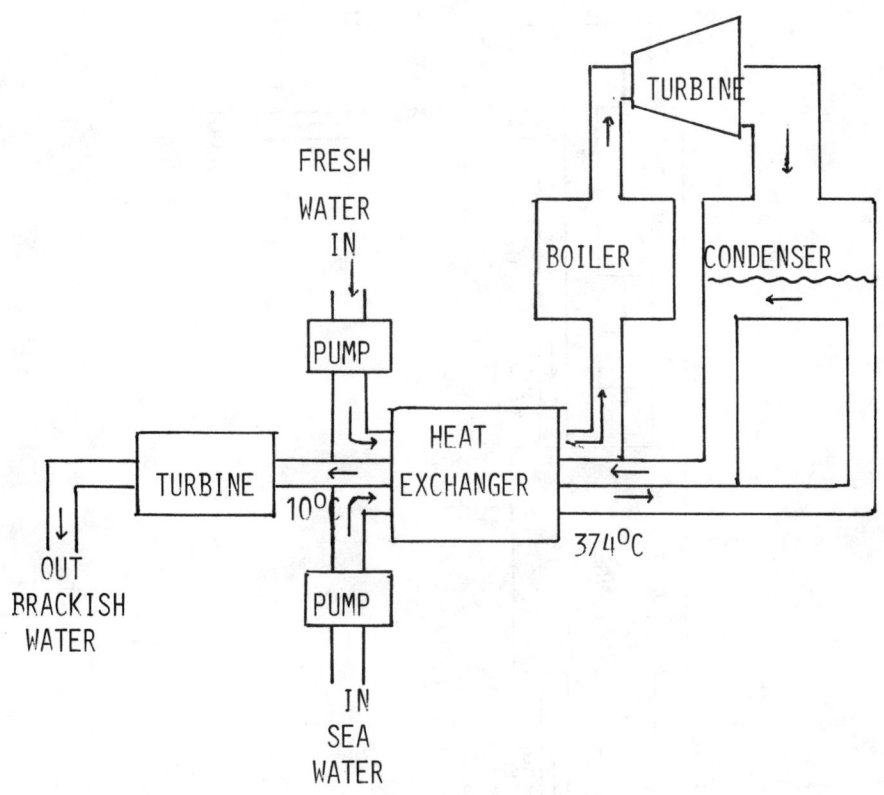

Figure 1. Steam Turbine

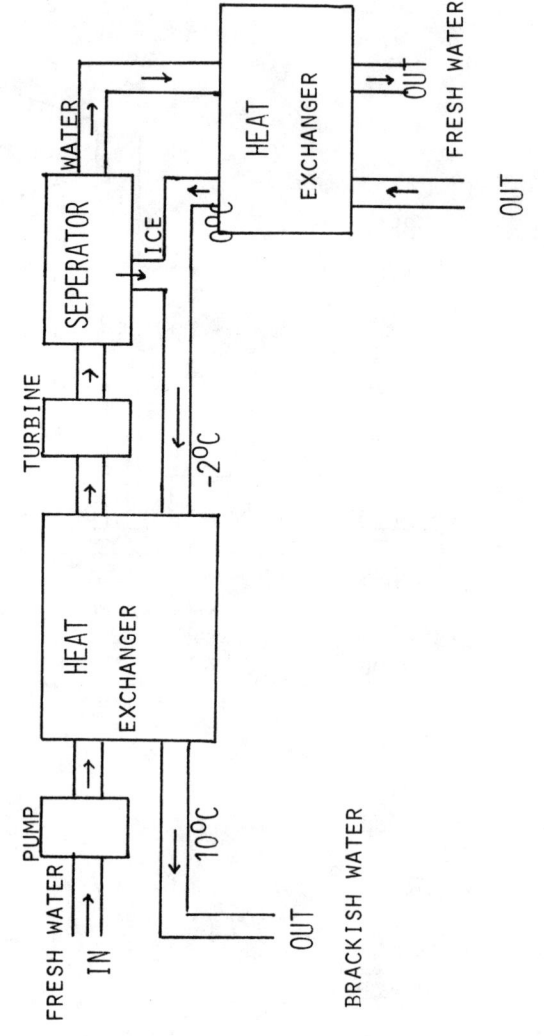

Figure 2. Freezing Machine

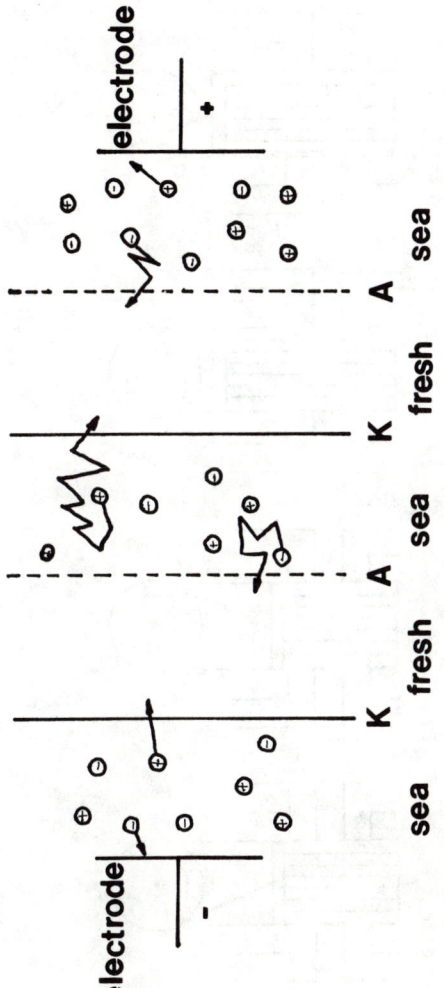

Figure 3. Concentration Cells

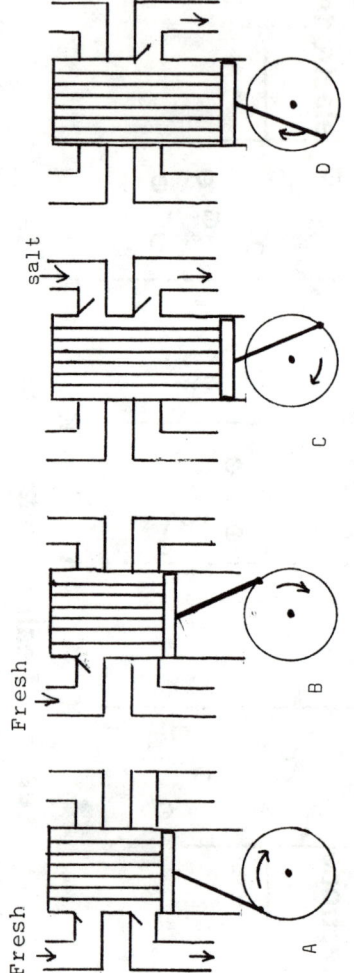

Figure 4. Mechanochemical Engine

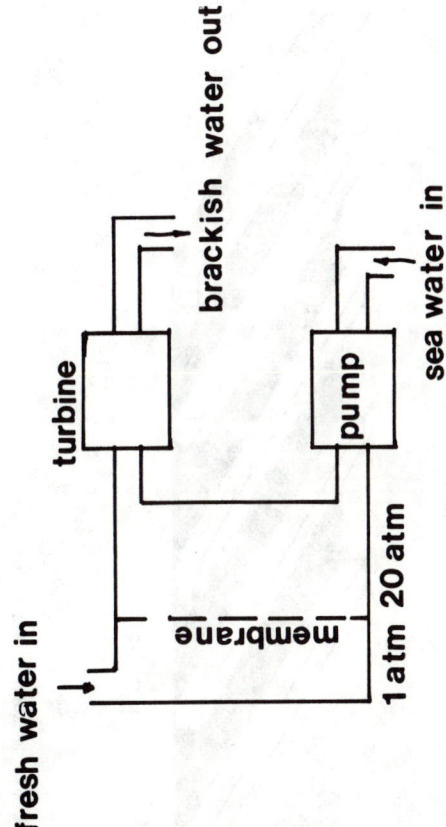

Figure 5. Osmotic Turbine

Figure 6. Membrane Structure

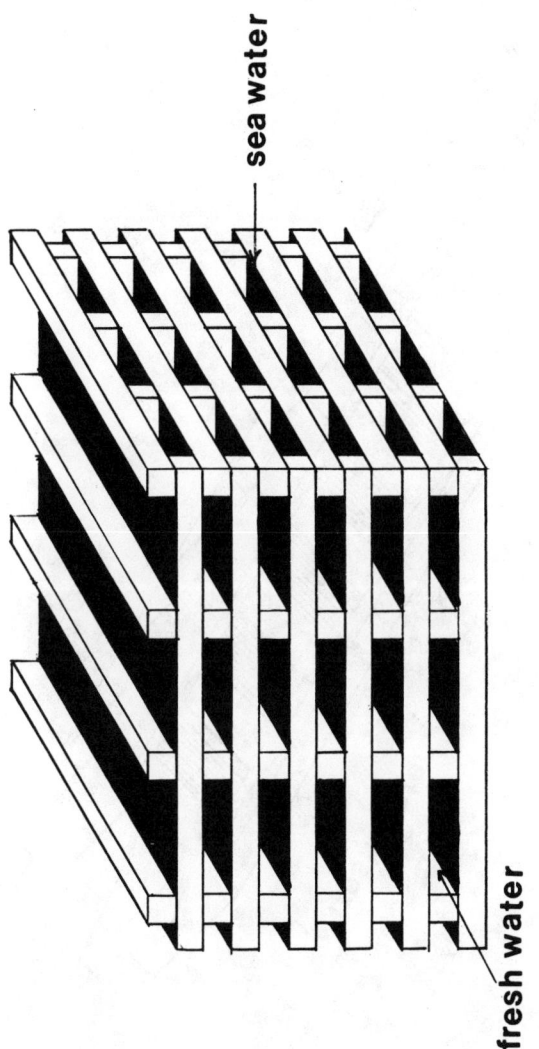

Figure 7   Part of a Pile Segment

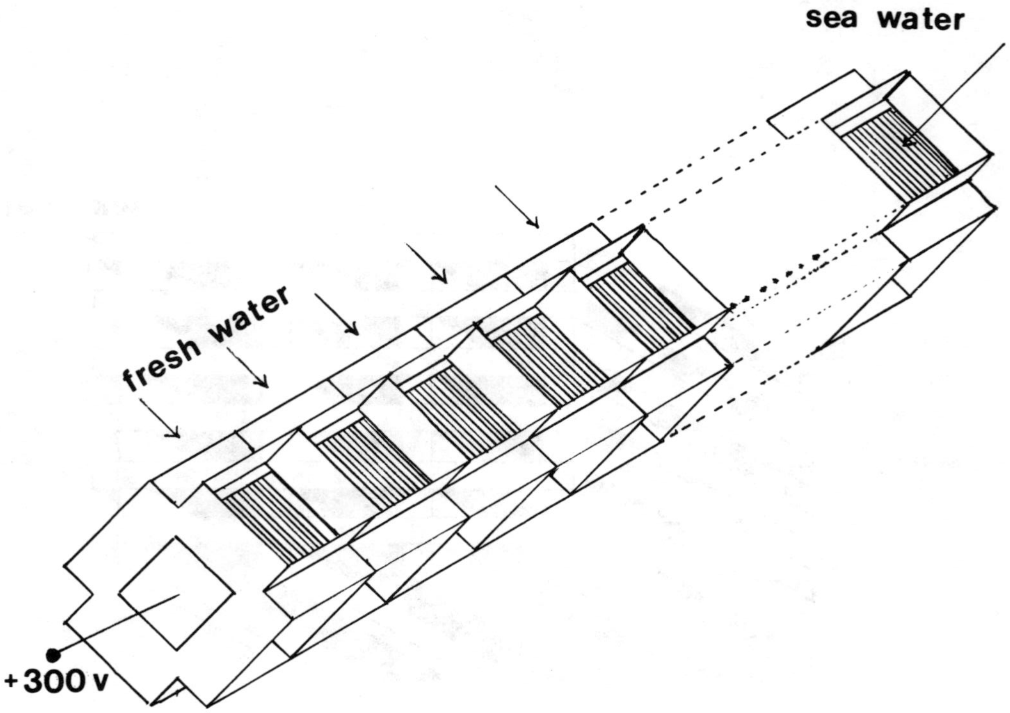

Figure 8. Membrane Pile

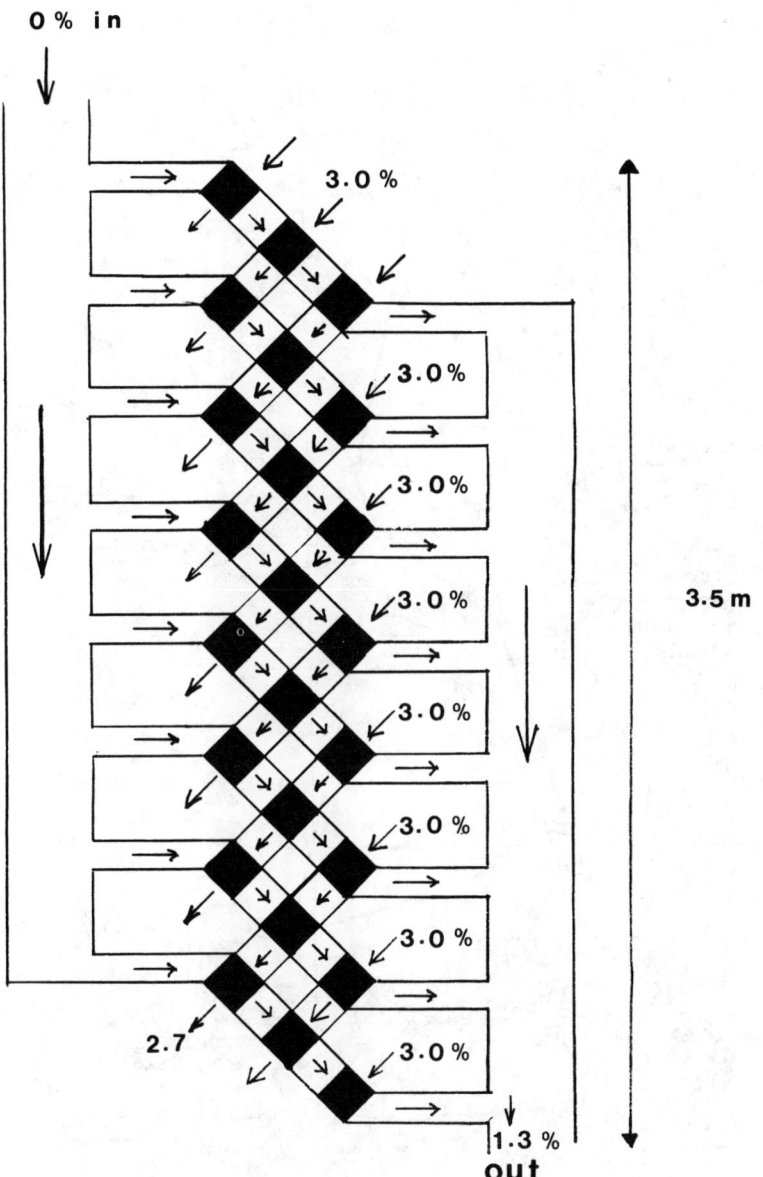

Figure 9. 20 kW Module

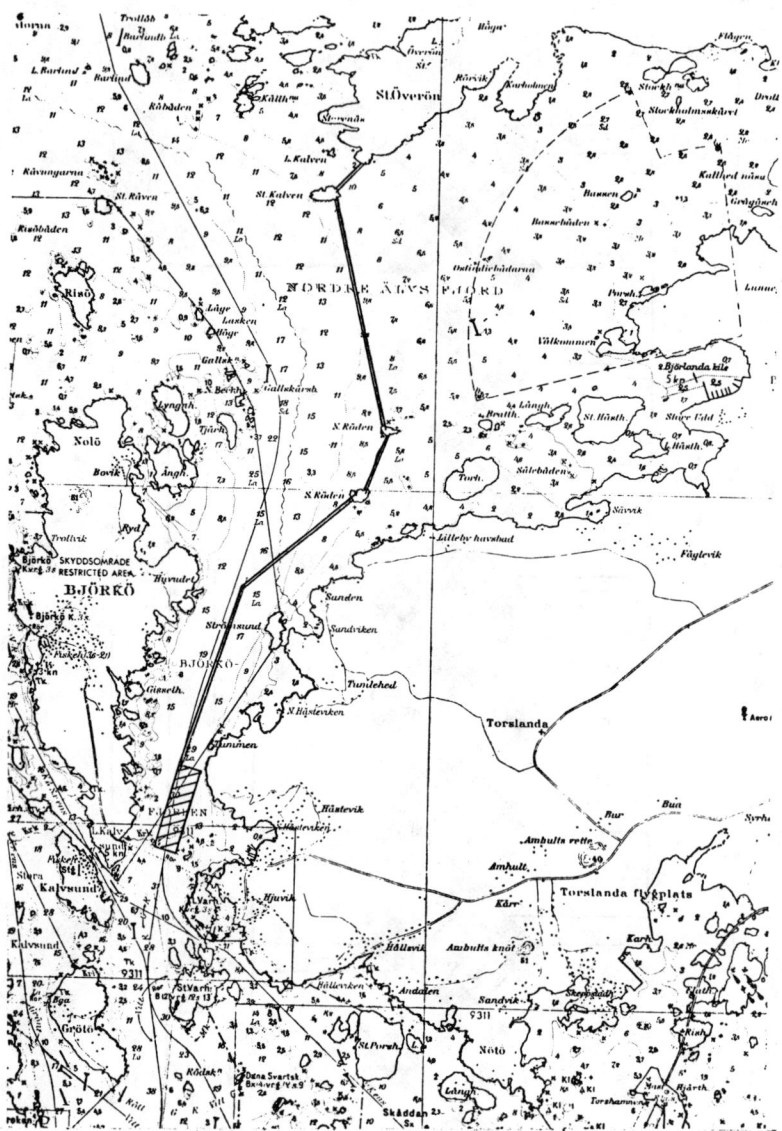

Figure 10. The Power Plant Between Björkö and Hisingen

Heat Extraction from a Salt Gradient Solar Pond

F. ZANGRANDO and H. C. BRYANT
Department of Physics and Astronomy
University of New Mexico
Albuquerque, New Mexico, USA

ABSTRACT

A salt gradient solar pond is an efficient, low cost solar energy collection and long range storage system for low temperature heat. A full-scale demonstration salt gradient solar pond has been in operation at the University for over two years. The object of our research is to establish operational parameters, selection criteria for the materials to be used, cost, performance, as well as to study the physical behavior of doubly-diffusive systems exposed to the environment. Although the current research is primarily geared to space heating (which varies seasonally) and industrial process heating (which poses a constant demand), crop-drying, water desalination, cooling and electricity production are possible applications of the solar pond which are receiving attention elsewhere. In the second year of successful operation the large storage layer has reached a temperature of 93°C; this is a record high for sodium chloride ponds with storage. Heat has been successfully extracted from the 175 m^2 pond on a daily basis since November 4, 1977. The amount of energy extracted corresponds to the heating requirements of a 185 m^2 house in Albuquerque. Current estimates indicate that the pond will meet this demand, dropping to a minimum temperature of about 30°C in early February 1978.

A. INTRODUCTION

A body of water collects a large amount of energy from the sun throughout the year yet, in general, the water temperature remains below ambient as heating produces a convective circulation which transports the energy absorbed to the surface, where it is quickly dispersed.

A few naturally salty lakes such as Lake Wanda [1], Los Roques [2], Hot Lake [3] and Lake Mahega [4] display an unusual behavior: the temperature of the water at the bottom is substantially higher than the surface temperature. Studies of these lakes have shown a non-uniform vertical distribution of salts, commonly established by salt leaching at the bottom of the lake while a river (or the sea) supplies fresh water (or low salinity brine) to the surface. In this manner a natural salinity gradient is established and maintained, preventing convection because the increase in density due to the dissolved salts offsets the effect of thermal expansion due to local heat absorption. The density gradient in turn allows for a temperature gradient to develop, positive downward yet stable, and for characteristically higher temperatures at the bottom of the lake. Temperatures of 60-70°C are not uncommon. Since water is a poor conductor of heat and it is also opaque to infrared, the salinity gradient allows the solar flux, which is mostly visible and ultraviolet, to penetrate deeply, effectively "trapping" the energy within the lake's lower regions.

B. HISTORICAL DEVELOPMENT

Natural occurrence of solar lakes was reported for the first time in 1902; Von Kalecsinsky not only described a phenomenon which had puzzled him greatly but also proposed the establishment of artificial solar lakes to harness the energy of the sun for home and industrial purposes [5]. In 1948 the same idea was expressed by Dr. Rudolph Block to the Israel National Research Council and, about ten years later, research on solar ponds began in Israel [6-9]. At that time, solar ponds were considered solely for the production of electricity; as they were not competitive with the then plentiful and cheap conventional fuels, the research was discontinued.

Awareness of an energy shortage brought a change in attitudes regarding the applications of solar ponds; instead of electricity generation, the main thrust was directed to low temperature applications. Proposed uses include: space heating and crop drying [10], water desalination [11, 12], space cooling [13] and industrial process heat. Theoretical studies and predictions for the behavior of solar ponds have been performed [9, 10, 14, 15, 16] and some experimental models, mostly laboratory ones, have been monitored and analysed [17-20]; the bulk of this research involves salt stabilization, very shallow gradient regions and no storage convective layer to provide increased heat storage capacity. Extensive economic feasibility studies, comparisons and predictions have also been performed for various economically favorable applications of solar ponds [21,22] and the idea of power production has been revived [13] although solar ponds are apparently not yet competitive to conventional fuels in this application [23].

C. THE PRINCIPLE OF A SOLAR POND WITH STORAGE

Section A describes how a salinity gradient can effectively prevent convection in natural lakes. If below the region of strong gradient a large convective layer is added, this region will act as storage, insulated from the atmosphere by the gradient above. This storage region would make the solar pond relatively insensitive even to weekly fluctuations of the insolation and ambient temperature, while facilitating heat removal procedures since this region would be large, homogeneous and convective. This idea was presented by Rabl and Nielsen [10] as a means to provide seasonal storage for winter home heating. They suggested separating these two regions with a partition: if the partition were stiff, fresh water could be used in the storage region without problems due to buoyancy effects; if the partition were of thin plastic, water at the maximum gradient concentration would be used. As the cost of a stiff partition is high and the installation is not trivial, we opted for a salty convective layer and no partition at all. In this case the solar pond contains no physical partition but is still divided in two regions: the upper one (gradient) in which the salinity content changes from no salt at the surface to a maximum concentration at the "boundary"; the lower one (storage) which extends from this "boundary" down to the floor of the pond and which is homogeneous at the maximum concentration. See figure 1.

D. OVERVIEW OF OPERATING PRINCIPLES AND EXISTING FACILITIES

Until now we have described the research performed on salinity gradient stabilized solar ponds. A second approach to the problem of preventing convection of water is viscosity stabilization [24], in which the addition of organic polymers or other substances to water will "gel" the pond and maintain it statically stable. A third approach has been investigated by a group from the Lawrence Livermore Laboratory: the shallow solar pond [25]. This approach is very different from the previous two as thermal convection of the water is not prevented. The ponds contain a shallow layer of water which is exposed to the sun during the day and is covered with a transparent material which retards loss to the environment; at night, the heated water must be moved and stored in a large, insulated reservoir.

The following discussion will be limited to the salinity gradient concept. To our knowledge, the following are the only facilities currently in operation or in planning stages; they are divided according to the salt used to establish the gradient and stabilize the pond against convection.

a) Ponds using sodium chloride (and long term storage):
The solar pond at the Ohio State University was built in the Summer of 1975; estimated area, 200 m^2; depth, 2.5 m [26]. The one at the University of New Mexico (UNM), which will be described in detail within the next sections, was built a few months later and is of comparable size: area, 175 m^2; depth, 2.5 m. A third pond, covered by a greenhouse, has also been in operation since 1975 in Wooster, Ohio; area, 155 m^2; depth, 3.6 m [27]. A large solar pond is currently under construction in Miamisburg, Ohio and will be used to supply heat for a municipal swimmingpool and adjacent shower building; area, 1400 m^2; depth, 3.5 m [28]. A small demonstration pond is planned for Des Moines, Iowa [29]; surface area will be approximately 60 m^2.

b) Ponds using magnesium chloride:
Experimental work has been revived in Israel with the construction of a 2000 m^2 pond in Sodom, near the Dead Sea. This pond is only 0.8 m deep as long term storage has not been researched there. A second large pond should supply cooling and hot water for a hotel near the Dead Sea [13]. Plans to turn part of the Dead Sea itself into a solar pond have also been submitted [30].

E. POSSIBLE SALTS

Any salt as well as any natural brine can be used to establish a solar pond; the criterion for selection of a particular salt should be based on the following factors:
1. It must be safe to handle and environmentally sound; disposal problems associated with toxicity and ground water contamination must be carefully evaluated.
2. It must be cheap and readily available; the proximity to a large

supply of a particular salt will weigh favorably on the choice.
3. It must not reduce the optical transmission characteristics of water; salts which enhance organic growth, those which form colloidal suspensions or simply color the water should be avoided.
4. The solubility should be strongly temperature dependent in order to reduce or eliminate diffusion; a saturated pond would have a stable gradient with no diffusion.

Sodium and magnesium chlorides satisfy the first three criteria; other salts which have excellent temperature dependence do not satisfy the remaining necessary conditions. Having chosen to minimize initial costs, we are presently using NaCl, despite its almost constant solubility.

Salts which satisfy criteria 1, 3, 4 might be ideal as they would allow the pond to be in an entirely saturated state and therefore eliminate problems with diffusion. This fact alone may render them economically justifiable as there would be no replenishment necessary and no maintenance requirements. A detailed study of possible salts is planned for next year at the University of New Mexico.

F. CONSTRUCTION AND INSTRUMENTATION

The solar pond at UNM was built in the Fall of 1975 with dimensions: top diameter, 15 m; depth, 2.5 m; bank angle, 34°; average collecting area, 105 m^2; total capacity, 230 m^3; average storing capacity, 130 m^3. Figure 2 is a schematic diagram of the pond. The pit was excavated to about one half the desired depth and the dirt removed raised the banks above the level of the surrounding ground for drainage. The walls were then smoothed and compacted to prevent possible liner perforation, as no insulation separates the liner from the walls. Although the sloping walls result in a large loss in volume compared to vertical walls, the former do not require retaining work. The circular shape of the pond was mainly chosen to facilitate mathematical modelling of its behavior but it also maximizes the volume per wall area.

Thermocouples were buried at uniform depths in the ground below the pond and in the side walls to allow for monitoring of the temperature distribution around the pond (for locations see figure 2). Two motor-driven thermocouples scan the depth of the pond to monitor the internal temperature distribution. The density distribution in the pond has been measured until now by weighing samples extracted from different levels. As this method provides us only with discrete data points, we are in the process of switching to a sensitive conductivity probe which will allow collection of continuous data proportional to the local density. A black and white Eppley pyranometer has been installed near the pond to measure the total (direct and diffuse) radiation on a horizontal surface. The signal is displayed on a strip chart recorder with integrator. Until now all data obtained has been recorded on strip charts or manually. We have recently acquired a data logger to digitize all information received on a daily basis and produce a tape which will be fed directly to the computer for analysis.

G. MATERIALS USED

Pond Liner: Hypalon, 45 mils, 3 plies with two nylon mesh reinforcement between the plies. This is guarantedd for twenty years against "normal" weathering and has not been tested by the factory above 77°C. Our experience is that it softens at 100°C, but appears to remain strong enough for the purpose. It is our impression that a thinner liner, ie. 30 mils thick, would have performed equally well. Maximum operating temperature so far has been 93°C and the liner has performed satisfactorily until now. We have not encountered problems associated with weeds growing through the liner, as reported by other research teams using Butyl rubber liners. The Hypalon liner was manufactured into a square of 22 m x 22 m with all seams glued at the factory. Since the pond was circular by choice while the liner was square by necessity, the excess material from the corners was later glued in the field to line a subsidiary shallow pond. These field seams have also performed well but have never been subjected to water temperatures above 50°C.
Evaporation Ponds Liner: black polyethylene, 8 mils thick, with no reinforcement. These sheets were installed directly on dirt and sand in the Summer of 1977 and have been adequate for the purpose until present.
Pump and Filter: Commercial swimming pool equipment with a stainless steel sand filter, brass pump, 1 Hp. motor; maximum flow 66 gpm, 1.5 inch suction and discharge ports. No problems have been registered with the units themselves but at high operating temperature the plastic leaf strainer preceeding the pump has partially melted while the original PVC piping connecting the six-way valve to the filter unit developed a leak at the inlet to the filter. Both were minor problems.
Valves: gate, three-way and four-way valves, all brass. These allow use of only one pump to circulate water or brine to and from any reservoir, including circulation through the heat exchanger.
Piping: polyethylene hose is used for all cold water lines and for hot brine syphons where the pressure drop is small. Although suggested for temperatures below 70°C, successful syphoning has been performed up to 85°C. Multipurpose steam hose is used on all other suction and discharge lines carrying hot brine. This hose is overrated for the application but we could not find any other flexible hose which could withstand fluid temperatures around 100°C.
Diffuser Assembly: two plates of plexiglas, 30cm in diameter, 2cm thick, with a 3mm gap on a stainless steel support, driven on the bank of the pond by a synchronous motor and reduction drive to scan the depth of the pond at selected rates. This is used for selective injection of fresh water or brine into the gradient region as well as return of the salty brine to the convective layer, after circulation through the heat exchanger.
Temperature Sensors: copper-constantan thermocouples; 24 gauge wire was silver soldered and encapsulated in "shrink tubes." No failures have been registered yet.
Scanners: thermocouples attached to weights are suspended on fishing lines by means of a pulley arrangement. Vertical scans into the pond can be made at any selected position over the east-west or north-south diameter of the pond. The fishing line is rolled on calibrated, threaded drums driven by synchronous motors so that vertical position of the thermocouples is a known linear function of time. Scanning rates are 6 and 18 cm/min. Delimiting markers are attached to the lead wire and reverse the direction of motion by activating a switch. The horizontal temperature distribution is obtained by using the scanners simultaneously.

Wall Scanner: synchronous motor drive on a continuous loop to which a thermocouple is attached. The support is made of thin stainless steel tubing welded as a ladder, and rests on the bank of the pond. The drive and loop are mounted off-center to provide three distances from the wall: 5, 9, 15cm. The scanning rate is 6 cm/min and can be matched to the vertical scanners for comparison of vertical temperature distribution.
Heat Exchanger: at present, 5cm diameter brass tubes, 3.6 m long, in 4 parallel sections 38cm apart. This heat exchanger has been installed in the shallow subsidiary pond.

H. FILLING PROCEDURE

Several schemes are adequate to establish the gradient:
1. Natural Diffusion: a layer of fresh water (region II) is superposed onto a layer of high salt concentration (region I). The procedure would be to mix salt and water in the pond to produce the high salinity layer, then add fresh water above. If the supply of fresh water is large, the mixing at the boundary will enhance salt redistribution. The salt migrates to the surface by diffusion and with time, establishes a salinity gradient. This is of course the simplest but also the slowest process. See figure 3.1.
2. Redistribution: once the two aforementioned regions are established, water can be pumped from region II and injected into region I with a horizontal diffuser which scans the depth of the pond. The rate of redistribution will be dependent on the concentration gradient required. This is the method we propose and support for future establishment of the gradient. This method has been used very effectively to change the salinity gradient in our pond. It can be done automatically and does not require the operation to be performed all at once. See figure 3.2.
3. Stacking: after mixing a high concentration solution in order to establish the storage layer, batches of progressively lower salinity brine are "stacked" one over the other until fresh water is added to the surface. This is the method we used originally; it is successful but has two drawbacks. One is that a mixing tank is required; two, the layers are exposed to wind action throughout the filling process and mixing can occur beyond the desired amount. This therefore requires continuous operation if climatic conditions are unfavorable. See figure 3.3.

The solar pond at UNM was filled in a 28 hour period starting November 27, 1975. A convective layer 20 cm thick containing 15% NaCl by weight was established in the pond. Above this, fifteen layers about 3 cm thick were pumped one over the other through a horizontal diffuser floating on the brine. Batches of brine were mixed in the subsidiary pond, starting at a 14% concentration, the correct volume was pumped into the solar pond and the brine in the mixing pond was diluted to a 13% concentration; the procedure was repeated until fresh water was added to the surface. At completion, the solar pond contained a gradient region 70 cm thick, for a total depth of 90 cm. The pond was not filled to capacity to allow experimentation with the gradient through the winter. In the Spring and Summer 1976 the pond was progressively filled to capacity by injection of high salinity into the convective layer.

I. LIGHT PENETRATION

As discussed in section A, water, with or without dissolved salts, is opaque to long wavelengths; this characteristic, which prevents the convective layer from radiating the energy absorbed also prevents a portion of the sunlight impinging on the surface of the pond from reaching the convective layer itself. Solar energy reaching the pond surface is partially reflected, absorbed and transmitted; absorption of the long wavelengths is an important effect as 27% of the total solar spectrum is absorbed within the first centimeter of water. The collection efficiency is therefore limited by the absorption of water in the gradient layer.

Rabl and Nielsen [10] have shown that the insolation can be approximated well by dividing the spectrum into four bands, each with an effective attenuation length. Thus

$$H(x) = H_s \zeta \sum_{n=1}^{4} \eta_n e^{-\mu_n x}$$

where H_s is the insolation at the surface, ζ the transmission coefficient, μ_n the effective absorption coefficient for a given range of wavelengths and η_n are constants.

We have found that the fraction of light remaining after passing through a thickness x of clear water can be expressed as accurately in the range 1 cm to 10 m by [31]:

$$f(x) = a - b \ln x$$

where x is the depth of water in centimeters, a = 0.73 and b = 0.08. In figure 4 we compare these two curves by plotting $f(x)$ and $H(x)/H_s$ with $\zeta=1$. Experimental data obtained at the UNM pond agrees approximately with these predictions. A solar cell scan without temperature compensation is superposed on figure 4 and is presented for illustration purposes only. We should note that little data has been collected on this subject so far but plans to improve the accuracy and frequency of our records with the installation of a scanning pyranometer (instead of a solar cell) should materialize shortly.

J. WATER CLARITY

The surface of the solar pond is cleaned using two different procedures: the surface is flushed on a continuous basis using a low rate fresh water feed and a slow draw-off syphon which collects the run off into the evaporation ponds for salt recycling; after severe dust storms the entire surface layer is pumped, filtered, then returned to the pond. Most materials that are blown into the solar pond either float or settle rapidly to the bottom, with the occasional exception of some leaves which remain in suspension for a longer time. If a convective layer is present at the surface, these leaves will sink to the base of this layer, marking the boundary with the gradient region below. Once at this level, the leaves appear to remain there until

enough salt has crystallized on their surface and makes them sink to the
bottom; this has never been a serious problem. As there are several decid-
uous trees in the vicinity of the pond, we have used a floating net, fixed
in the pond, to immediately collect leaves and other floating debris through
the natural circulation of the surface water in times of breeze or even
strong winds (at these times the surface is swept efficiently and collection
of all debris at the banks is possible).

The debris that have settled to the bottom do not affect the transmission
nor inhibit the absorption of the liner; no attempt has therefore been
made to remove this settlement. On occasion, leaves appear to have colored
the water in the convecting region; this does not affect the collecting
efficiency of the pond since we do want total absorption in this region.
If this effect were so strong as to absorb all the transmitted light at the
top of the storage layer, it could create a problem as a new gradient would
be established, inhibiting heat extraction; we have observed a temporary
condition of this nature once, in the Spring of 1976. The effect disappear-
ed in a short time, probably due to eventual settling of the leaf pigment
(see figure 5, curve of June 2, 1976).

In an uncovered pond, there seems to be no reason to have a black liner; in
fact, a light-colored liner might reduce the tendency for localized convection since it would decrease the amount of heat absorbed at the walls. As
for the bottom of the pond, it is inevitably covered with brown debris which
mask the original color of the liner.

To eliminate turbidity caused by biological growths in the upper region of
the pond, a copper sulfate solution is sprayed over the entire pond two or
three times a year. Two pounds of copper sulfate diluted in warm water are
sufficient to remove all visible organisms and clear the pond in three days.
Samples collected from the pond show very low turbidity and, if the solar
angle is high, one can see details of the bottom very clearly.

K. ECONOMIC EVALUATION

Excluding land, the construction and installation costs to duplicate a pond
such as this one, with an average collecting area of 105 m^2 (1130 ft^2) are
estimated to be as follows:

1.	Excavation	$ 900
2.	Hand Labor	400
3.	Liner, Hypalon, 30 mils, reinforced	1500
4.	Salt, 40 tons *	1400
5.	Pumps, piping and heat exchangers	1500
		$5700

*this is an amount of salt sufficient for successful operation of the solar
pond in the residential heating mode.

This amounts to $54/$m^2$ ($5/$ft^2$), which is below the total cost ot any other
comparable solar heating installation. While items 3, 4, 5 should be

approximately constant for any location or application, items 1 and 2, as well as land, will vary widely for different locations and could be partially eliminated if excavating machinery and hand labor were already available at the site. Costs per unit area would also decrease for larger solar ponds as the unit construction and liner costs would drop sharply and the edge losses would be proportionally smaller, improving the pond's efficiency.

Previous cost estimates [10, 21, 22] are confirmed by the extrapolation we have made on the basis of costs incurred building and operating the UNM facility. Our predictions [32] are that a solar pond supplying heat for one house compares favorably with electric heating while a large pond supplying heat to a cluster of homes or apartment buildings can be competitive even with gas heating. These comparisons were made on basis of resource and equipment expenditures, excluding maintenance requirements.

L. TEMPERATURE RESPONSE OF THE SOLAR POND AND THE GROUND BELOW.

The UNM pond was originally filled in November 1975 with water at 20°C. Shortly after filling, the surface of the solar pond became a 5 cm thick sheet of ice which persisted for over one month; temperature scans were not taken during this period. Figure 5 shows several temperature scans taken from January 1976 to August 1977, depicting the temperature distribution in the solar pond and in the ground below. While the thickness of the insulation layer was maintained between 60 and 110 cm throughout this time, the thickness of the convective layer was increased from 20 to 150 cm by August 1976.

The temperature difference in the curves for August 8, 1976 and 1977 clearly indicates that the solar pond is far from reaching steady state; this slow response can be attributed mainly to the ground, which approaches steady state very slowly, approximately as $1/\sqrt{t}$. Following Rabl and Nielsen [10] and considering the ground under the pond as a semi-infinite solid with a skin depth of 2 m, the temperature of the ground is only at about 3/4 of its steady state value after two years of operation. This effect can also be seen in figure 6 which depicts the temperature distribution under the solar pond as recorded by the arrays of thermocouples burried in the ground on the East-West axis. The approximate isotherms obtained indicate that there is no appreciable ground water movement under the pond.

The daily temperature response of the solar pond in both a sunny and a cloudy day is shown in figure 7. As one can see, the daily change in ambient temperature and insolation affects the surface temperature as one would expect, while the effect on the convective layer is greatly minimized. In both cases, the temperature of the convective layer reaches a daily maximum at about 17:00, a daily minimum at about 8:00. In a reasonably sunny day (August 22), the convective layer temperature varies by 1.3°C from daily maximum to minimum, with a net gain of 0.9°C over the 24 hours. In a cloudy day (August 11), in which the insolation is well below the period average, the convective layer maintains its temperature within a range of 1.1°C, with no net gain for this day. The sharp drop in surface temperature around 18:00 is due to rain, characteristic of this time of the year. Comparison of the two convective layer temperature curves indicates a very similar response and verifies that daily rather than hourly average temperatures and insolation are quite satisfactory for an accurate modelling of the solar pond.

Figures 8 and 9 show instead the response of the pond to long periods of low insolation and/or low average ambient temperature. Figure 8 portrays the behavior of the convective layer temperature in January 1977, while the pond was dropping to its yearly minimum of 31°C. As we have seen previously, a day's reduction in insolation below average does not entail a drop in pond temperature; in this case the duration is seven days and still no great effect is registered. Instead, at consistently low ambient temperatures (two weeks below 0°C) the net temperature drop is doubled, from 0.2 to 0.4°C/day. At this time of course, the surface of the pond was completely frozen and the amount of light penetrating the pond was reduced. However, similar occurrences when the surface was not frozen reflect the same behavior. Figure 9 depicts similar circumstances during the period March-April 1977, when the temperature of the convective layer was well on its way up.

Rabl and Nielsen [10] calculated the steady state temperature for a pond of infinite lateral extent and consequently no edge losses; in the UNM pond edge effects cannot be neglected. To compensate for this condition in the derivation, one can introduce a term which is proportional to the difference between pond temperature T and ground temperature T_g. Since the temperature of the undisturbed ground remains approximately constant throughout the year and is equivalent to the average ambient temperature $\bar{T}_a$, this ground loading term will appear in both the time dependent and independent equations as: HL $\tilde{T}$ and HL$(\bar{T}-\bar{T}_a)$, where HL is the heating requirement of the ground per unit area-degree second and is equal to 1 W/m^2-°C for the UNM pond. This value is determined from the data collected; the total heat load of the ground is comparable to the conductive loss through the insulating layer and amounts to about 60 W/m^2.

Let us consider that all the energy reaching a depth ℓ_i is absorbed there and is balanced by the heat flow through the gradient, the heat flow to the ground and the heat extracted. Following the notation of Ref.10, the heat equation is modified to

$$\bar{H}(x=\ell_i) = K_w \frac{\partial \bar{v}}{\partial x}\bigg|_{x=\ell_i} + HL(\bar{T}-\bar{T}_a) + \frac{\Psi}{A}(T_r-\bar{T}_a),$$

where the bar refers to average values, $\bar{H}$ is the radiation reaching the convective region (at depth ℓ_i), K_w is the thermal conductivity of water, $\bar{v}$ is the temperature of the water at a depth x, Ψ is the average home heating requirement per degree second, A is the area at depth ℓ_i and T_r is the room temperature.

Solving the time independent heat equation while expressing the radiation reaching a depth x as $\bar{H}(x)=\bar{H}_s\zeta$ (a-b ℓnx), the solution for the average steady state temperature at depth x becomes

$$\bar{v}(x) = \bar{T}_a + (\bar{T} - \bar{T}_a)\frac{x}{\ell_i} + \frac{\bar{H}_s \zeta}{K_w} b \times \ln\left(\frac{\ell_i}{x}\right)$$

while the equation for the mean temperature of the convective layer becomes

$$\bar{T} = \bar{T}_a + \frac{\bar{H}_s \zeta [(a+b) - b \ln \ell_i] - \frac{\Psi}{A}(T_r - \bar{T}_a)}{\left(\frac{K_w}{\ell_i} + HL\right)}$$

Using data collected at the UNM pond and setting $\Psi=0$, the prediction for the steady state, average convective layer temperature is 81°C. By considering the heat extraction necessary to supply one house in Albuquerque (4.75×10^7 Joules per degree-day for 2416 °C-days), the predicted average convective layer temperature drops to 67°C. As can be seen in figure 10, the solar pond is rapidly approaching the predicted value for the no-load condition; as the average convective layer temperature was 65°C this fall, the 1977 cycle should be reproduced next year since heat extraction from the solar pond is currently under way.

For the two consecutive summers of operation, the UNM pond has reached the maximum yearly temperature within the same week: 71°C on August 7, 1976 and 93°C on August 6, 1977. To our knowledge this is a record high for NaCl ponds with large storage capacity. The minimum temperature was 31°C on January 30, 1977. The recorded lag between convective layer temperature and insolation is therefore 1.5 months for the two summers and 1.3 for the one winter. As instabilities occurred in both years at the maximum temperature (see section M for details), it is premature to say that the UNM pond falls short of the theoretical prediction: two months.

The sharp drops in pond temperature for the period April-August, 1976 are not due to climatic conditions but rather to the addition of cold brine to the bottom in order to raise the total level of the pond.

M. STABILITY OF THE GRADIENT

For a fluid to be in static equilibrium, the density gradient must be positive downward, thus

$$\frac{d\rho}{dx} > 0,$$

where x is positive down. In the salt gradient solar pond, the density is a function of the salt concentration C and the temperature T; this condition can therefore be expressed as

$$\frac{d\rho}{dx} = \frac{\partial \rho}{\partial C} \cdot \frac{\partial C}{\partial x} + \frac{\partial \rho}{\partial T} \cdot \frac{\partial T}{\partial x} > 0$$

As $\frac{\partial \rho}{\partial T} < 0$, it may be written as

$$\frac{\partial \rho}{\partial C} \cdot \frac{\partial C}{\partial x} > \frac{\partial \rho}{\partial T} \cdot \frac{\partial T}{\partial x}$$

However, in a doubly-diffusive system such as the solar pond, where the salt acts as the stabilizing element, it is not sufficient to satisfy the static condition to insure stability, as the departure from equilibrium is not necessarily to aperiodic modes as in a homogeneous fluid. Instead, it may be to overstable modes of increasing amplitude[33, 34]. Consider a particle of fluid at a certain depth x, perturbed by a small disturbance to $x-\Delta x$. As heat diffuses much more rapidly than salt, this element of fluid will become colder but will maintain most of its salt. The density will increase and the restoring force will therefore be greater than previously (overstabilizing); the particle will overshoot the position $x+\Delta x$ by, say, Δ. At $x+\Delta x +\Delta$ the element will become hotter but will not gain much salt; it will therefore be less dense and again, the restoring force will be greater than previously. In this manner a "small" disturbance will create an overstable oscillatory mode with increasingly larger amplitude.

Weinberger states the stability condition in a solar pond [9] as

$$\frac{\partial \rho}{\partial C} \cdot \frac{\partial C}{\partial x} > f \frac{\partial \rho}{\partial T} \cdot \frac{\partial T}{\partial x},$$

with $f= (\nu + \alpha')/(\nu + \alpha'')$; ν is the kinematic viscosity and α', α'' are the diffusivities of heat and salt. We are presently investigating the validity of this form as the derivation considers the established temperature gradient but does not account for the energy absorbed during the daily heating. Results of the study will be presented in a subsequent paper.

The solar pond has been operated at consistently low salinity gradients to establish the stability limit experimentally. Records of convective zones appearing in the otherwise non-convective region show local density gradient values below 1.4×10^{-3} gm/cm^4 for previous temperature gradients of up to 1°C/cm. These conditions have been registered a few times during the past two years; under these circumstances the shape of the temperature distribution changes in a few days with small convective layers appearing in the regions of lowest gradient. It has been our experience that a small and localized instability, bounded by strong gradients, disappears spontaneously in a short period of time. Otherwise, a scanning redistribution of brine (see section O for details) will correct this condition in one or two days.

On occasions, we have also recorded a breakdown of the insulating layer when the gradient should have been stable. This occurs only after a strong and lasting windstorm and the results appear as a complete destruction of the non-convective region which is replaced by a succession of small convective layers. We do not yet understand the mechanism by which wind can act as a catalyst for instabilities in regions where direct action could not be possible. Comparison of data and experiences with the solar pond

project at Ohio State University show that this is a condition peculiar to
the UNM pond only. It would therefore seem that the bowl shape as well as
the less inclined walls of the UNM pond are the major contributors to this
occurrence. Our conjecture is that an oscillation of the entire pond can
be set up by sufficiently strong and lasting winds and that this disturbance
is great enough to start convective cells at the walls even in the presence
of a stable gradient. Once it is disturbed enough, overstable modes could
be maintained, causing the convective cells to spread towards the center of
the pond, given sufficient duration of the wind disturbance.

Two more effects can create a disturbance of the gradient at the walls.
Daily wall heating due to the absorption of radiation by the liner forms
small convective cells; this condition appears to involve only a small
boundary layer at the wall itself and the effect subsides every evening.
On the other hand, recent laboratory experiments performed at the University
of Cambridge [35] show the possible formation of horizontal layers in a
doubly-diffusive system induced by the presence of an insulated sloping
boundary. As the density must remain constant horizontally while the curves
of constant temperature and salinity must be perpendicular to the boundary,
vorticity is generated at the wall.

We then have three conditions which might disturb the gradient at the walls:
the mechanical perturbation of the wind itself, the convection caused by
daily wall heating and the vorticity induced by the slanted walls. It is
noteworthy that whatever the cause, the entire body of the pond appears to
be affected only in the presence of strong winds. While these occurrences
have not hindered the operation of the solar pond, they have provided us
with a considerable amount of data to study the behavior of a salinity
gradient exposed to daily heating and cooling, as well as other climatic
conditions.

N. SURFACE CONVECTIVE LAYER

As the surface of the UNM pond is exposed to the atmosphere, rapid night
cooling induces an even stronger destabilizing temperature gradient in the
upper region of the pond; wind stress on the surface also favors mixing
and the surface invariably becomes convective for a depth of about 10 cm.
Wind action would be minimized by an array of pipes floating on the surface
of the solar pond. If, in addition, a small amount of hot brine from the
convective layer were to be circulated through these pipes each evening, a
higher surface temperature would be maintained during the rapid cooling
period and convection could be prevented.

An obvious but costly alternative to this procedure would be to shield the
solar pond with a plastic, air inflated bubble. This cover would decrease
the solar input into the pond but it would also prevent the large evaporation characteristic of arid regions such as New Mexico. This would maintain a higher surface temperature, decreasing the heat loss by conduction
through the gradient. A cover would also prevent dirt from blowing into
the solar pond, minimizing cleaning requirements. As stated, it is a costly
measure which would benefit the small residential pond, but would not be

reasonable for either a low-cost pond or a very large one. Having chosen to minimize materials cost, we accept surface convection and maintain its size small by syphoning from the base of this layer and introducing fresh water on top. For periods of strong wind storms such as the spring, the scheme presented at the beginning of this section will be tried.

0. MAINTENANCE OF THE GRADIENT

As the working principle of a salt stabilized solar pond requires a salinity gradient to be present, diffusion of salt from the high concentration region (storage layer) to the low concentration region (surface) takes place continuously. The tendency is therefore towards destruction of the gradient, unless salt is removed from the surface and replenished into the storage layer. This is accomplished by flushing the surface of the solar pond with fresh water (or low salinity brine), collecting and evaporating the runoff and subsequently reinjecting the concentrated brine into the convective layer.

A different method has been proposed by Tabor [8], denoted as "the falling pond method". In this case hot brine is extracted from the convective layer and passed through a flash evaporator to produce concentrate and steam. In this manner, water is extracted from the brine and the steam can be used for electricity production. The concentrate, of smaller volume and higher salinity than the brine extracted originally, is returned to the bottom of the pond lowering the total level; fresh water is then added to the top of the pond, to bring the level back to the original position. If the drop in volume matches the diffusion rate, there is no net migration of salt and the gradient is maintained. This method requires initially a very large amount of salt and is therefore economically favorable only if salt is available at low cost.

In the event convection appears in the otherwise non-convecting insulation region, the gradient can only restore itself if the vertical dimension of these layers is less than 4 cm; otherwise redistribution is necessary. We have developed a method to correct these instabilities which has proven to be successful and generally expedient: a scanning injection of fresh water or high salinity brine into the anomalous region. If the gradient above the instability is strong, fresh water from the surface of the pond is pumped through the scanning horizontal diffuser. If the gradient below is weak, high salinity brine from the convective layer is used. The mixing created by pumping brine into the solar pond, plus the internal circulation due to the difference in temperature and density between the brine at a particular level and the brine injected, do not further convection; instead they generally reestablish a linear gradient within one day. In the few instances of complete gradient breakdown, hot brine from the convective layer has been syphoned to the auxiliary basin, mixed with more salt and pumped into the solar pond with the diffuser scanning from the top of the instability down to the convective layer, increasing the salinity of the entire pond. This procedure has been observed to drop the temperature of the convective layer by a few degrees only. The flow rate through the diffuser is now 110 liters/minute while the vertical scanning rate ranges from 30 sec./cm to

2 min/cm through the insulating layer, depending on the amount of salt necessary.

P. ENERGY BALANCE AND EFFICIENCY

A schematic diagram of the percent energy distribution in the solar pond, based on data collected at the UNM facility, is shown in figure 11. The predicted steady state efficiency for a solar pond of infinite lateral dimensions and a gradient of 1 m is between 20 and 30%, depending on the operating temperature [9]; this compares favorably with the distribution on figure 11 as the UNM pond is not yet at steady state.

Temperature gradients from the boundary of the storage layer to: a) the insulating layer, b) the walls and c) the floor of the pond are plotted in figure 12 for reference. These are computed on a daily basis from the temperature scans in the solar pond, as well as the data collected from the thermocouple arrays buried under the pond (for locations, see figure 2). Temperature gradients into the ground show fairly smooth curves, a yearly cycle and extremes matching the trend of the convective layer. Note that in the month of January and for part of February, the gradient into the ground below the pond was reduced to zero or slightly negative values, returning heat to the pond. The temperature gradient in the insulating layer instead has no apparent trend except for higher gradient values equidistant times from the pond's maximum. In June-July, due to maximum insolation, the pond is heating at its maximum rate while in September-October, when the temperature of the solar pond is still high, the ambient temperature drops sharply; both these conditions increase the temperature gradient. A few of the sharp peaks are related to instabilities in the gradient; these peaks occur because the actual insulating depth is reduced abruptly as a convective layer provides only a reduction of light transmitted but does not account for any insulation.

The efficiency of the solar pond could be increased by the installation of a reflecting wall on the north bank. Depending on the geometry and orientation, the wall could even double the input into the solar pond (by image), at a time when direct solar input is low. This project will be undertaken next year, after we have collected operational data on heat extraction from the solar pond as it stands now, in order to compare performances.

Q. HEAT EXTRACTION

The UNM solar pond was designed to supply the hot water and space heating requirement of a 185 m^2 house in Albuquerque; this amounts to approximately 33 MWH per year of useful heat. As no building of this approximate size is available near the facility, we have decided to model a standard 185 m^2 all frame-stucco, flat roof, single story house which is occupied by a five-person family. This house is heated exclusively by natural gas, with the exception of a rarely used fireplace. During the summer and early fall we monitored gas consumption used for hot water and cooking, to determine what fraction of the total winter use would actually supply heating. At present, only that fraction utilized for space heating is extracted from the solar pond; this amounted to 5.94 x 10^{10} Joules in the 1976-77 heating season. Figure 13 shows a comparison between the monthly gas consumption of this home and the 1976-77 monthly heating requirement in degree-days. A base of 10 MCF/month (the amount used for hot water and cooking) should be subtracted from the total to present the actual heating requirement.

On every morning of the heating period the gas meter is estimated to the nearest 10 cubic feet of gas and the portion used for heating is calculated. This in turn is converted directly into energy, using the conversion from MCF to therm supplied by the gas company with each monthly bill (approximately 8.8) and and equivalent amount of energy is extracted from the solar pond. Inefficiencies of the floor furnaces are not taken into account as we estimate them to be roughly equivalent to heat losses in the pipes, if the solar pond were used for direct heating. Because of the large thermal inertia of the pond, a delay of one day between the demand of the house and the withdrawal of that energy should be negligible; in this way, we can measure the pond's effectiveness and applicability to residential heating.

Heat is extracted by direct circulation of hot brine from the convective layer into a heat exchanger. This method was chosen not only because it is the most efficient in terms of heat transfer but also to analyse the effect of localized extraction of brine from the solar pond and to monitor possible corrosion problems inside the heat exchanger. The brine is pumped out of the convective layer, circulated through the heat exchanger, then returned to the convective layer through the diffuser; several arrangements of inlet/ outlet positions will be tried. At present brine is extracted from the bottom and injected 30 cm below the convective layer boundary to provoke perturbations of this boundary by enhanced convection. Throughout this first month of daily extraction and despite the deliberate disturbances, no detrimental effects have been noticed. Tests performed periodically since May 1977 were equally successful, even at the highest pond temperature (least stability).

The heat exchanger was placed in the auxiliary pond for two reasons: the effects of salt on the outside of the heat exchanger can be monitored and the heat extracted can be used to enhance evaporation of the surface runoff. While this setting represents approximately the condition of a heat exchanger at the bottom of the solar pond, it also renders the exchanger accessible for analysis of exterior corrosion and scaling problems.

As we are primarily interested in the response of the solar pond to large, periodic and localized fluid extraction and injection, characteristics and efficiency of the heat exchanger are secondary at this time; the relevant parameter is the amount of heat extracted from the solar pond. For reference, the heat exchanger is described in section G while the pumping rates have been around 110 liters/min. Daily heat extraction began on November 4, 1977 and totals 3.1 x 10^9 Joules at this time, with an average thermal output power of 66 KW for the duration of heat extraction. Figure 14 shows a superposition of the convective layer temperature for the two years of operation; the temperature in now 51°C. If the heat extraction continues to be as successful as it has been until now, the pond temperature should not drop below last year's minimum and that curve should be duplicated with a slight shift.

R. CONCLUSIONS

The UNM solar pond has been operated successfully for two consecutive years and sodium chloride has proven adequate for stabilization of the gradient to temperatures above 90°C. Operational controls have been established for gradient maintenance. Heat has been extracted on a daily basis without causing any detrimental effects and this operation will continue throughout the winter in amounts equivalent to the heating requirements of a 185 m^2 house in Albuquerque. The behavior of the UNM pond, although it is not yet at steady state, follows theoretical predictions. Although more operating time is necessary to establish the reliability of the system, observations made to date indicate that the solar pond is an efficient, low cost solar energy collection and storage system, adequate to supply residential heating.

Further research is needed in the following areas:

1. Salts with strongly temperature dependent solubilities, suitable for solar pond operation.
2. Mechanisms causing occasional gradient instabilities.
3. Improved mathematical model of the pond.
4. Accurate light absorption data.
5. Performance and cost effectiveness of a reflecting wall.

S. ACKNOWLEDGEMENTS

This research was supported in part by the State of New Mexico Energy Resources Board (ERB-161, ERB-76-202) and is presently under contract to the Energy Research and Development Administration/ Department of Energy (EG-77-5-04-3977).

The authors wish to express their appreciation to Javad Rassouli for his dedicated work.

REFERENCES

1. Wilson, A.T., Wellman, H.W., "Lake Wanda: an Antartic Lake", Nature, vol. 196, p. 1171 (1962).

2. Hudec, P.P., Sonnefeld, P., "Hot Brines on Los Roques, Venezuela", Science, vol. 185, p. 440 (1974).

3. Anderson, G.C., "Some Limnological Features of a Shallow Saline Meromictic Lake", Limnology and Oceanography, vol. 3, p. 259 (1958).

4. Melack, J.M., Kilham, P., "Lake Mahega: a Mesotropic Sulfatochloride Lake in Western Uganda", African Journal of Tropical Hydrobiology and Fisheries, vol. 2, p. 141 (1972).

5. Von Kalecsinsky, "Uber die Ungarischen Warmen und Heissen Kochsalzseen als Natuelische Warme Accumulatoren", Annals der Physik, vol. 4, p. 408 (1902).

6. Tabor, H., "Large Area Solar Collectors for Power Production", Solar Energy, vol. 7, 4 (1963).

7. Tabor, H., "Solar Ponds", Electronic and Power, September 1964.

8. Tabor, H., "Solar Pond Project", Solar Energy, vol. 9, 4 (1965).

9. Weinberger, H., "The Physics of the Solar Pond", Solar Energy, vol. 8, 2 (1964).

10. Rabl, A., Nielsen, C.E., "Solar Ponds for Space Heating", Solar Energy, vol. 17, 1 (1975).

11. Hirschmann, J.R., "Salt Flats as Solar Heat Collectors for Industrial Purposes", Solar Energy, vol. 13, 4 (1970).

12. Tabor, H., "Solar Ponds as Heat Source for Low-temperature Multi-effect Distillation Plants", Desalination, vol. 17, p. 289 (1975).

13. Granot, N., Director of the Building Center of Israel, private communication (1977).

14. Usmanov, Y.U., et al., "Solar Ponds as Accumulators of Solar Energy", Geliotekhnika, vol. 5, p. 49 (1969).

15. Eliseev, V.N., et al., "Theoretical Investigation of the Thermal Regime of a Solar Pond", Geliotekhnika, vol. 7, p. 17 (1971).

16. Eliseev, V.N., et al., "Determining the Efficiency of a Solar Salt Pond", Geliotekhnika, vol. 9, p. 48 (1973).

17. Chepurniy, V., Savage, S.B., "An Analytical and Experimental Investigation of a Laboratory Solar Pond Model", ASME Publication 74-WA/SOL-3 (1974).

18. Saulnier, B., et al., "Experimental Testing of a Solar Pond", BRI-R119, Brace Research Institute (1975).

19. Jain, G.C., "Heating of Solar Ponds", The Paris Congress on Solar Energy, July 1973.

20. Nielsen, C.E., Rabl, A., "Operation of a Small Salt Gradient Solar Pond", ISES meeting, Los Angeles, California (1975).

21. Styris, D.L., et al., "The Non-convecting Solar Pond, an Overview of Technological Status and Possible Pond Applications", BNWL-1891, Battelle-Northwest Laboratories (1975).

22. Styris, D.L., et al., "The Nonconvecting Solar Pond Applied to Building and Process Heating", Solar Energy, vol. 18, p. 245 (1976).

23. Drumheller, K., et al., "Comparison of Solar Pond Concepts for Electrical Power Generation", BNWL-1951, Battelle-Northwest Laboratories (1975).

24. Shaffer, L.H., "Viscosity Stabilized Solar Ponds: Phase 1", CEM-4200-572, Center for the Environment and Man, Inc., November 1976.

25. Dickinson, W.C., Neifert, R.D., "Parametric Performance and Cost Analysis of the Proposed Sohio Solar Process Heat Facility", UCRL-51783, Lawrence Livermore Laboratory (1975).

26. Nielsen, C.E., "Experience with a Prototype Solar Pond for Space Heating", ISES meeting, Winnipeg, Canada (1976).

27. Badger, P.C., et al., "A Prototype Solar Pond for Heating Greenhouses and Rural Residences", ISES-American Section meeting, Orlando, Florida (1977).

28. Bowser, M., private communication (1977).

29. Hull, J., private communication (1977).

30. Assaf, G., "The Dead Sea, a Scheme for a Solar Lake", Solar Energy, vol. 18, 4 (1976).

31. Bryant, H.C., Colbeck, I., "A Solar Pond for London?", Solar Energy, vol. 19, p. 321 (1977).

32. Zangrando, F., Bryant, H.C., "Operation and Maintenance of a Salt Gradient Solar Pond", Helioscience Institute Conference, Palm Springs, California (1977).

33. Turner, J.S., Buoyancy Effects in Fluids, Cambridge Press (1973).

34. Huppert, H.E., "Thermosolutal Convection", for publication in the Journal of Fluid Mechanics.

35. Linden, P.F., Weber, J.E., "The Formation of Layers in a Double-diffusive System with a Sloping Boundary", Journal of Fluid Mechanics, vol. 81, 4 (1977).

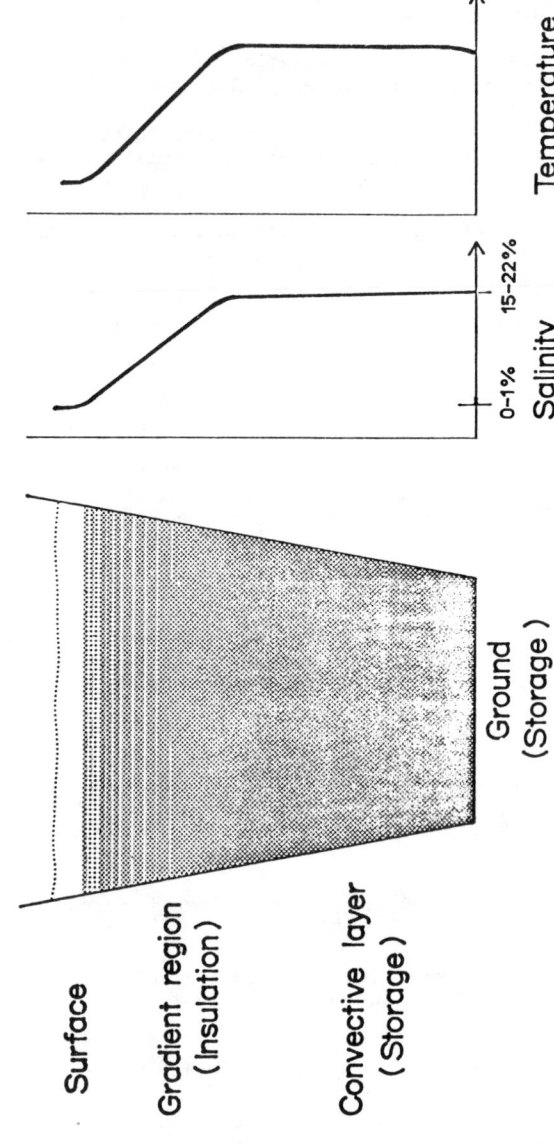

Figure 1. The Salt Gradient Solar Pond With Storage

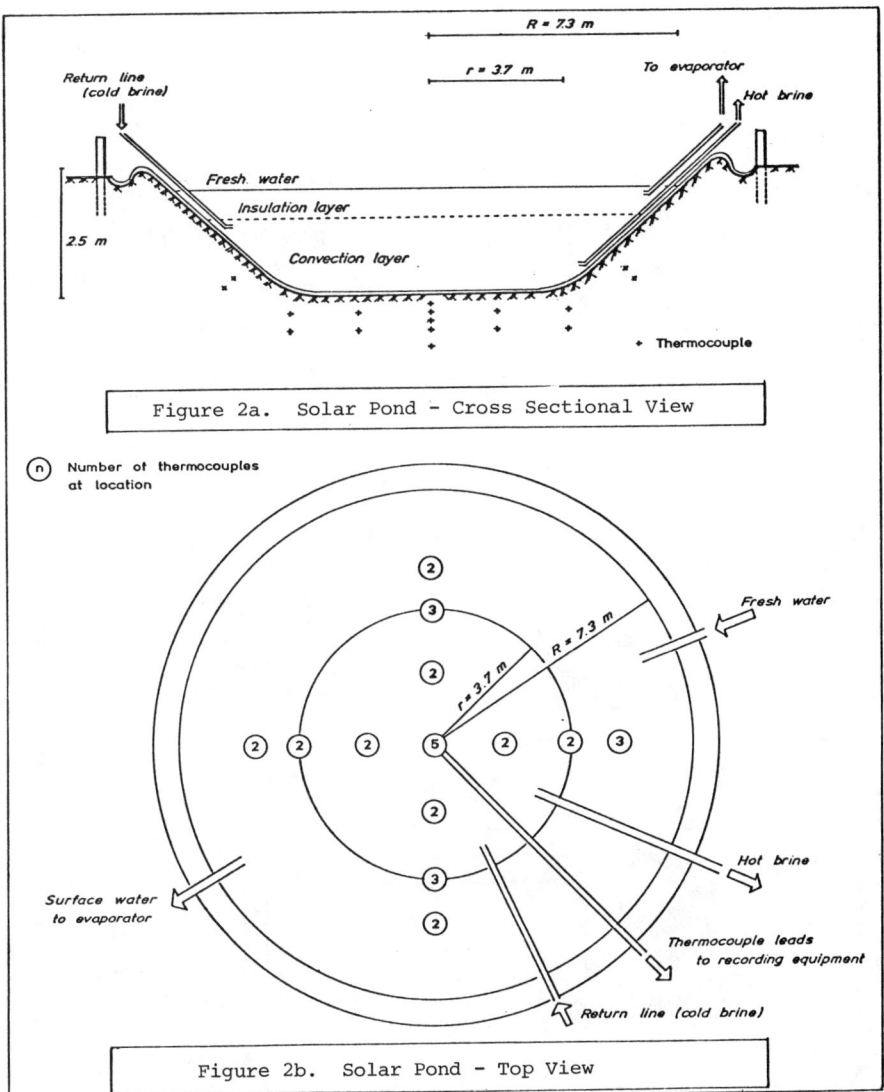

Figure 2a. Solar Pond - Cross Sectional View

Figure 2b. Solar Pond - Top View

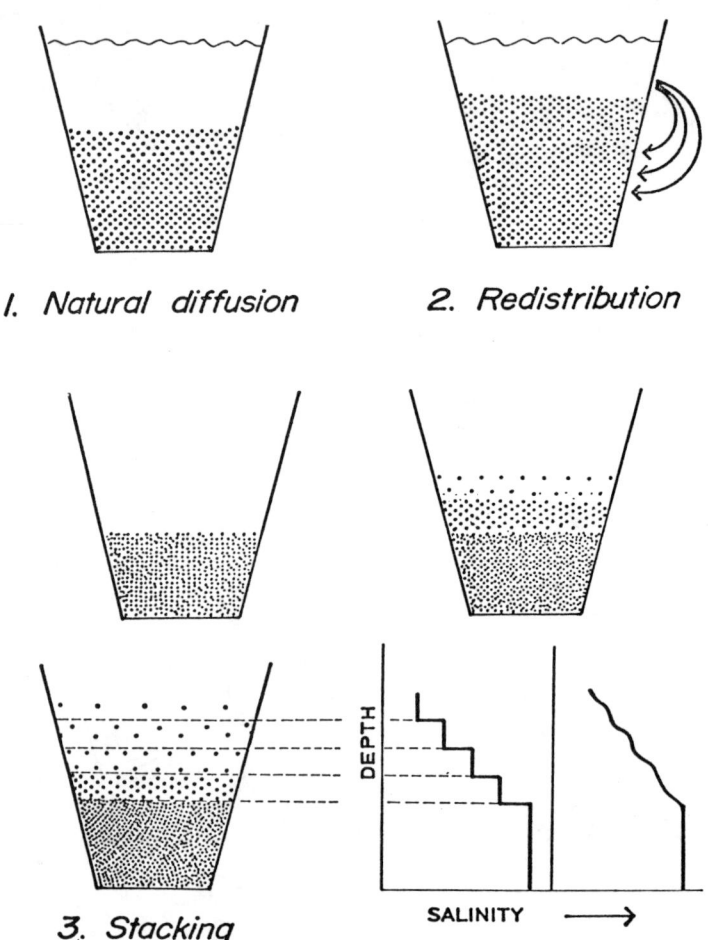

Figure 3. Methods to Establish the Gradient

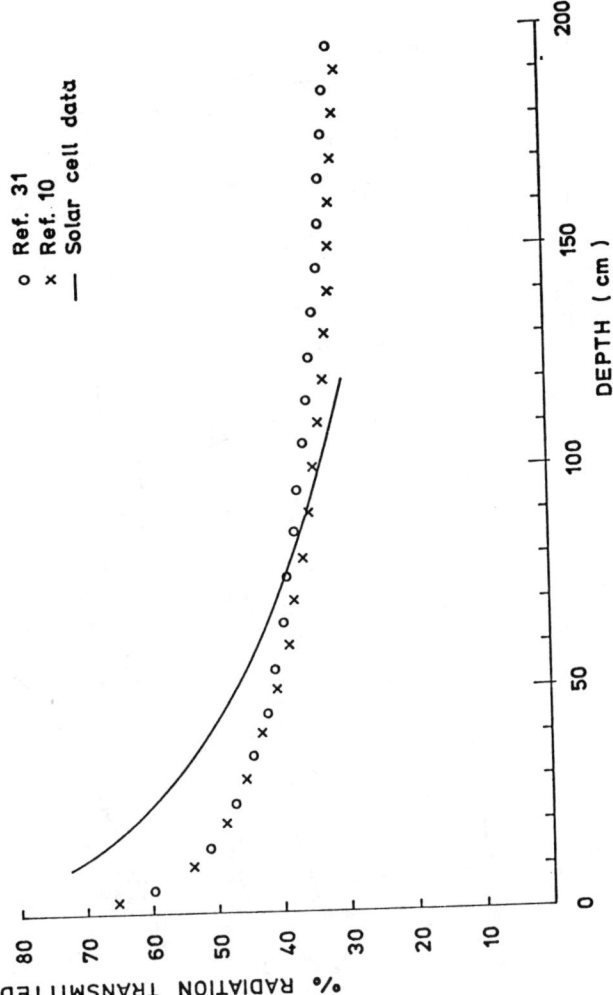

Figure 4. Fraction of Light Remaining at Depth X

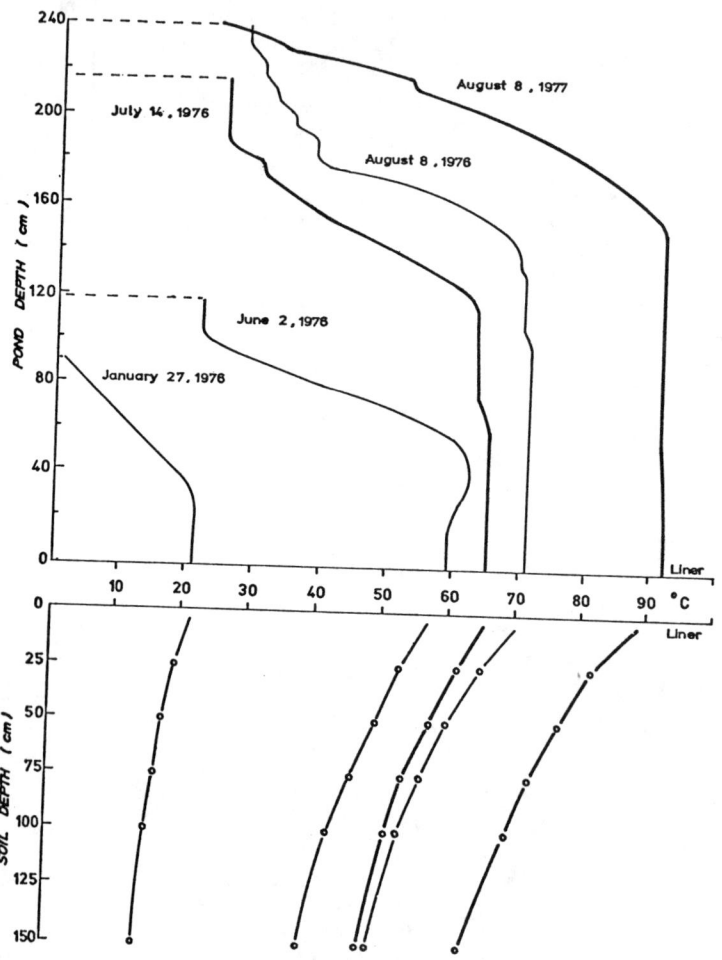

Figure 5. Temperature Porfile of the Solar Pond and The Ground Below (Center)

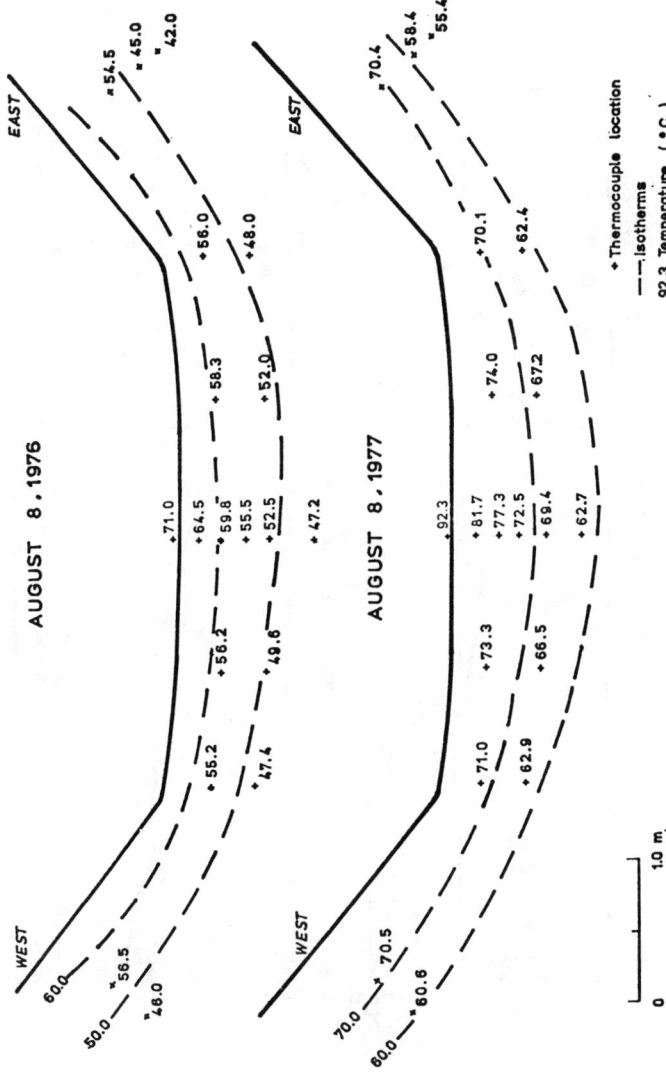

Figure 6. Temperature Profiles of the Ground

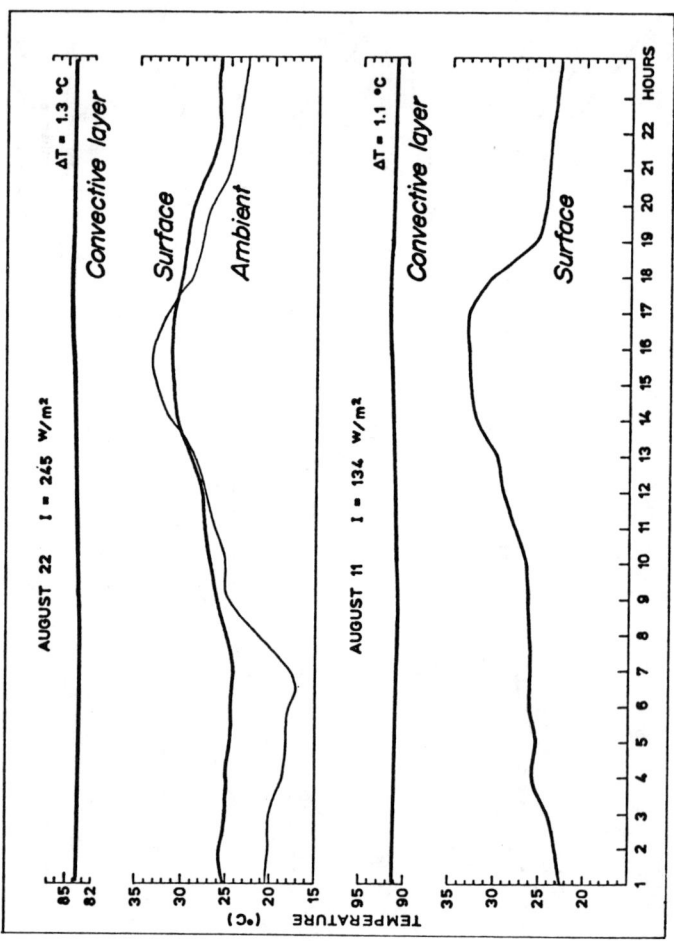

Figure 7. Daily Temperature Response

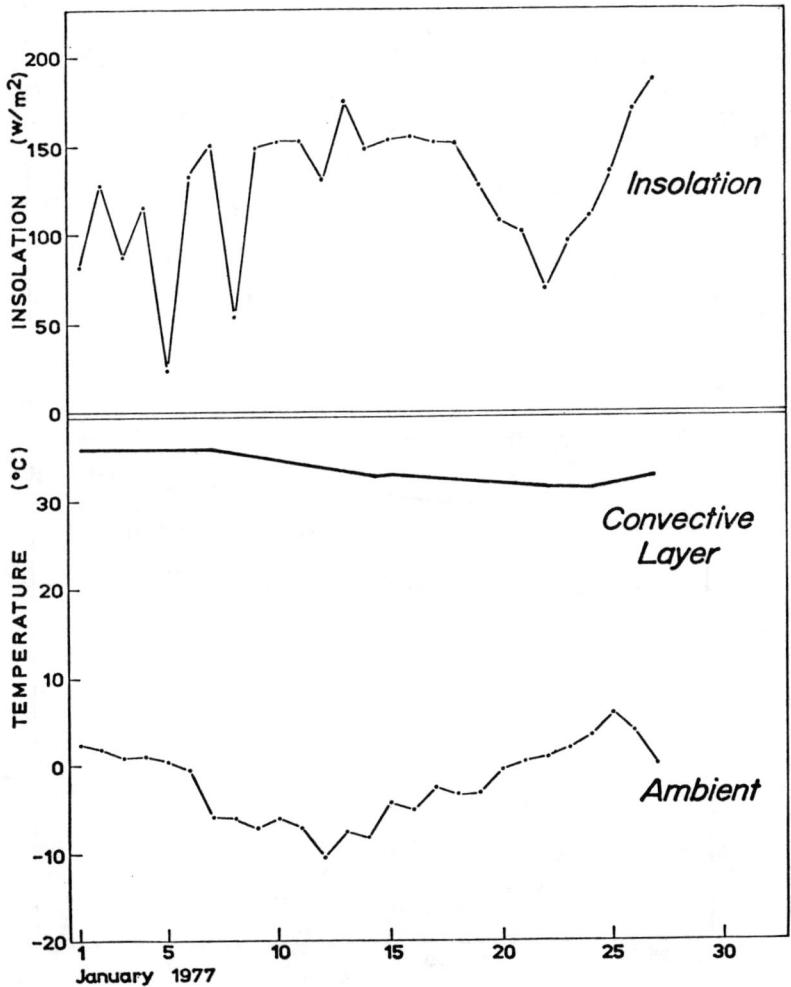

Figure 8. Temperature Response of the Convective Layer

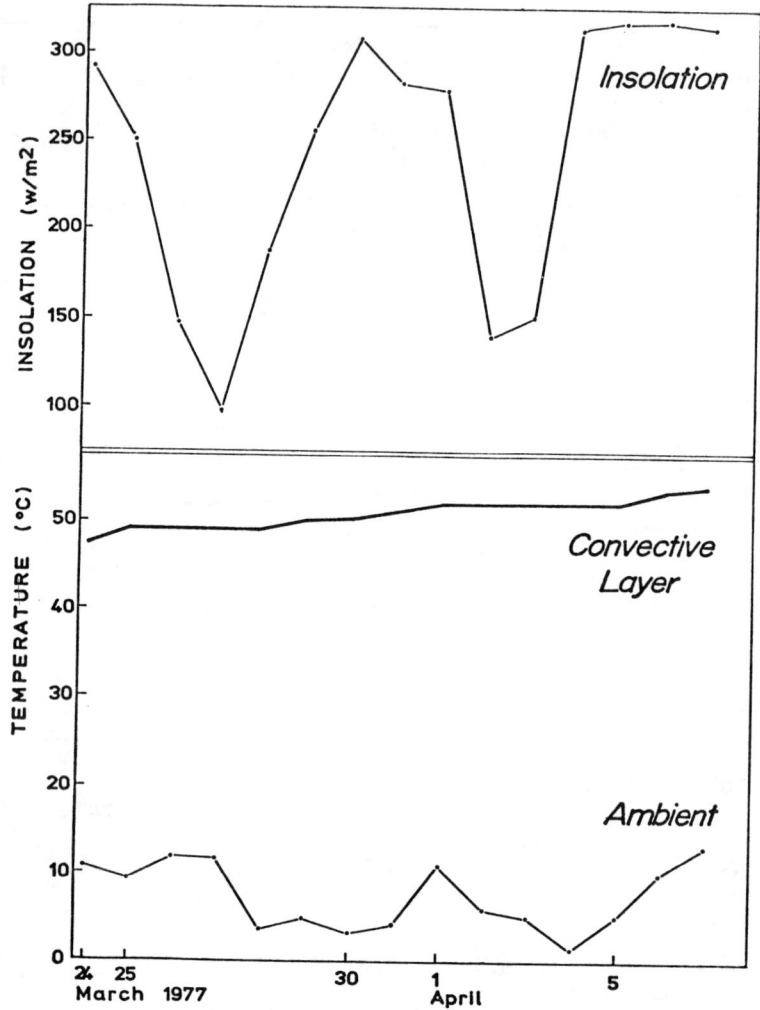

Figure 9. Temperature Response of the Convective Layer (2)

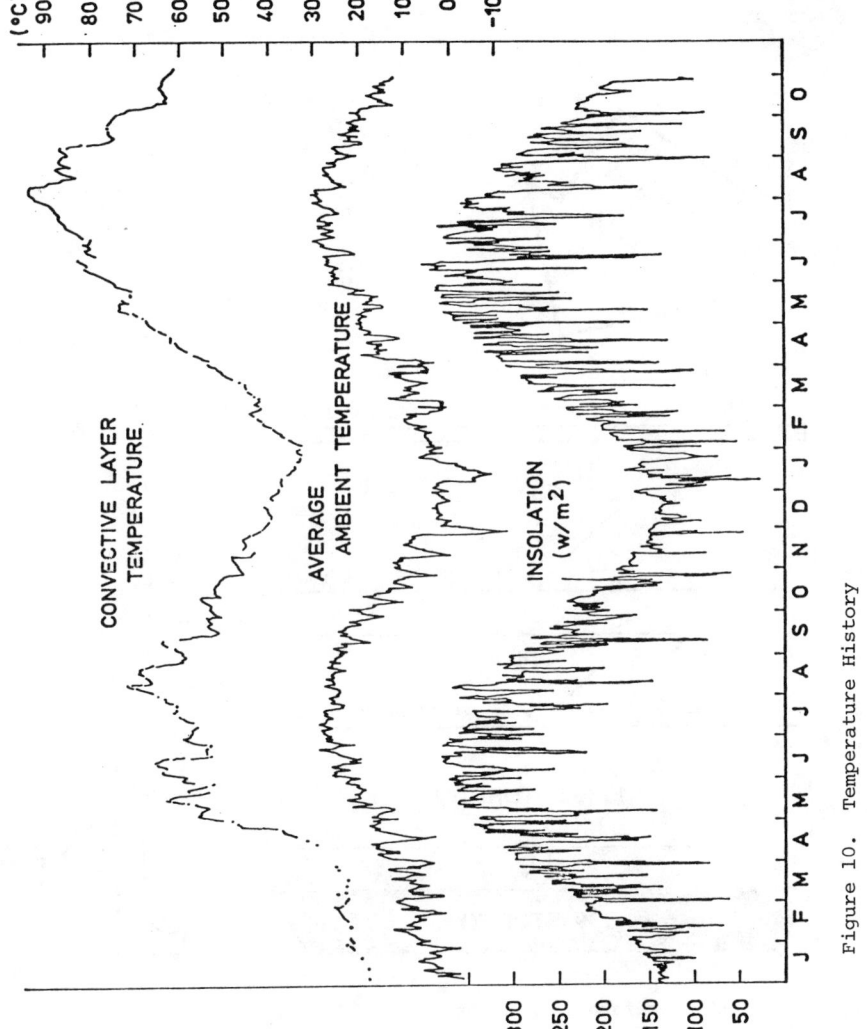

Figure 10. Temperature History

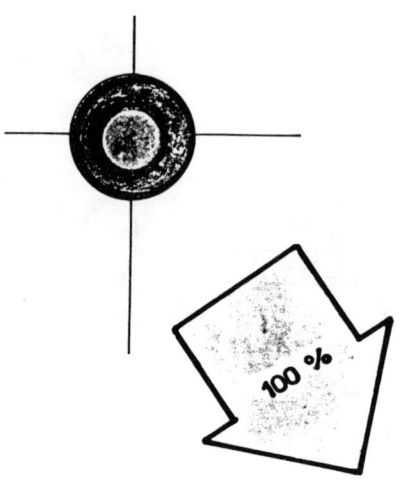
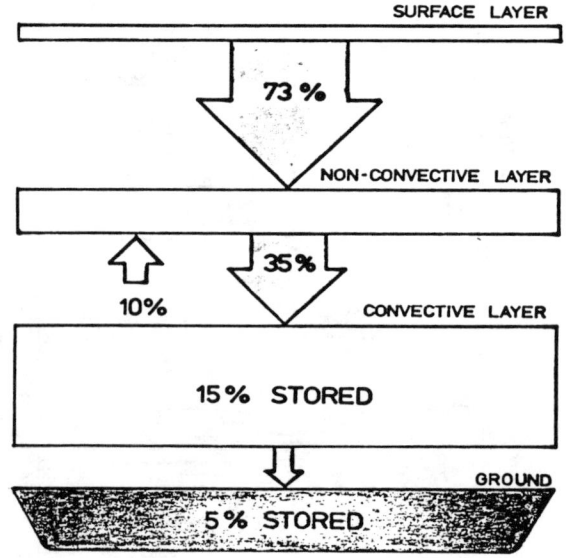
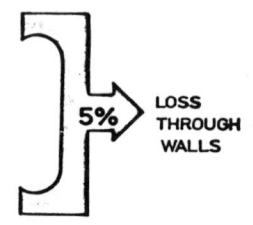

Figure 11. Diagramatic Heat Balance

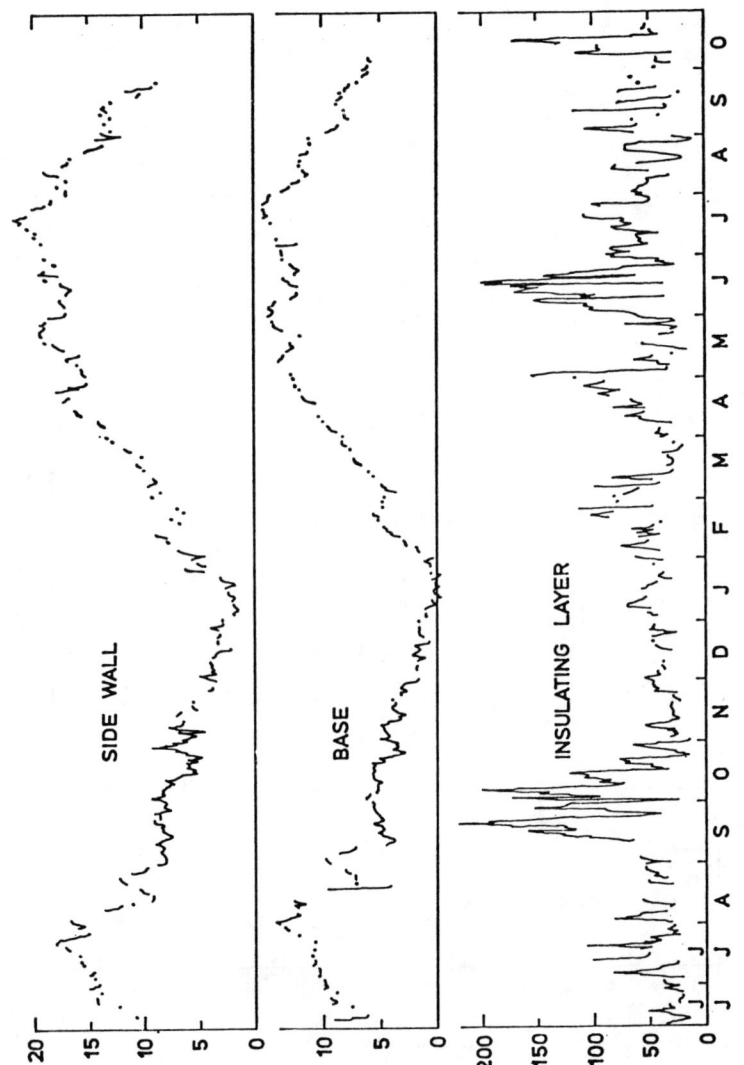

Figure 12. Temperature Gradients in °C/M

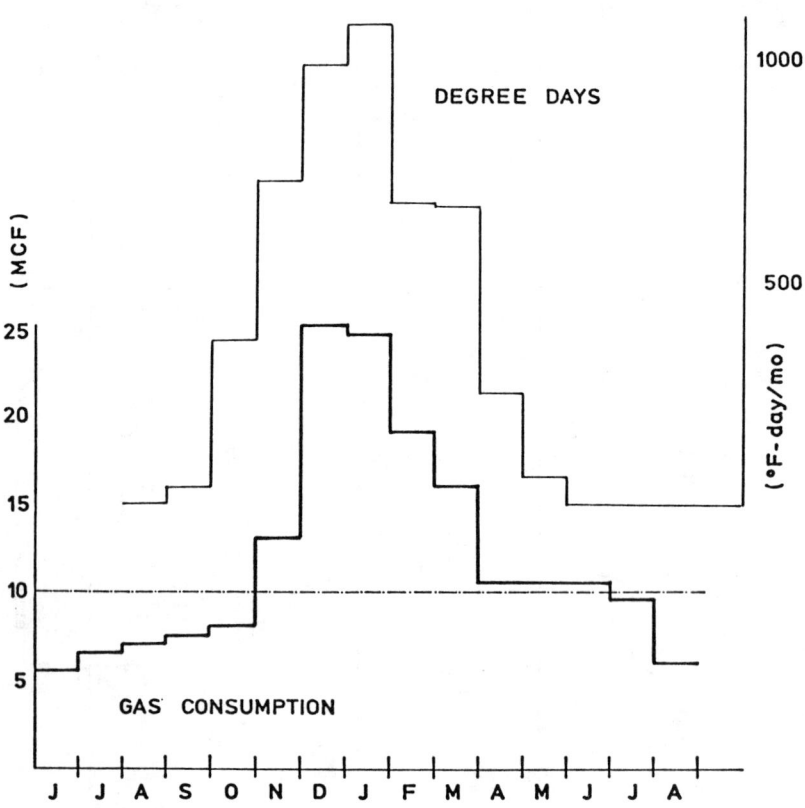

Figure 13. Heating Requirements

2966

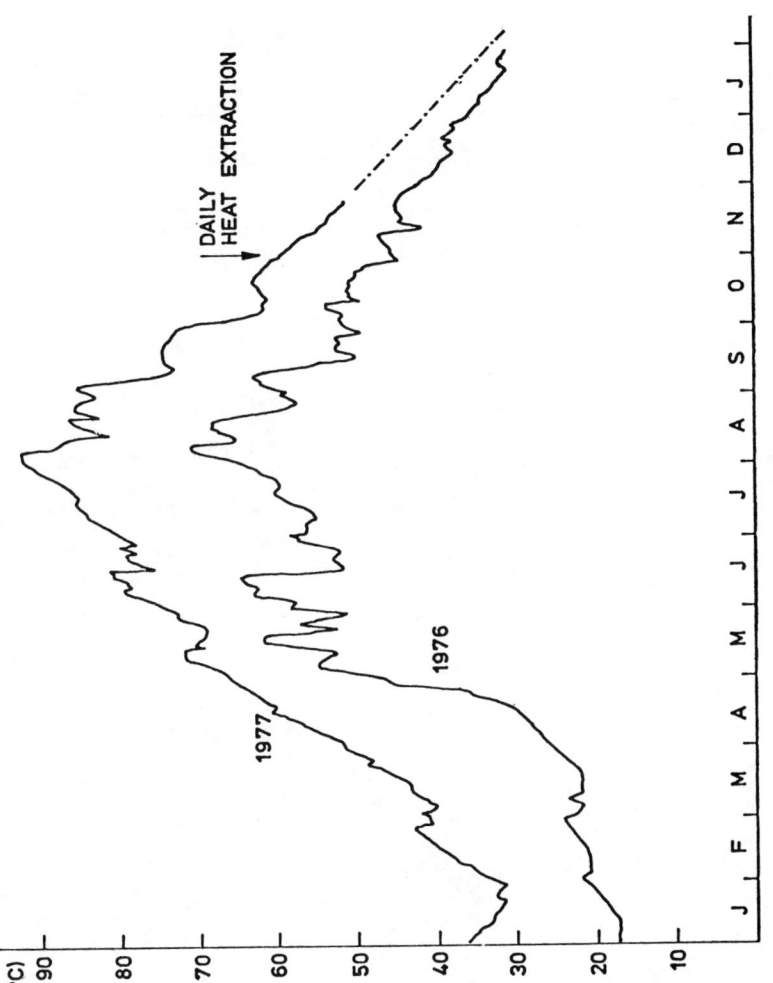

Figure 14. Projected Temperature Response

TJ
163.15
M5
1977
v. 6

APR 19 1979